Fundamentals and Applications of Ion Exchange

NATO ASI Series

Advanced Science Institutes Series

A Series presenting the results of activities sponsored by the NATO Science Committee, which aims at the dissemination of advanced scientific and technological knowledge, with a view to strengthening links between scientific communities.

The Series is published by an international board of publishers in conjunction with the NATO Scientific Affairs Division

A	**Life Sciences**	Plenum Publishing Corporation
B	**Physics**	London and New York
C	**Mathematical and Physical Sciences**	D. Reidel Publishing Company Dordrecht and Boston
D	**Behavioural and Social Sciences**	Martinus Nijhoff Publishers Dordrecht/Boston/Lancaster
E	**Applied Sciences**	
F	**Computer and Systems Sciences**	Springer-Verlag Berlin/Heidelberg/New York
G	**Ecological Sciences**	

Series E: Applied Sciences – No. 98

Fundamentals and Applications of Ion Exchange

Edited by

Lorenzo Liberti

Senior Researcher
Water Research Institute
National Research Council of Italy

Associate Professor of Industrial Chemistry
Faculty of Engineering
University of Bari
Bari, Italy

John R. Millar

Formerly Senior Research Scientist
Duolite International Inc
Redwood City, CA 94062, USA

1985 **Martinus Nijhoff Publishers**
Dordrecht / Boston / Lancaster
Published in cooperation with NATO Scientific Affairs Division

Post-Conference Proceedings of the NATO Advanced Study Institute on Mass Transfer and Kinetics of Ion Exchange, Maratea, Italy, May 31-June 11, 1982

Library of Congress Cataloging in Publication Data

Main entry under title:

Fundamentals and applications of ion exchange.

(NATO ASI series. Series E, Applied sciences; no. 98)
"Post-conference proceedings of the NATO Advanced Study Institute on Mass Transfer and Kinetics of Ion Exchange, Maratea, Italy, May 31-June 11, 1982"-- P.
"Published in cooperation with NATO Scientific Affairs Division."
1. Ion exchange--Congresses. I. Liberti, Lorenzo. II. Millar, John R. III. NATO Advanced Study Institute on Mass Transfer and Kinetics of Ion Exchange (1982 : Maratea, Italy) IV. North Atlantic Treaty Organization. Scientific Affaris Divison. V. Series.
QD561.F825 1985 660.2'9723 85-18842
ISBN 90-247-3229-8

ISBN 90-247-3229-8 (this volume)
ISBN 90-247-2689-1 (series)

Distributors for the United States and Canada: Kluwer Boston, Inc., 190 Old Derby Street, Hingham, MA 02043, USA

Distributors for the UK and Ireland: Kluwer Academic Publishers, MTP Press Ltd, Falcon House, Queen Square, Lancaster LA1 1RN, UK

Distributors for all other countries: Kluwer Academic Publishers Group, Distribution Center, P.O. Box 322, 3300 AH Dordrecht, The Netherlands

Printed in The Netherlands

PREFACE

"Ion exchange", as Dr. Robert Kunin has said, "is a unique technology since it occupies a special place in at least three other scientific disciplines - polymer chemistry, polyelectrolytes and adsorption." It may also lay claim to being one of the most widely used industrially. From its origins in water treatment and the sugar industry, through hydrometallurgical applications as diverse as the treatment of plating wastes and the tonnage production of uranium, to the present-day production of ultrapure water for the microelectronics industry, the recovery of valuable materials from sewage effluents and pollution control, the uses of ion exchange are legion.

As a result, it is well-nigh impossible to prevent infiltration by the real world of even the most academic of conferences on the subject. It came as no surprise to the Scientific Board of the NATO Advanced Study Institute on "Mass Transfer & Kinetics of Ion Exchange" that one third of the lecturers, and one half of their advanced students, were from Industry, nor that the two round-table discussions, which specially featured industrial applications and future requirements, were well attended and enthusiastically debated.

This volume presents a collection of the contributions from the audience which followed the (already published) principal lectures at each of the five sessions and which, so far, have only been published in abstract. These ranged from theoretical exegesis to down-to-earth practicality and covered a wide range of topics inspired by the original idea of the Institute without necessarily being restricted in content by the academic restraint of its formal title.

In the two and a half years since these papers were originally delivered the ion exchange world has seen a number of changes, to be sure, but these have been mainly of economic origin. The state of the art, as might be expected of a mature technology, has advanced but little; the theory, which in ion exchange has usually lagged behind practical application, remains much as it was discussed, in congenial company and amid beautiful surroundings, at Acquafredda di Maratea.

Lorenzo Liberti | John R. Millar

TABLE OF CONTENTS

ION EXCHANGE TOWARDS THE TWENTY FIRST CENTURY

C. Calmon

511 Ewing Street
Princeton, NJ 08540 USA

TOWARDS DEFINITION OF THE ION EXCHANGE FIELD

Before consideration is given to the future development in ion exchange, the term "Ion Exchange" must be understood to encompass many aspects of separation. The classical text on "Ion Exchange" by Helfferich defines "ion exchangers" but not the term "ion exchange". Also, many chapters on the ion exchangers have nothing to do with ion exchange, as can be seen from the following:

1. Ligand exchange - deals with molecular exchange.
2. Electrodialysis - deals with the membranes which act as selective barriers and conductive media.
3. Catalysis - deals with the transformation of organic molecules.
4. Sorption - involves the sorption of organic molecules e.g. through ion exclusion.
5. Liquid ion exchangers - are not solid insoluble products, but these can be incorporated into polymeric matrices to form extracting resins.

From the above it is evident that the text deals with the utilization of ion exchangers for many functions often not involving ion exchange. So the topic for discussion should be polymeric matrices with functional groups.

Basically ion exchange is a separation process which utilizes functional or active sites on so-called inert matrices. Both from the fouling of ion exchangers by high-molecular weight organics and the difficulties involved due to the interaction of certain organics with the polymeric matrix, it appears that there is no such thing as an absolute inert matrix. So even the definition for an ion exchanger as an inert and insoluble solid is rather limited. trace solubility of ion exchange resins is a problem in the manufacturing of microcircuits.

The removal of colloids and the filtering out of organisms by certain ion exchangers with high porosities and the recent development of microporous membranes with polar groups for removal of very fine particulates in solution expand the use of entities with ion exchanger groups. In view of the above can enzymes, biochemicals, and microbiocells immobilized on polymers be excluded from ion exchange texts? In addition there are electro and photo conductive resins in which polar groups attached to polymeric matrices. So the field has to be expanded beyond ions and beyond exchange.

The reason for the limited views of ion exchange until recently was the fact that the early work in ion exchange was done for water treatment. The application involved the exchange of damaging or toxic ions for inocuous ions or the partial or total removal of ionic species. The contributions of the early water treatment companies and the ion manufacturers were:

1. They showed how to prepare inert polymeric matrices.

2. How to attach polar or active sites to the matrices.

3. The application of these to water and wastewater treatment and to resource recovery.

However, the leaders in the field today are academicians and in small companies developing products for the chromatographic and biochemical fields. Examination of Mathur´s (et al) book, "Polymers as Aids in Organic Chemistry" (Academic Press) and Hodge´s and Sherrington´s book, "Polymer-Supported Reactions in Organic Synthesis" (John Wiely and Sons), will show the tremendous progress or advancements being made by the synthetic organic and biochemist in this field.

Naturally, there will arise many specialized divisions in this field. The question will be how cross fertilization within the field can take place. In the future the biochemical and pharmaceutical fields may outstrip the water treatment industry, resulting in large companies to set up divisions to research the development of the new reactive polymers and determine what can be done with these.

PRESENT STATUS

Soon after the development of the modified natural products and the condensation polymers as supports for polar groups, their shortcomings became evident. The introduction of the styrene-divinylbenzene matrices improved the supports and gave matrices with well defined compositions.

During the next fifteen years (1945-1960) the emphasis outside applications was on evaluating exchangers with various polar groups.

The next step, prompted by the fouling of the highly basic ion exchangers when dealing with high molecular weight organics, led to the modification of the matrices themselves e.g. the macroporous resins, open structured celluloses, and the acrylics as binding sites. So not only did the polar group become of importance in ion exchangers but also the matrix to which it is attached.

FUTURE DEVELOPMENT

The saying "When one gets the urge to prophecy, it is better one goes to bed until the urge passes over", is now well established. However, there is another old saying, "Future events cast their shadows long in advance", which gives a base for making some guesses as to future developments. About 20 years before reverse osmosis swept the world, an experiment at University of Florida with cellulose acetate showed its potential for desalination. The experiments and developments in the academic world are the "shadows" of future events. So let us find some "shadows".

In recent work with ion exchangers it has been shown that selection of an ion exchanger is affected by the inert groups that are attached to the polar group.

The difference in properties of Type I anion exchangers, which have trimethyl groups on the nitrogen, and Type II resins, which have an ethanol group on the nitrogen in place of one methyl group, is well known.

The selectivity for sulfate-nitrate of a strong base resin is reversed to nitrate-sulfate if the methyl groups on the nitrogen are changed to ethyl groups.

An arm of $CH_2 \cdot NH \cdot CH_2$ separating a PO_3H_2 from the aromatic matrix makes it possible to remove traces of calcium and magnesium from brine feeds used in the chlor-alkali electrodialytic cells.

Now three components define the properties of ion exchangers -

1. Type of polar group.

2. Structure of the matrix.

3. Inert groups on the polar groups and other groups adjacent to it.

CHELATES

For the removal of toxic metal ions, recovery of metal ions from the low grade ores and purification of plating baths, chelates are coming into use. A great deal of work in synthesizing and evaluation of chelates is going on in universities.

Ion Exchange membranes

Improvement in the ion exchange membranes for use in the caustic-chlorine cell is resulting in the elimination of the mercury cell.

Use of exchangers at high concentrations

Continuous counter current ion exchange, (CCIX), units have not been successful due to certain operating practices. However, today fifteen (CCIX) large plants are operating for the recovery of ammonium nitrate and water in fertilizer manufacture. In some cases the ionic concentration runs in thousands of ppm. Small CCIX units are operating in purification of rinse waters and recovery of metal ions in plating companies. Small countercurrent units for laboratories are appearing also. The knowledge gained from these will lead to more CCIX system applications as more concentrated solutions, e.g. in hydrometallurgy, are used in recovery of metal ions from poorer ores.

Ion exchange systems can be polluting if the regeneration effluents have no value and if disposed in sewers, ponds or streams. Ion exchange may become in the future due to this aspect only as scavengers after R.O. or E.D. The sirotherm process with improvement may overcome the problem of pollution due to ion exchanges.

For improving kinetics fine ion exchange beads are used in shallow beds.

NEW DIRECTIONS

In order to keep alert to future potentials in ion exchange, the following disciplines should be followed:

1. Chromatography.
2. Polymer research along with applications to
   a. reagents as aids in organic chemistry
   b. organo-metallics
3. Immobilization of enzymes and other biochemicals.
4. Immobilization of microbiocells.

Chromatography

The media, tools and techniques in chromatography offer opportunities for applications the same as above in ion exchange. Ion exchange chromatography has been utilized in rare earth recovery and separation of various radioactive species. Horvath has shown the separation of one hundred organic acids in a drop of urine by means of HPLC with an ion exchanger. Today the separation of fructose from glucose by means of a cation exchanger utilizing water as the eluting agent is becoming a large industrial process. Ion chromatography for determining ions in ppb quantities have become standard tools in nuclear power plants.

Polymeric Reagents in Organic Chemistry

Today organic and biochemists are preparing very complex functional groups attached to polymeric matrices. These researchers will develop highly selective exchangers by attaching more organic molecules which give specific tests for metal ions.

Organo-Metallic Polymers

The Petroleum industry and various academicians are developing an array of organo-metallic polymers for catalytic work. The matrices and possibly group sites may develop into more stable ion exchangers.

Immobilization of Biochemicals

The immobilization of enzymes, begun by academics in the 1950's, is now utilized in Japan on a large industrial scale. The pickup or destruction of organisms, viruses, pyrogens and other metabolites in water may be a future possibility through specific groups or biochemical attachments.

Immobilization of Microbiocells

For many organics there developed microorganisms for their breakdown to CO_2 and to other inocuous chemicals. Already a pilot plant is evaluating immobilized specific microbiocells for reducing COD due to organics. Will we see in the future immobilized microbiocells for pretreatment of water prior to ion exchange or membrane processes to prevent fouling? Will the same technique be used for destroying the halomethane precursors such as humic acids? No doubt it will come in the near future. The advantage of immobilized microbiocells is that they can be made into a continuous system.

Computerization of ion exchange operation will become more common. The development of more exact and varied sensors will make possible more accurate control in automated systems on small and large scale processes. Processes which are today only laboratory curiosities will become industrial processes as control via the microprocessing will make control more automatic. Stable ion exchangers will be required.

In the future, the choices will be many and more complex so that more specialized individuals will be necessary to develop processes dealing with ion exchange in a broader sense than previously thought of.

The Use of Some Coordinating Copolymers in Hydrometallurgy

Michael J. Hudson

Department of Chemistry, University of Reading, Whiteknights, Reading RG6 2AD, UK

This paper will concentrate its main attention upon two reagents:

1. A polymeric N-dithiocarboxylate
2. A polymeric thiol

A poly [N-(dithiocarboxylato)] polymer (PIED) reacts with metal ions such as iron, cobalt, nickel, copper, zinc and silver to produce stable compounds in which the metal is co-ordinated to the sulphur atoms in the N-dithiocarboxylate groups. The capacity of the polymer PIED for a particular metal is a function of the sulphur content, pH, complexing agents, and other metals which are present in solution. The polymer rapidly takes up the metals (eg. Co^{2+} at pH 6.5 has a $t_{\frac{1}{2}}$ of one and a half minutes). Above a pH of 10 the polymer dissolves but is reprecipitated on the addition of acid. The stable metal-polymer complexes can be used for metal extractions over a wide pH range eg cobalt (II)-PIED can be used to extract silver from a solution containing cobalt and silver. Consequently, the metal salts are recommended for metal extraction. It is not possible to elute the metals using standard reagents such as cyanide but the cobalt-PIED is reduced by hydrogen at 1400 K to give cobalt metal although the polymer backbone is destroyed. The nickel(II)-PIED compound can be quantitatively oxidised to nickel(IV)-PIED which is stable in air and in which all of the nickel ions are oxidised to nickel(IV).

A second investigation with the polymer PIED concerns its use as an extractant for rhodium and iridium towards the end[2] of the Platinum Group Metals refining process (i.e. after solvent extraction of Au, Pd, Pt). The results indicate that the co-polymer is surprisingly stable in hydrochloric acid (5 molar) and that good extractions of rhodium and iridium can be achieved (99.9%). The polymer has a high capacity for both rhodium and iridium and the rates of uptake of rhodium and iridium on the polymer are acceptable ($t_{\frac{1}{2}}$), for (Rh and Ir ca. 5 min). The metal-polymer complexes have also proved to be easy to decompose either by ashing or by chemical treatment e.g. with hydrogen peroxide. The chemical treatment with a hydrogen peroxide, sulphuric acid mixture is the preferred route as separation of concentrated solutions of rhodium and iridium sulphates is then possible.

There have been many attempts to prepare polymeric reagents which are capable of selectively extracting metal ions such as those of silver. A thiol polymer was used to extract ions of Ag(I) and Cu from separate and mixed solutions. The polymer[3] has a capacity of 1.86 mol kg^{-1} of Ag(I) with respect to the dry polymer from silver nitrate solution and 0.88 mol kg^{-1} of Cu from copper(II) chloride solution. In a mixed Ag(I) and Cu(II) nitrate solution only the Ag(I) is extracted. The polymer does have low capacities for many other ions including Fe(II), Fe(III), Co(II), Ni(II) and Zn(II). The time ($t_{\frac{1}{2}}$) for the silver uptake on half the sites of the polymer is well under five minutes at pH 5. The silver can be eluted from the polymer-silver complex using KCN (0.1 M). The time ($t_{\frac{1}{2}}$) for half of the silver to be eluted is under five minutes. The regenerated polymer can be used for repeated separations without loss of capacity. When copper ions are complexed by the polymer and there is evidence that some, at least, of the ions are reduced to Cu(I). The time ($t_{\frac{1}{2}}$) for copper extraction is approximately five minutes and the copper is probably eluted as a Cu(I) species.

The thiol resin can be used for separation of precious metals from base metals. The resin has, for example, no affinity for Fe(II), Fe(III), Co(II) or Ni(II). The capacities of the resin in IMHCl are for Au(III) (720 mg/g resins), Pd(II) (240 mg/g resin), Pt(IV) (240 mg/g resin), or Rh(III)

(60 mg/g resin). There is some selectivity between the precious metals themselves e.g. Au(III)>Pd(II)∿Ag(I)>Pt(IV)>Rh(III). The separation ratios on sample batch separations are also encouraging in certain cases Pd(II) / Cu(II) (α = 48); Au(III) / Rh(III) (α = 94); Ag(I) / Cu(II) (at pH 1.0, NHO_3) (α = 2,750). Selectivities for Pt(IV) and Rh(III) are, however, low but differential elution is being investigated. In the copper complex of 2-S-methyl-1,3,4-thiadiazole-5-thiol (MTT) which is a low molecular weight analogue the copper is present entirely as Cu(I). The bonding in this and the corresponding Ag complex is thought to involve the 2-S-methyl and 5-thiol sulphur atoms which gives rise to polymeric species with linear S-M-S stereochemistry. The bonding in the polymer-metal compounds is thought to be similar.

References

1. Giwa C.O., and Hudson, M.J. Hydromet. (1982), 8, 65-75.
2. Hudson, M.J. and Thorns, J.T. Angewandte Chemie, (accepted for publication) 1983.
3. Hudson, M.J., and Shepherd, M.J. Hydromet. (1983), 9, 223-234.

THE CCS-ION EXCHANGE CONTACTOR

Kurt Meuli

CHRIST LTD, Aesch, Switzerland

The successful application of an ion exchange process depends on the properties of the ion exchange resin on one hand and the ion exchange device on the other hand.From the point of view of a water treatment engineer, the improvement of the characteristics of the ion exchange resin has always been given priority over the development of the ion exchange device.The success of reverse osmosis, in recent years, has kept the drive to improve ion exchange equipment low.Even today,most advantages gained by choosing the proper resin are often lost through inappropriate equipment.

For the most important application of the ion exchange process, the production of demineralized water, the changing cost structure and the increasing importance of environmental constraints promote the demand for improved water treatment systems.A well designed, technically advanced ion exchange device may push the limits of application towards higher salinities and well into the areas which reverse osmosis processes have taken over in recent years.

A typical example with comparable costs is the demineralization of 150 m3/h (or 3600 m3/d or 1'240'000 m3 per year) of a water with a TDS of 6 eq/m3 (50% HCO_3^- and 10% Na^+) Fig.1.It represents for the actual European cost parameters the economic limit of application between ion exchange and reverse osmosis.The total costs were calculated for four different processes applied to this demineralization problem (Fig.2).

Process 1 CHRIST-continuous ion exchange system (CCS)
Ion exchange in a quasi-continuous contactor with internal resin recirculation (moving bed reactor).Every process step, especially loading and regeneration, takes place in separate zones, optimally adapted to the ion exchange process.

Process 2 Counter-current regenerated ion exchange process (RX)
Ion exchange, carried out in two parallel trains,alternately loaded and regenerated in countercurrent mode.

Process 3 Reverse Osmosis (RO)
Reverse osmosis with an extremely high permeate yield.

Process 4 Reverse Osmosis (RO)
Reverse osmosis with a normal permeate yield.

This comparison (Fig.3) shows that the total costs are very similar at lower rawwater costs.This similarity however diminishes considerably with an increase in the raw water price.The advantage of RO process 3 with a high permeate yield, due to its low internal consumption of water, over RO process 4 with a lower conversion factor, then of course becomes evident.

An extremely low consumption of water which is equivalent to a highly concentrated waste water is an important characteristic of the CCS-contactor.

When a bed of normal depth of loaded ion exchange resin is regenerated with a slight excess of regenerants, bell-shaped elution curves arise (Fig.4).The maximum eluate concentration is strongly related to the operating capacity of the resin as well as the concentration of the regenerant.The average effluent concentration however lies far below this value.If a highly concentrated effluent is the desired product, then the concentration may be increased through fractionation followed by a recycling of the fractions with low concentrations or by increasing the bed-depth indefinitely. This latter situation corresponds to the continuous exchange between the two phases which uniformly and countercurrently flow through a contactor (Fig.5).Per time unit, the water stream must transport at least an amount of regenerants equivalent to the operating capacity of the ion exchange resin, flowing in the opposite direction.

The result of this ideal countercurrent mode of operation is an eluate curve in which at least the maximum eluate concentration of a fixed-bed system is obtained (Fig.6).With the CCS-contactor it is possible to realize a countercurrent ion exchange operation with practically an ideal behaviour.

Description of the CCS-contactor (Fig.7)

Raw water fed into the bottom of the loading section 1 flows upstream through the resin bed until certain quality criteria are met in the effluent.At this point,the flow of water is interrupted, loaded resin is drawn off at the bottom of section 1 and transferred into pulse section 3 which determines the volume of the resin to be transported.The resin in the pulse section is then backwashed to remove fine suspended solids through section 2.The resin collected in the pulse section is moved by means of water into the regeneration loop 4,replacing an equivalent amount of fully regenerated and rinsed resin out of rinse section 5 into the loading vessel 1.Immediately after completion of the resin transport, the exhaustion section 1 is put into service again, while at the same time regenerant is introduced into regeneration loop 4 and treated water is drawn through section 5 for the purpose of rinsing of the resin.Depending on the design and the ionic load of the raw water, there are two to ten such pulses per hour.

A treated water storage tank after the CCS ion exchange unit permits quasi continuous operation, thus eliminating the need for a parallel unit.

The CCS-system is an improved version of the Higgins system with the difference that the loading section works in the upstream mode with a portion of the resin in the fluidized state.A uniform water quality is achieved due to the fact that small segments of the resin are moved in rather short intervals and that the ion exchange beds in each zone contain several of these increments at varying states of loading or regeneration at any given time.This mode of operation permits to achieve a steady state situation with the typical counter-current characteristics:

- very small leakage which is practically independent of the variation of flow and salinity
- high regeneration efficiency due to long contact time and strict control of the sequence of process steps
- regeneration levels of 110 to 130% of theory
- high operating capacities with values of 60 to 80% of the total capacity can be achieved due to the fact that only the resin with the highest loading is drawn off for regeneration
- very low consumption of rinsing water
- very low overall water consumption
- high concentration of regeneration effluent
- very insensitive to high loads of suspended solids due to short cycles and frequent backwashing
- low space requirement since only one chain is required and the regeneration column can be kept very narrow

With the help of the equilibrium diagrams (Fig.8),typical for the softening of water, the behaviour in the contactor during loading and regeneration and the conditions for the maximum effluent concentration may be derived.All mixtures which are not in equilibrium are represented by a point outside of the equilibrium line. The point B corresponds to the mixture of raw water and loaded ion exchange resin, namely the state at the water inlet and resin outlet of the loading zone.On its way to point A through the loading zone, the water meets increasingly higher regenerated resin. At point A, the treated water leaves the loading zone, after having been in contact with completely fresh regenerated resin. The loaded ion exchange resin with its loading rb is transferred to the regeneration column, where it comes, at first, in contact with brine.This mixture is represented by the point D.Every point in the operating line DC corresponds to the mixture in a cross section of the regeneration column.At point C, regenerated resin,still loaded to the degree ra, is transferred back to the loading zone after being in contact with fresh brine. The positions of the operating lines with respect to the position of the equilibrium lines,characterize the ion exchange efficiency of both zones.With the help of the Mc Cabe-Thiele method, a step-by-step construction between the two lines may be drawn.The number of steps corresponds to the number of theoretical ion exchange stages required for the given process.

The bed height required for a theoretical ion exchange stage or a transfer unit must be determined experimentally with respect to the actual operating conditions.

The Mc Cabe-Thiele method is a very useful instrument to quantify and to visualize the ion exchange process in moving bed reactors meeting the following prerequisites :

- a uniform countercurrent flux of both phases, in the loading as well as in the regeneration zone
- the resin must be removed from the bed of one zone and added to the bed of the other zone as a layer of uniform thickness
- the portions of resin have to be small compared to the total volume of the zones

The experience has shown that these conditions are fulfilled by the CCS-contactor.The first full-scale CCS-plant was installed at Hoechst in Frankfurt, and has practically been in continuous operation since its start-up in 1974.This plant, consisting of aCCS-cation contactor and a CCS-anion contactor, with a CO_2-removing device inbetween, produces 200 m3 of water per hour with a conductivity of less than 5 μS/cm from a raw water containing between 6 and 7 eq/m3.

Even if this plant, in the first place, was not designed for a maximum concentration of waste water, it would still approach our concept of an ideal concentration device, with its internal consumption of water of about 3%. These results, combined with the tolerance of the CCS-contactors for suspended particles in the raw water, and the high temperature range of the resin, extend the field of application of ion exchangers when compared to the conventional fixed-bed ion exchange plants and RO-systems considerably.

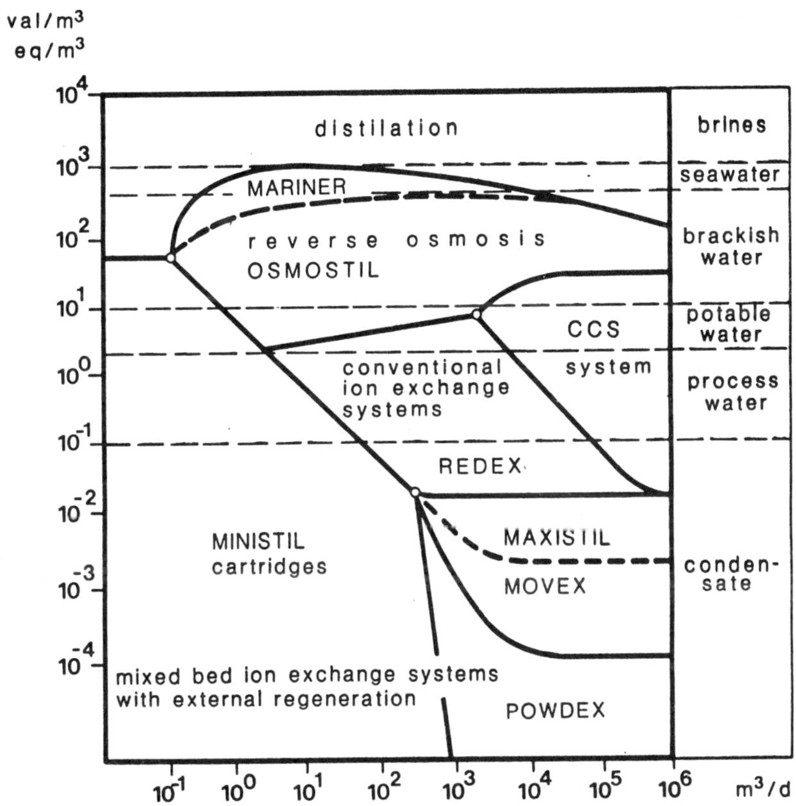
val/m3
eq/m3
10^4
10^3
10^2
10^1
10^0
10^-1
10^-2
10^-3
10^-4
distilation
brines
seawater
MARINER
brackish water
reverse osmosis
OSMOSTIL
potable water
CCS
system
process water
conventional ion exchange systems
REDEX
MAXISTIL
MINISTIL cartridges
condensate
MOVEX
mixed bed ion exchange systems with external regeneration
POWDEX
10^-1
10^0
10^1
10^2
10^3
10^4
10^5
10^6
m3/d

Fig 1

150 m3/h (1"240'000 m3/year), 6 eq/m3 (50% HCO_3; 10% Na)

PROCESS		CCS	RX	RO	RO
variant		1	2	3	4
Plant costs	SFr 10^3	1696	1350	1950	1880
Product water yield	%	97	87	90	75
Degree of desalination	%	99	99	85	95
Demand of surface	m2	50	110	75	70
Demand of space	m3	450	1000	225	210
Chemical consumption:					
- Hydrochloric acid HCl 30%	kg/m3	1,1	1,36	-	-
- Sulphuric acid H_2SO_4 96%	kg/m3	-	-	0,14	0,10
- Sodium hydroxide NaOH 30%	kg/m3	0,93	1,22	-	-
- Hexametaphosphate	kg/m3	-	-	0,01	0,01
Energy consumption	kWh/m3	0,13	0,18	1,20	1,40
Internal water consumption	m3/m3	0,03	0,15	0,11	0,33

Costs per year

Annual instalment 15% (eg.10 years depreciation and 8% interest on the inv.capital)

- 15% of plant costs	SFr 10^3	254	202	291	281
- 15% of space costs (SFr 350/m3)	SFr 10^3	24	53	12	11
Attendance (5 persons)	SFr 10^3	375	375	375	375
Resin loss (7%)	SFr 10^3	6	9	-	-
Membrane replacement (20%)	SFr 10^3	-	-	188	179
Maintenance (3% of the inv.cap.)	SFr 10^3	56	51	83	79

Costs per m3

(150 m3/h, 95% load: 1"240'000 m3/year)

Plant	SFr/m3	0,205	0,163	0,235	0,227
Space requirement	SFr/m3	0,019	0,045	0,010	0,009
Attendance	SFr/m3	0,300	0,300	0,300	0,300
Energy (SFr. -.10/kWh)	SFr/m3	0,013	0,018	0,120	0,140
HCl 30% (SFr 20/100 kg)	SFr/m3	0,200	0,250	-	-
H_2SO_4 96% (SFr 37/100 kg)	SFr/m3	-	-	0,056	0,051
NaOH 30% (SFr 15/100 kg)	SFr/m3	0,140	0,180	-	-
Hexametaphosphate (SFr 3/kg)	SFr/m3	-	-	0,033	0,040
Resin loss	SFr/m3	0,005	0,007	-	-
Membrane replacement	SFr/m3	-	-	0,152	0,144
Maintenance	SFr/m3	0,045	0,041	0,067	0,064
Total costs except water consumpt.	SFr/m3	0,927	1,004	0,973	0,975
with costs for water consumption:					
Raw water costs SFr 0.20/m3	SFr/m3	0,933	1,034	0,995	1,041
Raw water costs SFr 1.00/m3	SFr/m3	0,957	1,154	1,084	1,305
Raw water costs SFr 2.00/m3	SFr/m3	0,987	1,304	1,195	1,635

Fig 2

	RAW WATER COSTS		
	SFr 0.20/m3	SFr 1.00/m3	SFr 2.00/m3
CCS	0,933	0,957	0,987
RX	1,034	1,154	1,304
RO	0,995	1,084	1,195
RO	1,041	1,305	1,635

Fig 3

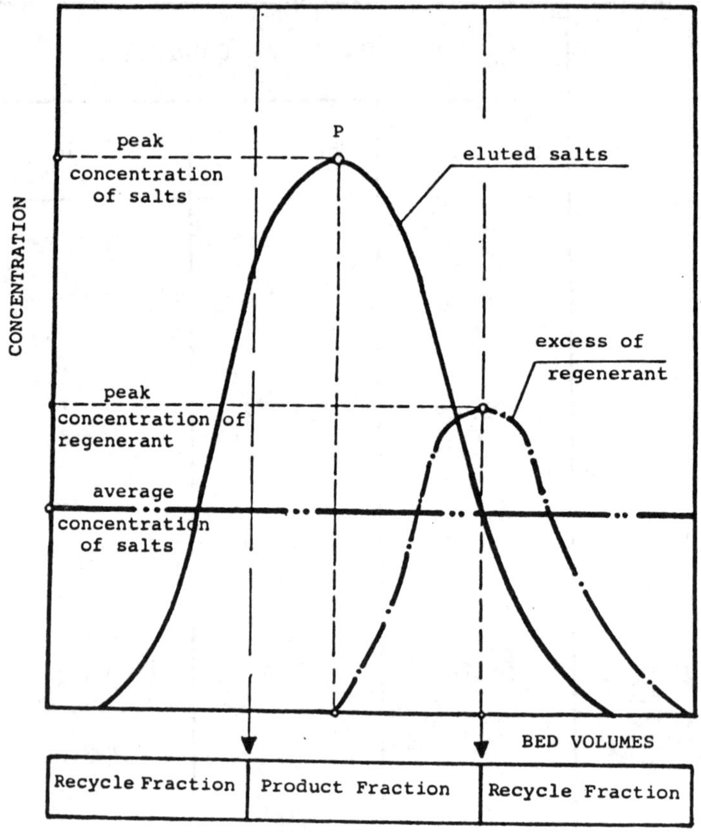
P
peak
concentration
of salts
eluted salts
CONCENTRATION
excess of
regenerant
peak
concentration of
regenerant
average
concentration
of salts
BED VOLUMES
Recycle Fraction
Product Fraction
Recycle Fraction

Fig 4

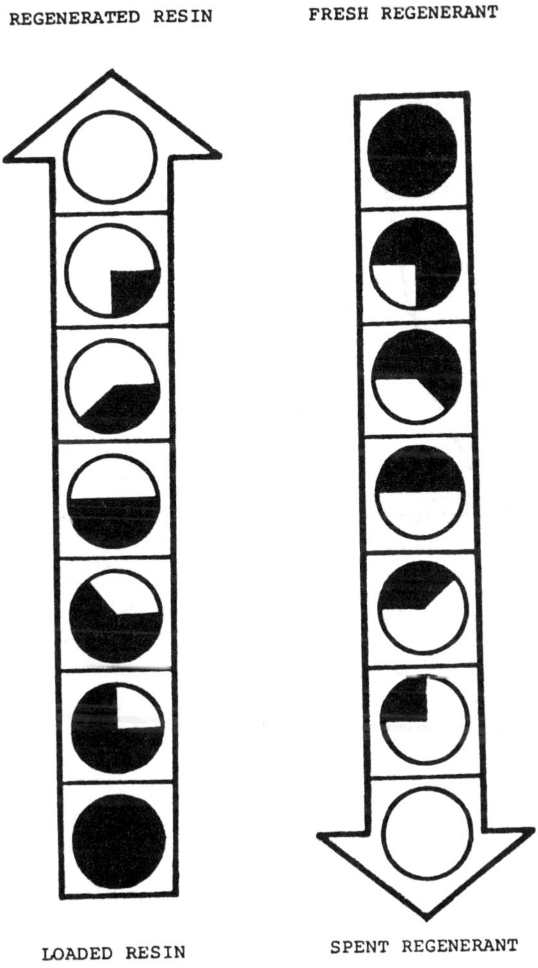
REGENERATED RESIN
FRESH REGENERANT
LOADED RESIN
SPENT REGENERANT

Fig 5

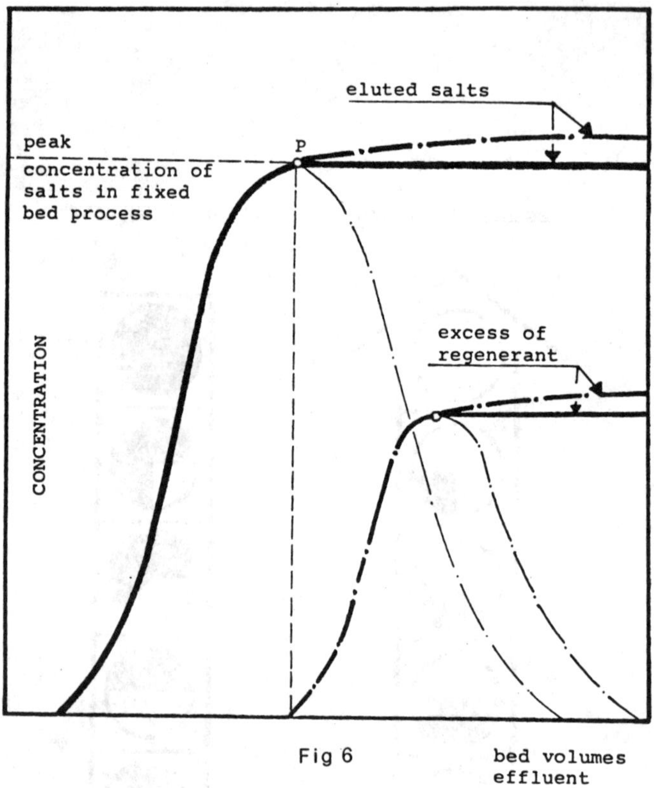

eluted salts
peak
P
concentration of
salts in fixed
bed process
excess of
regenerant
CONCENTRATION
bed volumes
effluent

Fig 6

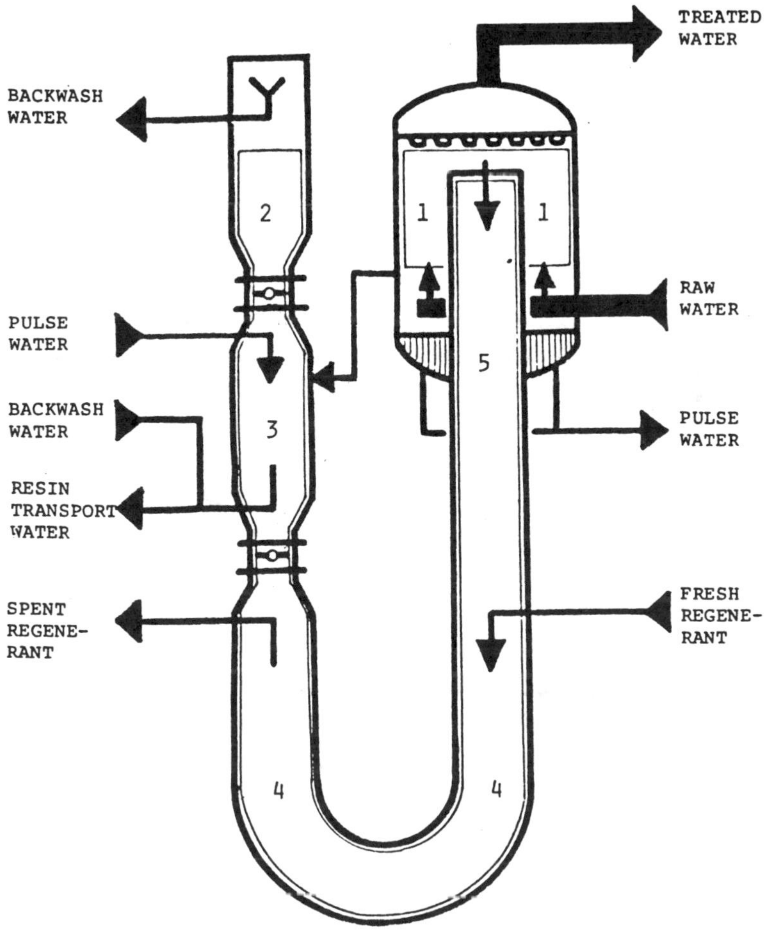
TREATED
WATER
BACKWASH
WATER
2
1
1
RAW
WATER
PULSE
WATER
5
BACKWASH
WATER
3
PULSE
WATER
RESIN
TRANSPORT
WATER
SPENT
REGENE-
RANT
FRESH
REGENE-
RANT
4
4

Fig 7

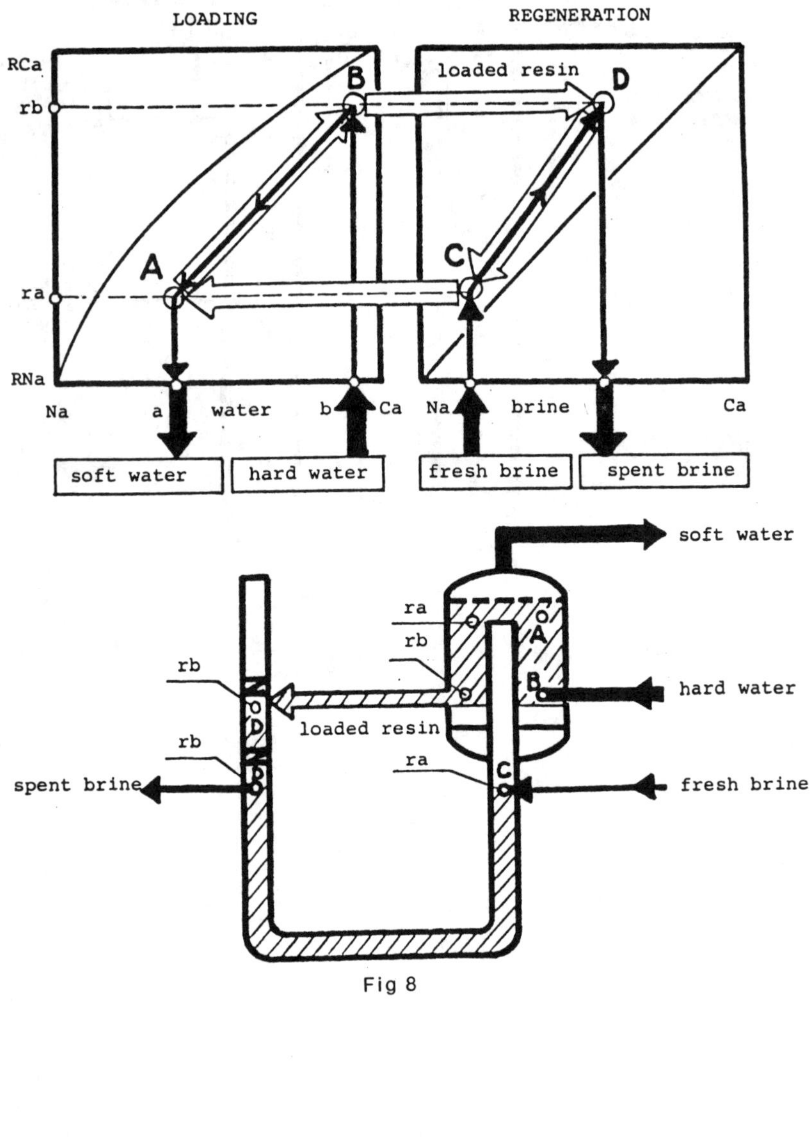
LOADING
REGENERATION
RCa
rb
ra
RNa
A
B
C
D
loaded resin
Na
a
water
b
Ca
Na
brine
Ca
soft water
hard water
fresh brine
spent brine
soft water
ra
rb
A
B
hard water
rb
D
loaded resin
rb
ra
C
spent brine
fresh brine

Fig 8

OUTLOOK AND DESIRES OF ION EXCHANGE RESIN TECHNOLOGY DEVELOPMENT IN THE NUCLEAR FIELD

F.L.Rizzo

Ansaldo Impianti,La Spezia,Italy

INTRODUCTION

Due to the radioactivity discharge problems in nuclear power plants,any method allowing to mantain the process water characteristics,at the same time minimizing the water discharge,would result in a considerable improvement for the cost-benefit analysis.

For powdex condensate systems,water quantity required for convetional backwash operation is approx.50 m^3 (0.36 m^3/ m^2 for an exchange surface of 140 m^2).For a mixed bed condensate system the equivalent discharge is 250 m^3.

In order to reduce these quantities,different methods have been proposed:

- Powdex systems:'backwash recovery system',i.e.,to treat the filter backwash by an exhaust powdex and to recycle this water;
- Polishing systems:a) to recover and store part of post-regeneration water,to be reused for the backwash operation before the following regeneration

  b) to recover into the hot well the waters of resin transfer to the service station and the pre-service washing waters.

Both methods improve the situation of the radwaste system, although the plant may become overloaded from both an economical and operation point of view.

PURITY LEVEL OF WATERS

To understand how the water chemistry may influence the nuclear plant radiological activities, it may suffice to remember that the pollution level on the piping surfaces is directly proportional to the main concentration of heavy metals in the waters, either under ionic or particle form. This concentration shall be maintained at a level of some ppb and a few fraction of μ Siemens in normal operation conditions to guarantee an acceptable ambient activity increase.

According to experience, the yearly increase attains 150 μmR/h for plants whose water is kept at about 15 ppb with a conductivity below 1 μS/cm. This rate can hardly be ensured, particularly on account of the fact that even the smallest transient produces peaks exceeding normal values by 20 to 30 times. The feed water purity level, should be brought to values below 1 ppb. However, this can hardly be obtained for the time being, mainly on account of all draining lines flowing together downstream the cycle water of the treatment due to thermal reasons.

Some washing techniques now available allow to remove pollution from lines thus preventing its exponential increase in time.

However, even if these washing systems take away the crud deposits, they also reactivate surfaces, which become more active sources of metal ions, thus canceling the benefits of passivation attained in normal operating conditions.

The quantity analysis to ascertain convenience of such treatments is, therefore, difficult and uncertain and not so widely used so far.

The treatment systems involve, in addition, transfer and storage of suspensions, whose decantation form highly radioactive cakes and gives therefore rise to a large number of objections under the plant construction point of view. To appraise influence upon design, it will suffice to think that suspensions must be transferred through constant grade pipings to prevent local storage pockets and that pipings must be laid into special screening concrete trenches.

Concerning the treatment of wastes, the water evaporation systems must be manufactured with special alloys, such as incoloy or inconel, to withstand corrosion in the ambient conditions in which they are to operate.

Likewise, all the filtering and cementation systems must be automatized as much as possible and provided with duplications and separations of components not to stop operation in case of maintenance. So, huge costs are to be expected for these plants.

On the other hand, good quality of waters is the only guarantee, particularly in BWR reactors, for low activity levels and, to facilitate management and maintenance operations, the only real condition permitting to comply with the so called "ALARA" concept for the whole plant.

TRENDS IN THE BWR AND PWR DESIGN

Let us give now some activity rates that we have accertained in the potentially active systems in commercial plants of 1000 MW. Table 1 gives the activity rates of a BWR, table 2 those of a PWR. Table 3 gives for both types an appraisal of the main quantities of resins contained and the design quantities discharged every year from plants of this kind.
It appears that quantities are such as to pose the question if regeneration may be convenient or not, considering that the value of a regeneration plant may be compared to the cost of the resins discharged in a decade.

TABLE 1 - BWR

	Ident. letters	Design capacity		Nº	Type	Inlet activity	Outlet activity
Treatment system for reactor primary water	RWCU	GPM "	125 250	2	powdex (1) (2)	2.04 μCi/cc 3.06 "	0.27 μCi/cc 0.31 "
Treatment system for fuel pool water	FPFD	"	1650	2	powdex	0.045 "	0.005 "
Demineralization syst. for recovered waters	WD	"	500	3	L.M. (3) (4)	1.1 E-01 " 2 E-04 "	1.1 E-03 " 2 E-06 "
Filtration system for active cakes	TBF	"	250	3	L. TD		1.1 E-03 "
Main condensate treatment system	TC	"	26000	6	LM	0.03 "	0.0035 "

1) With RWCU flowrate equal to 1% of feedwater flow
2) With RWCU flowrate equal to 2% of feedwater flow
3) Filtrate from travelling belt filters
4) Distillate from evaporators

TABLE 2 - PWR

	Ident. letters	Flowrate	N°	Type	Inlet activity	Outlet activity
Primary circulation water treatment system	CVCS	75 GPM (nor.) 120 " (max.)	2 1	LM CATIONIC	1.4 µCi/cc 0.5 "	0.5 µCi/cc 0.15 "
Boron regeneration system	BTRS	75 "	4	LM	"	"
Boron treatment system	BRS	120 " 35 "	2 1	LM ANIONIC	1.4 "	0.15 "
Recovery water treatment water	RW	100 "	2	LM	0.0001"	0.00001 "
Pool water treatment system	SFP	265 "	2	LM	0.001 "	0.0001 "

TABLE 3

Reactor	System	Contained resins	Discharged resins	Type
BWR	RWCU	80 kg *	5100 litri/anno	powders
	FPFD	220 kg *	4000 litri/anno	"
	RWFD	20428 litri	7000 "	granulated
	TBF	variabile	13800 "	diatom earths
	TC	78400 litri	26000 "	granulated
PWR	CVCS	2550 litri	1700 "	granulated
	BTRS	10188 "	2550 "	"
	BRS	2550 "	2550 "	"
	RW	1700 "	2550 "	"
	SFP	1700 "	1700 "	"

* Value for system designed at 2% FW flow

ECONOMICAL ASPECTS

In the BWR, the trend is to use pulverized resins for both the reactor and fuel pool water purifying systems, so that no regeneration is made.
In the condensate and radwaste treatment systems, where mixed beds are used, resort to regeneration and evaporation is, instead, preferred.
Conversely, in the PWR, the tendency is to reject always, perhaps on account of a more difficult situation due to the presence of boron.
As it appears, by summing up the values of table 3, the radioactive resin consumption/year is about 17,000 litres for the BWR, excluding the diatom earths of the travelling belt filters. In order to obtain this value, it should be taken into account that in normal operation the amount of resins actually discharged from RWCJ and FPFD systems is the one shown in table 3, while the discharge from RWD and TC systems is only one quarter of the shown design value.
Conversely, for the PWR, the radioactive resin consumption/year value obtained is about 11,000 litres.
The difference is due to the fact that the resins from the cycle condensate treatment are usually conveyed to the radwaste treatment, in the BWR, and only exceptionally, in case of losses, to the steam generator in the PWR. From this point of view, the management costs are penalized in the ratio of about 1.5 to 1.
However, it is worthwhile to remember in this connection that, in plants of this size, where the costs for production shortages exceed one billion/day, such a penalization is almost negligible.
On the other hand, the BWR penalization (compared with PWR) arising from the need of conveying all the resins to the radwaste system is compensated by less problems in the condensate treatment system and/or by a longer duration of the demineralizers operating cycles.
This situation is originated by the particular BWR water chemistry, which doesnt't require any chemical conditioner. In this case, the condensate treatment system operates in a neutral environment (pH = 6.6 ÷ 7.5) and therefore it can easily maintain the required chloride (200 ppb) and silica (5 ppb) concentrations inside the reactor water.
Conversely, the PWR water chemistry needs the utilization of conditioners like ammonia and hydrazine, in order to rize pH up to 8.8 ÷ 9.2. All that imples the need of keeping the sodium exhaust

cationic resin active groups under 0.3% during the regenerating phase, contemporarly regenerating 95.5% of the anionic resin active groups under OH form.
Since the sodium is always the first element released by the spent resins, whatever is the cause of depletion, it results clear that, if cationic resins are under hydrogenous form, it will be necessary to regenerate them more frequently because of the ammonia depletion. All that implies the use of highly sophisticated techniques, in order to obtain the whole resins regeneration, including both cationic and anionic resins. This result can be obtained by two methods:

1) Displacement of the sodium fixed on the cationic resin by another cation (Calex and Ammonex processes);

2) Improvement of the cationic and anionic resin separation (Trisbed, Ambersep and Seprex methods).

But naturally, all these technologies penalize the chemistry of PWR cycle water more than a greater quantity of resins.

POSSIBLE IMPROVEMENTS

The possible improvements on these plants are:

1) Treatment system very flexible in flowrate in order to enhance treatment to a by far larger extent during the start-up stage, when the quantity of corrosion products is greater, including drains (not treated in normal conditions) and prevent accumulation of corrosion products in the reactor.

2) Systems permitting to recirculate the same waters repeatedly.

3) Use of anticorrosion additives acceptable for reactors.

4) Methods for re-utilization and storage of the waters used for transfer of resins during the regeneration stage.

5) Large use of stainless materials or alloys not containing cobalt (in the place, for instance, of stellite) or other elements which may generate long life radionuclides.

6) And, first of all, introduction of mechanical or magnetic systems as pre-filters. These systems allow, in fact, for quick regeneration without increasing plant complexity, and for reduction of filtering load and pollution of resins downstream.

I should like to point out, in this connection, that in the Alto Lazio power plant, the heat cycle will be so arranged that a part of the water separated in the separators/heaters is conveyed back to the nuclear boiler without crossing the condenser , i.e. without going through the condensate treatment system; likewise, the condensates of the last H.P. preheater will be sent to the reactor without any previous treatment. This means that about 32% of the total circulating water flowrate does not undergo treatment at powers exceeding 70%.

On account of the high temperatures involved, an improvement might be obtained by installing an electromagnetic ball filter on the pumping forward line of drains.

In fact, according to some studies on this subject, the iron oxydes at the temperature in question (200°C) consist mainly of magnetite or γ-ferrite having both a ferromagnetic behaviour.

The ball filter was developed by KWU and has been used already in both the conventional and nuclear power plants.

It consists of an austenitic steel reservoir filled with balls of steel or other magnetic material. The height of the ball layer is about 1 metre. The filter is surrounded by copper wire coils fed with rectified three-phase current. The separation factor depends to a large extent upon the magnetic field intensity and upon the inlet concentration.

It decreases slowly, as the inlet concentration is reduced, from 98% / $Ci > 2$ ppm to about 80% / $Ci < 0.02$ ppm.

Other variables influencing the separation factor are the filtration temperature and velocity. The higher is the temperature the greater is the separation factor.

To clean the filter, it will suffice to stop inlet of water for a while, demagnetize the balls for a few seconds and rinse.

The whole operation will take no more than one minute.

Electromagnetic filters are available having capacities ranging between 100 and 1200 t/h.

I want to emphasize two very important facts, in this connection:

a) On account of the very short regeneration time, no stand-by unit is required. In fact, treatment can be suspended for 60 seconds.

b) The treated water can be used also for regeneration purposes.

These filters are highly efficient even in case of removal of non magnetic metal impurities. This may be explained by the fact that under the action of the axial magnetic field, the filter

balls become small magnets with their North/South axes in the same direction of the water flow to be treated. The caught impurities tend to settle along the lines of force of the small magnets thus forming porous interstices like the bristles of a brush whic ensure a sort of mechanical filtration also for non-magnetic metal particles.
The filters are only one of the alternative methods we may resort to for plants which cannot use resins on account of their ranges of working temperatures.
With a view to reducing the installation costs, it might often be necessary to renounce to the circuit portion having the task of reducing water temperature for demineralization purposes and of increasing it before mixing the demineralized water with the cycle water in order not to impair the total heat cycle efficiency.
To operate the BWR primary water demineralization system, whose value is in the range of some billions, a sophisticated system is required upstream, including canned pumps operating at temperatures nearing the saturation temperature and regeneration and non-regeneration heat exchangers (duplicated and separated from each other) and designed for seismic conditions with sophisticated quality controls. The composite drawings may give an idea of the complexity of such a system. Its value exceeds by far the one of the demineralization system it serves.
If the resins had the possibility of working at temperatures of about 250°C all this might be avoided.
However, we can but rely on chemists hoping that they can develop new sophisticated resins capable to operate to higher temperatures.
Waiting for development of the high temperature resin technology, I think that only the prefilters can give a valid contribution.
For additional information on this subject, reference should be made to the studies made by A.I. in the scope of the R&S contract between A.I. and CNEN for industrial promotion.
These methods, such as the magnetic filters, have not a very high efficiency but they allow to treat considerable flowrates so that a compensation exists between treated percentages and efficiencies.
Another objective to be hoped might be the development of regeneration methods for powdered resins which on account of their double function of filters and demineralizers are very advantageous for purification of low ionic contents waters such as those of the large open pools.
I think that new techniques should be studied and verified, using mixed inverse osmosis and resin systems. In fact, the cost of

the osmosis system is independent, if the suitable membrane is chosen, from the salt contents and it has not a management cost linked to the regeneration frequency of resins. This allow to reduce the quantities of water to be evaporated and cemented as well as to spare resins.
Even if their introduction into the plant cycle requires additional studies and tests, I think that the mixed systems in question may be profitable solutions to be associated with the demineralizers used at present.

CONCLUSIONS

So far, I have spoken about what we can do now to improve performance of the purification systems in the power reactors, but what can be said for the future? Even assuming that nuclear development in Italy may be accomplished in the next two decades, I do not believe that the commercial plants will be different from the water thermal ones. Therefore, keeping into account the outputs of these machines and that construction of one of them would take at least 10 years, I think that, in consideration of the economical and technical resources of our country, no more than ten plants may be built in the next twenty years.
This means that this technology is mainly a test bench for the resin producers as concerns quality and capability of production rather than a desirable supply market.
As our task on this team of experts can only be that of a child playing with complex toys without realizing how they are made inside, I resume in brief our wishes: "Give us systems capable of keeping our waters below the metal contents of one ppb, producing a small quantity of rejection water and resins, requiring small maintenance and suitable to operate up to 250°C, if possible!"

PURIFICATION OF EFFLUENTS OF ACID COPPER AND NICKEL PLATING GALVANIC PROCESSES WITH CONVENTIONAL CATION EXCHANGE RESINS. COPPER AND NICKEL RECOVERY.

Roberto Stortini

Ausind, Montedison Group

ABSTRACT

The paper discusses the use of cation exchange resins in depuration processes of dynamic rinse water from acid copper and nickel electroplating.
Operations are performed by isolating rinse water from each plating process. The purpose is to obtain by regeneration of the exhausted exchanger pure and concentrated solutions of one metallic ion (Cu^{++} or Ni^{++}) which may be recycled into the main electroplating tank following a slight correction of properties and thus recovered with financial advantages.
The formation of slurry as well as involved environmental and financial inconveniences are thus eliminated. The use of weakly acid (carboxylic) and strongly acid (sulphonic) exchangers is compared and advantages and disadvantages thereof are outlined.
The use of cationic sulphonic exchangers, resulting from countercurrent regeneration, appears to be more rational in consideration of related opportunities in the production of de-cationized water.

INTRODUCTION

The disposal of sludges resulting from the purification of galvanic effluents involves environmental and economic drawbacks :

- environmental due to the dangers caused by the possible limited dissolution of sludges

- economic due to the value of the products present in sludges
The problem is the same whether sludges derive from batch waste treatments or from regeneration eluates in recycle demineralization plants. In the latter case, the rinse water is almost totally recovered.
The most convenient solution of the problem consists in the prevention of sludge formation, and involves the adoption of processes making it possible to recover concentrated solutions of processing products.
The composition of these recovery solutions is very similar to that of the plating tank where they may be recycled after appropriate correction.
This is possible, but it requires the rinse water from different processes of the same galvanic plant be individually segregated.
The following is a comparative analysis of results obtainable in the recovery of Cu^{++} (Ni^{++}) from acid copper (nickel) plating rinse water by using the following conventional cation exchangers :
- weakly acid, carboxylic type
- strongly acid; sulphonic type

WEAKLY ACID CARBOXYLIC CATION EXCAHNGERS

Because of their low acidity, carboxylic exchangers cannot decationize acid solutions such as those resulting from the dynamic washing of galvanized products submitted to acid copper (nickel) plating, or in any case exchange cations from neutral salts in hydrogen cycle.
Consequenlty, before use, the exchanger has to be partially conditioned in the salt form by means of NaOH.
The polluting ion removal from the acid solutions resulting from dynamic rinse baths is performed through the following reaction :

$$\begin{matrix} RCOO^-Na^{+} \\ RCOO^-Na^{+} \end{matrix} + Me^{++} = \left\{ \begin{matrix} RCOO^- \\ RCOO^- \end{matrix} Me^{++} \right\} + 2\,Na^+$$

For the same reason mentioned above, the pH value of treated solutions must range between 4 and 5. Under those conditions, exchange capacity are used in the order of 3 equivalents per liter of carboxylic resin. Such results are possible when operating according to the "merry-go-round" technique.
The exhausted exchanger regeneration is performed by means of concentrated acid solutrions (ex., 15% H_2SO_4), and makes it possible to obtain eluates that may be easily used again in the plating tank

after some small composition adjustments, if required.
The advantages resulting from this procedure are the following :

- high exchange capacity and, consequently, long operating cycle
- high regeneration efficiency -a typical feature of the exchanger. Because of this reason, concentrated ion solutions may be obtained with a slight excess of free mineral acidity- no special technique is required.

The fact is to be recalled that if the carboxylic exchange resin system is the only one provided for the waste water treatment, at each cycle the solution from the dynamic rinsing leads an increased Na^+ ion concentration equivalent to the amount of removed Me^{++} ions. Therefore, it is necessary to design the recycling system so that there is a continuous blowdown and softened water make-up in order to prevent sodium salinity build-up.
This problem does not arise when a central deionization system is provided supplying various users with deionized water.
In this case, effluents from carboxylic filters flow to the central deionization system.

SULPHONIC CATIONIC EXCHANGERS

The solutions to the problem of sludge disposal according to the previously mentioned terms of reference may also be ensured through the use of sulphonic cation exchangers operating in H^+ cycle. In this case, the effluent water from the filter is decationized; therefore, an anion filter is sufficient to obtain deionized water which may be recycled to dynamic rinsing. Where a central deionization system is in operation, decationized water may flow to it as it does not increase the cation load, and be recycled to dynamic rinsing, in this case too.
With the sulphonic exchange, it is not possible to have as high operating capacities as with the carboxylic exchanger due also to a lower regeneration efficiency.
Therefore, in order to obtain solutions that may be directly recycled, after a small adjustment of their characteristics, if required, regeneration techniques have to be adopted allowing for maximum efficiency and requiring more accurate technologies with a consequent constraint on operating capacity, as in the case of countercurrent regeneration. Results are here reviewed following countercurrent regeneration with low regeneration and high efficiency. The optimal level is considered 50 g H_2SO_4 (100%) per liter of exchanger The resulting exchange capacity is equivalent to 25 g Cu^{++} per li-

ter of exchanger.
With regeneration levels exceeding 50 g H_2SO_4 (100%) the recovered effluent fraction contains residual free acid in a too high concentration, and recovery may be less easy, mainly in the case of acid copper plating.
With lower regeneration levels a better regeneration efficiency and less free acidity in the recovery fraction are obtained. In the latter case, the recovery fraction would be too limited to make and adequate selection of the regeneration eluate possible.
Graph 1 shows the exchange capacity pattern according to different regeneration levels.
Graph 2 shows the typical pattern of countercurrent regeneration with a regeneration level equivalent to 50 g H_2SO_4 (100%) per liter of exchanger in a 15% solution.
The exchanger in the H^+ form is exhausted with a solution having a composition similar to the one of the effluent solution from a dynamic rinsing of galvanized products submitted to acid copper plating.
The composition of this solution is the following :
$CuSO_4$ = 2 meq/l
H_2SO_4 = 0.2 meq/l
corresponding to the composition of the effluent from a dynamic rinsing following an adequate stagnant rinsing.
The exhaustion process is interrupted when in the effluent water the Cu^{++} concentration reach the value of 0.1 ppm.
Cu^{++} concentration in the recovered fraction is equivalent to 47 g/l, while free acidity is equivalent to 17 g H_2SO_4/l.
The recovery fraction is collected during the slow washing phase, and is equal to 0.5 bed-volumes.
Copper in non recovered effluent fractions is equivalent to 4% of used exchange capacity.
These fractions are recycled in the equalization tank placed before the exchanger.
Even in the case of acid nickel plating an adjustment may be required in the composition of the regeneration effluent fraction recovered before it is recycled in the plating tank.
Required adjustments may essentially concern the pH value.
If the pH value is lower than standard, it is adjusted through the addition of Ni^{++} basic carbonate.
More specifically, in the process of reelution of nickel from the exhausted resin, it is possible to use regeneration levels exeecing the optimal value for acid copper plating.
This procedure makes it possible to operate at higher capacity, and to recover a larger fraction of the regeneration effluent.

Table 1

Typical composition of acid copper and acid nickel platings

Copper sulfate baths	$CuSO_4 \cdot 5H_2O$	:	200–250 g/l
	H_2SO_4	:	10–65 "
	Cu^{++}	:	50–65 "
	pH	:	0–1
	Addition agents	:	according to formulation
Nickel sulfate baths	$NiSO_4 \cdot 6H_2O$	:	330 g/l
	$NiCl_2 \cdot 6H_2O$	:	45 "
	H_3BO_3	:	37 "
	Ni^{++}	:	85 "
	Addition agents	:	according to formulation

Fig. 1

VARIATION OF THE EXCHANGE CAPACITY WITH REGENERANT LEVEL

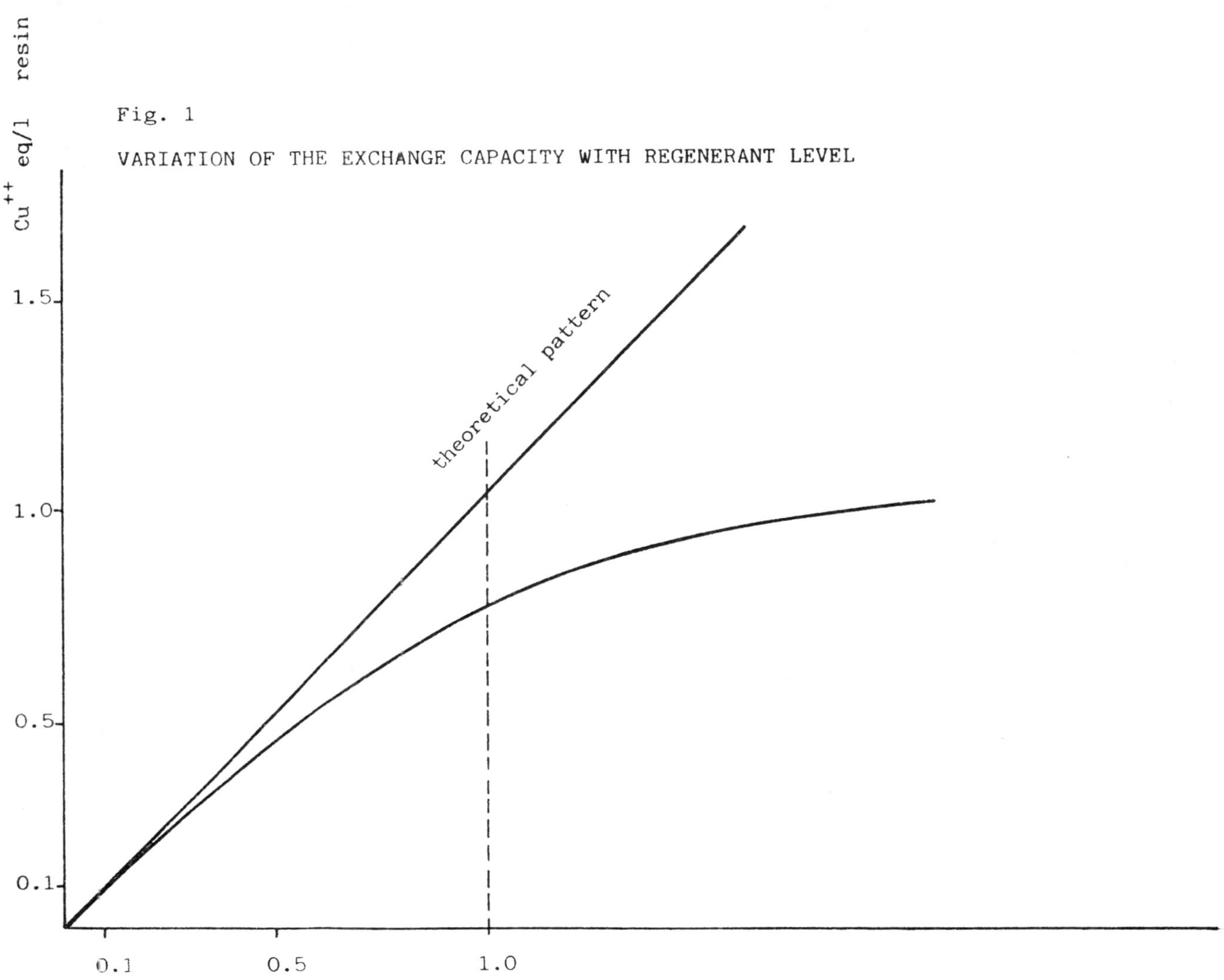

Fig. 2

COPPER ELUTION BY COUNTERFLOW REGENERATION

REGENERANT LEVEL : 50 g H_2SO_4 (100%) PER LITER OF RESIN

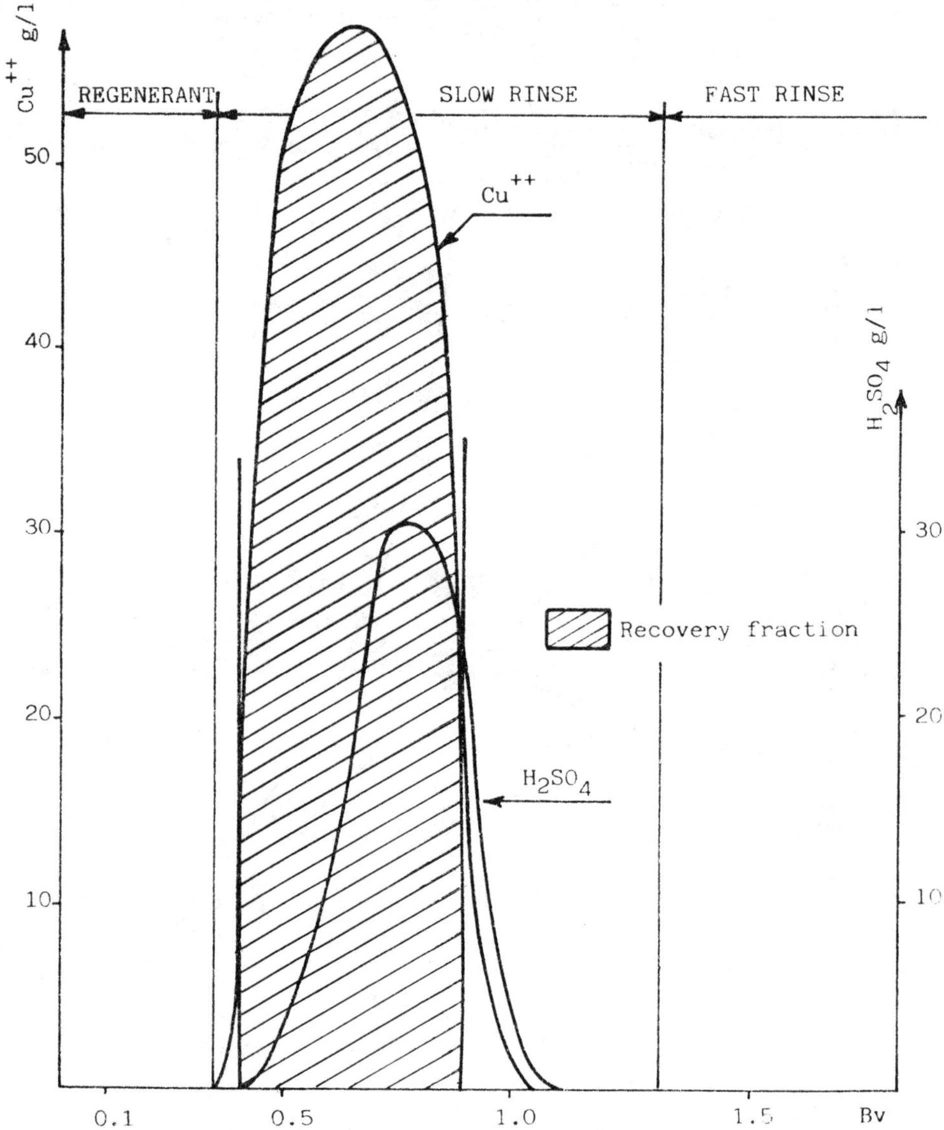

ANION RESIN KINETICS AT HIGH FLOW RATES

J.R. EMMETT

Consultant

INTRODUCTION

The mixed bed deionisation process is extensively used for the final purification of boiler feed water in modern power stations. This application is known as Condensate Polishing and the process is characterised by low influent impurities, a desire for the maximum effluent purity and the use of very high flow rates. Under these conditions the lifetime of cation resins, particularly macroreticular cation resins, creates no difficulties but anion resin performance gradually deteriorates and replacement is required at approximately four to five year intervals.

The loss of anion resin performance not only involves a gradual loss of capacity but also a fall-off in kinetic performance. A simple test is therefore required which is pertinent to mixed bed column conditions, not only to enable the resin with best kinetic ability to be selected in the first place, but also to judge when the resin performance has deteriorated to a point where it cannot be relied upon to give the required degree of protection to boiler plant should there be a sudden ingress of impurities into the feed system.

With this objective in mind, pioneering work was carried out by the Central Electricity Generating Board, Midland Region, Scientific Services Department. Their findings were published at the International Water Conference Pittsburgh 1981 (1). The results indicated that there was an appreciable difference between the anion resins tested and it was noted that a resin giving good kinetic performance was one specially graded for use with three component mixed bed systems using an intermediate inert layer for which the larger bead size anion resin had been removed. With the original work, resin particle size was not a control parameter. As a consequence, the Spring House Research Laboratory of Rohm and Haas commenced a programme of work on similar lines to the original work but with both cation and anion resin particle size as strictly controlled parameters.

TEST PROCEDURE

Resin Preparation

A wide spectrum of commercial resins which included both gel and macroreticular was used in the test programme. All resins were given three regenerations, rinse and exhaustion cycles prior to final regeneration and rinse which was performed immediately before testing. During the conditioning cycles exhaustion was performed with two bed volumes of saturated sodium chloride at a flow rate of 4 BV/h. The rinse was with deionised water at 15 BV/h to a conductivity of 10 μS/cm.

The tests were carried out under mixed bed conditions and therefore the anion resin was mixed with an appropriate cation resin using a 2:1 cation-anion ratio with a total volume of 100 ml. To test anion resin kinetics under mixed bed conditions was considered important as the concentration per unit volume of OH^- sites on the anion resin is diluted by the cation resin. Conversely, the concentration of H^+ sites on the cation resin is diluted by the anion resin.

The regeneration of individual resins immediately prior to a test was carried out as follows :

Anion, 80 g/litre sodium hydroxide as 4% W/V solution

Cation, 80 g/litre sulphuric acid as 5% W/V solution

The regenerant flow rate was 5 BV/h in each case.

The above regenerant levels would impart the following estimated capacities to the resins used.

		Eq/litre
Cation 12% DVB	MR	1.036
Cation 10% DVB	Gel	1.00
Anion	MR	0.44
Anion	Gel	0.55

In broad terms, the hydrogen ion concentration in the test column, taking into consideration the 2:1 cation-anion mix, is of the order of 0.68 eq/litre and the hydroxyl ion concentration is of the order of 0.16 eq/litre.

Resin Particle Size Analysis

A complete particle size distribution analysis was determined on both the cation and anion resin used in each test. This was accomplished by means of an Hiac automatic particle size analyser. From the results of the size distribution analysis the harmonic mean size was calculated.

Test Rig

The flow diagram of the test rig is shown diagrammatically in Figure 1. Columns 1A, 1B are primary polishing columns for the laboratory deionised water while columns 2A and 2B are secondary polishing columns. This initial polishing of the laboratory deionised water ensured that the water for test purposes was of high purity with a conductivity always less than 0.059 µS/cm.

Column 3 accomodates the resin under test, the dimensions of the column being :

26 mm internal diameter

300 mm long

The column was jacketed and of borosilicate glass construction.

The preparation of the feed to the test column was carried out by metering a stock solution of sodium chloride (100 mg/litre as Cl) into the deionised water to give a 10 mg/litre chloride concentration entering the test column.

The total effluent flow passed through a conductivity cell and was then divided equally to the chloride and sodium analysers, details of which are as follows :

Chloride :	Orion Research low level chloride monitor, range 0 - 50 µg/litre Cl. Model 151701.
Sodium :	Orion Research Model 151102 Range 0.1 to 1000 µg/litre Na.

Conductivity : Foxboro 9210 conductivity monitor using 921 titanium conductivity cell. This instrument incorporated full temperature compensation.

Recording was incorporated for all instruments.

Test Operation

The conditioned and regenerated cation and anion components for any given test were mixed together and then rapidly introduced into the test column while draining from the base . This gave a good mixed bed.

The bed was then rinsed with high quality, polished, deionised water (conductivity $<$ 0.059) for 30 minutes at the test flow rate, prior to the introduction of the chloride feed. The conductivity at the end of the rinse period was noted and designated C T = 0.

The flow rate conditions were 200 ml/min resulting in the following resin specific flow rates :

120 BV/h to the whole bed

180 BV/h to the cation resin component

360 BV/h to the anion resin component

After the 30 minutes rinse period the sodium chloride injection into the feed system was commenced to give the 10 mg/litre influent chloride concentration.

Both the flow rate and the feed concentration were held constant throughout each test which was terminated when the chloride leakage reached 50 µg/litre as Cl.

Replicate analyses were run on several samples to indicate the reproducibility of the test procedure.

The tests were carried out at laboratory ambient temperature of approximately 20°C.

TEST RESULTS

The instruments incorporated in the test rig were all equipped with recording facilities. However, the readings of the three instruments were noted 15 and 30 minutes after the commencement of sodium chloride injection. These readings are given in Table 1. The average harmonic mean particle size for the combined cation and anion resin has been calculated from the two individual size analyses taking into consideration the cation-anion ratio of 2:1. Figure 2 gives a graphical presentation of the chloride leakage for the various resins tested. The curves are identified by their respective test number. The letter G signifies that the resin under test was of a gel structure. The remaining resins were of macroreticular structure.

EXAMINATION OF RESULTS

Since one of the main objects of the tests was to study the influence of anion resin particle size in relation to kinetic performance, it is logical to examine this aspect first. Figure 3 graphically illustrates the results obtained showing chloride leakage as a function of anion resin particle size. The linear relationship shown has been established by the root mean square method. Test 9 was not included in the calculation as the anion resin for this test was regenerated with 160 g/litre NaOH as compared with the standard value for the test series.

However, before accepting the relationship shown in Figure 3 as final, interpretation of the readings from the other instruments has to be taken into consideration. The incorporation of a sodium monitor into the test rig system might have been regarded as unnecessary when the object of the test was to measure anion leakage. In reality, it is now considered that the analytical observations from sodium measurement played an important part in the final interpretation of results and prevented an incorrect conclusion being reached.

Table 2 schedules the μg/litre concentration of chloride and sodium ion leakage in terms of equivalents/litre x 10^{-6}. It will be noted that the magnitude of sodium and chloride leakage is substantially stoichiometrically equal.

The ion exchange reactions which take place in mixed bed deionisation are represented by the following equations :

$$\text{a} \quad RH + Na^+ + Cl^- \rightleftharpoons RNa + H^+ + Cl^- \quad \text{(Eq. 1)}$$

$$\text{b} \quad ROH + H^+ + Cl^- \rightleftharpoons RCl + H_2O$$

$$\text{a}\quad ROH + Na^{+} + Cl^{-} \rightleftharpoons RCl + Na^{+} + OH^{-} \qquad \text{(Eq. 2)}$$

$$\text{b}\quad RH + Na^{+} + OH^{-} \rightleftharpoons RNa + H_2O$$

The anion kinetics which the experiments were designed to study are represented by Eq. 1b and 2a. Incomplete reaction of Eq. 2a would lead to sodium and chloride ion leakage in stoichiometrically equal quantities. However, the large excess of hydrogen sites in the mixed bed make equations 1a and 1b the predominant reactions. The degree of regeneration used on the cation resin was such that the sodium leakage should have been of the order of 1 to 1.5 μg/litre as Na.

Further proof that the leakage experienced was substantially neutral sodium chloride can be obtained from conductivity considerations. Figure 4 shows the calculated conductivity for sodium chloride and hydrochloric acid in the range under consideration.

Test results at 15 minutes for sodium ion leakage were in excellent agreement with the theoretical conductivity of sodium chloride and good agreement was found between chloride ion concentration and sodium chloride conductivity. Some of the measured conductivity readings fell on, or close to, the sodium chloride line. However, some of the measured conductivity points fell closer to the theoretical conductivity for hydrochloric acid indicating a degree of acid leakage. With a correctly working column the measured conductivity should have fallen close to the theoretical line for hydrochloric acid. It is notable that the approach to the acid conductivity occurs more at low leakage levels which correspond to smaller particle size.

In an attempt to explain the leakage of substantially neutral sodium chloride the ionic equilibria appertaining to mixed bed deionisation should be considered. The equilibria concerned are :

Cation resin equilibrium
$$K_{H}^{Na} = \frac{RNa}{RH} \times \frac{(H^{+})}{(Na^{+})} \qquad \text{(Eq. 3)}$$

Anion resin equilibrium
$$K_{OH}^{Cl} = \frac{R(Cl^{-})}{ROH} \times \frac{(OH^{-})}{Cl^{-}} \qquad \text{(Eq. 4)}$$

Dissociation of water
$$Kw = (H^{+})\ (OH^{-})$$

Electro-neutrality
$$(H^{+}) + (Na^{+}) = (OH^{-}) + (Cl^{-})$$

If, due to poor kinetic performance, the equilibrium of Eq. 4 is not achieved then the reaction of equation 1a cannot proceed to completion and residual sodium will be present in solution.

Equation 3 can be rearranged in terms of residual sodium concentration as follows :

$$Na^{+} = \frac{RNa}{RH} \times \frac{(H^{+})}{K_{H}^{Na}} \qquad \text{(Eq. 5)}$$

Under the test conditions using 12% DVB macroreticular resin, 80 g/litre of sulphuric acid regenerant would impart a capacity of approximately 1.036 equivalents, i.e. 55% of the resin sites would be in the hydrogen form and 45% would remain in the sodium form.

The sodium/hydrogen selectivity at extreme dilution has been established as a function of resin composition. The value at 45% sodium was found to be 2.2. Equation 5 can therefore be resolved to :

$$Na^{+} = 0.372\ (H^{+})$$

Clearly, leakage due to limitation in the kinetic performance of an anion resin will not give rise to sodium and chloride ion concentration being stoichiometrically equal.

It appears therefore that ion exchange considerations cannot explain the magnitude of the sodium leakage measured during the test series. It was therefore decided to examine the hydraulic conditions of the test column with specific respect to the wall effect when using a 26 mm diameter test column.

There is always a greater flow down the wall of a column than through the body of the material in the column. For this to be the explanation of sodium chloride leakage, both the cation and anion beads must be fully exhausted at the point of contact with the glass wall of the test column during the first few minutes after the commencement of sodium chloride dosing into the influent.

Confirmation that there was an initial period of undetectable leakage was obtained with Test 9, the chloride leakage of which is shown graphically in Figure 5. It will be noted that the graph lines for both the 80 and 160 g/litre regeneration levels indicated no measurable leakage for the first 3 to 4 minutes. During this period an influent flow of 200 ml/min and a concentration of 10 mg/litre as Cl would consume only 1.1 to 1.5% of the OH capacity imparted to the resin at the lower regeneration level. The commencement of leakage cannot, therefore, be due to the anion resin, as a whole, becoming exhausted.

Further confirmation that the wall effect cannot be neglected was obtained by reference to pressure loss in packed columns established in Chemical Engineering techniques.

Figure 6 shows the correction factor which has to be applied in assessing pressure drop across a packed bed due to the ease at which a fluid can flow adjacent to the wall. The average value of HMS of the combined cation and anion resins given in column 3 Table 1 was 0.777 mm, giving a ratio of particle diameter to tube diameter of 0.03. This requires a correction factor of 0.95 for pressure loss calculations indicating that the wall effect, whilst small, is not negligible.

What appears to be the case is that the development of analytical procedures which are accurate in the microgramme per litre range have revealed a wall effect which hitherto has gone undetected.

If the argument that leakage is a function of wall effect is correct, then the particle size of both cation and anion resin has to be taken into consideration. Figure 7 shows the sodium leakage (plotted as Cl to be comparable with Figure 3) as a function of the HMS of the combined resins. This relationship showed that the sodium leakage based on the combined resin HMS has similar characteristics to chloride leakage based on anion resin HMS. The standard deviations for both relationship are of the same order, 6.3 in the case of chloride leakage vs anion resin HMS and 6.97 for sodium leakage based on the combined HMS for both resins. The major difference is that the leakage curve for sodium leakage is displaced along the particle size scale as compared with the chloride leakage curve.

CONCLUSIONS

Small diameter laboratory test columns give rise to wall effects which create great difficulties when attempting to measure low level leakage with the mixed bed deionisation process. The use of larger diameter columns will obviously reduce the problem which in turn would mean that laboratory equipment would have to be geared up to handle much larger flow rates. An alternative solution may well be to use the smallest bead size for cation resin permissible with pressure considerations in glass columns. Resin bead size of 0.15 to 0.3 mm are available which would substantially reduce the wall effect.

Future work in this area should, as a control parameter, include the degree to which resins under test are regenerated. When examining anion resin kinetics it would be advantageous to fully regenerate the cation resin. This would require a regeneration level of approximately 400 g/litre for resins having up to 12% DVB crosslinking.

For the comparison of new anion resins consideration should be given to working in the milligram per litre range instead of the microgram per litre and compare resins by the slope of the breakthrough curve on exhaustion.

Consideration should be given to the desirability of studying the kinetics of anion acid neutralisation, Eq. 1b and anion salt splitting, Eq. 2a as separate entities. It is anticipated that the kinetics of salt splitting will be slower than the neutralisation reaction. With the condensate polishing process ammonia is used as a feed water conditioning agent to ensure a satisfactory pH value of the feed water in order to minimize corrosion. This is removed by the cation resin. As a consequence, the early stages of a mixed bed unit operation cycle are characterised by a high hydrogen ion content on the cation resin and therefore the anion reaction 1b will be predominant. However, later on in the treatment cycle much of the cation resin will be in the ammonium form and any ingress of impurities would lead to the anion salt splitting reaction, Eq. 2a becoming predominant. Therefore knowledge of the kinetics of both reactions is desirable.

Ray, N.J., Ball, M. and Coates, A.
Kinetic Performance Tests of Anion Exchange Resins.
42nd International Water Conference, Pittsburgh, 1981.

TABLE 1

Test	Resin H.M.S.			Chloride leakage µg/litre as Cl		Sodium leakage µg/litre as Na		Conductivity µS/cm		
N°	cation	anion	mean	T = 15	T = 30	T = 15	T = 30	T = 0	T = 15	T = 30
1	0.832	0.757	0.807	22.8	29.5	15	19.0	0.067	0.145	0.175
2	0.960	0.676	0.865	35.5	38.7	24	31.0	0.058	0.185	0.192
3	0.832	0.757	0.807	39.0	39.0	24	25.0	0.073	0.238	0.233
4	0.832	0.672	0.779	14.0	19.0	10	14.0	0.065	0.119	0.128
5	0.829	0.728	0.795	25.0	28.1	12	14.5	0.060	0.182	0.196
6	0.829	0.651	0.770	14.6	17.0	7	9.8	0.064	0.118	0.123
7	0.960	0.626	0.849	12.1	13.3	11	13.0	0.058	0.110	0.114
8	0.960	0.676	0.865	26.3	29.0	19	22.0	0.058	0.161	0.161
9	0.832	0.757	0.807	9.8	20.7	9	14.0	0.072	0.128	0.164
10	0.832	0.757	0.807	24.5	34.0	15	21.0	0.080	0.164	0.20
11	0.953	0.626	0.844	10.0	12.1	8.3	12.0	0.059	0.103	0.105
12	0.741	0.583	0.688	1.7	3.0	2.2	2.0	0.059	0.068	0.071
13	0.741	0.583	0.688	1.7	2.7	2.1	2.4	0.061	0.068	0.070
14	0.746	0.649	0.714	9.2	10.5	3.8	5.2	0.070	0.103	0.106
15	0.746	0.601	0.698	5.9	6.5	2.6	3.3	0.060	0.088	0.088
16	0.741	0.756	0.746	32.5	35.5	16.0	22.0	0.061	0.222	0.233
17	0.829	0.672	0.777	16.5	18.0	8.0	11.0	0.065	0.130	0.133
18	0.701	0.647	0.683	9.9	10.3	5.3	6.8	0.065	0.108	0.109

TABLE 2

Test N°	Leakage eq/litre x 10^{-6} T = 15		Leakage eq/litre x 10^{-6} T = 30	
	Cl	Na	Cl	Na
1	0.644	0.652	0.832	0.826
2	1.058	1.044	1.091	1.348
3	1.101	1.044	1.099	1.087
4	0.395	0.435	0.536	0.609
5	0.705	0.522	0.793	0.631
6	0.412	0.304	0.479	0.426
7	0.341	0.478	0.375	0.563
8	0.742	0.826	0.818	0.957
9	0.276	0.391	0.584	0.609
10	0.691	0.652	0.959	0.915
11	0.282	0.361	0.341	0.522
12	0.048	0.096	0.085	0.126
13	0.048	0.091	0.076	0.104
14	0.259	0.165	0.296	0.226
15	0.166	0.113	0.183	0.143
16	0.917	0.696	1.0	0.957
17	0.465	0.348	0.508	0.478
18	0.279	0.230	0.290	0.296

KINETICS TEST RIG - FLOW DIAGRAM

FIGURE 1

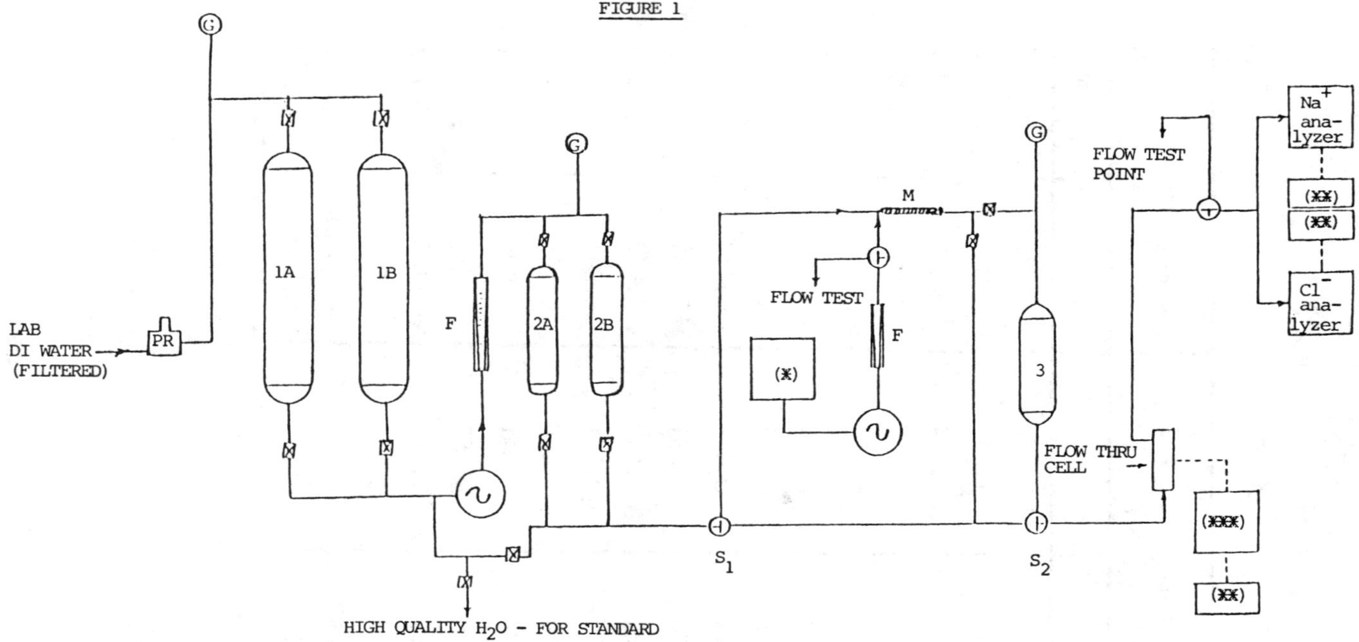

1A & 1B - PRIMARY POLISHING COLUMNS

2A & 2B - SECONDARY POLISHING COLUMNS

3 - RESIN TEST COLUMN

P.R. - PRESSURE REGULATOR

(G) - PRESSURE GAUGE

☒ - BALL VALVE, STAINLESS STEEL

(~) - POSITIVE DISPLACEMENT METERING PUMP

F - FLOW METERING

M - STATUS MIXER

(⊥) - 3-WAY VALVE, STAINLESS STEEL

(⊥) S_1 S_2 - 3-WAY SOLENOID VALVE, TEFLON BODY

(✱) - STOCK SALT SOLUTION BOTTLE - COLLAPSIBLE, <u>NOT</u> OPEN TO ATMOSPHERE

(✱✱) - RECORDER

(✱✱✱) - DIGITAL MΩ MONITOR

FIGURE 2

CHLORIDE LEAKAGE VERSUS TIME

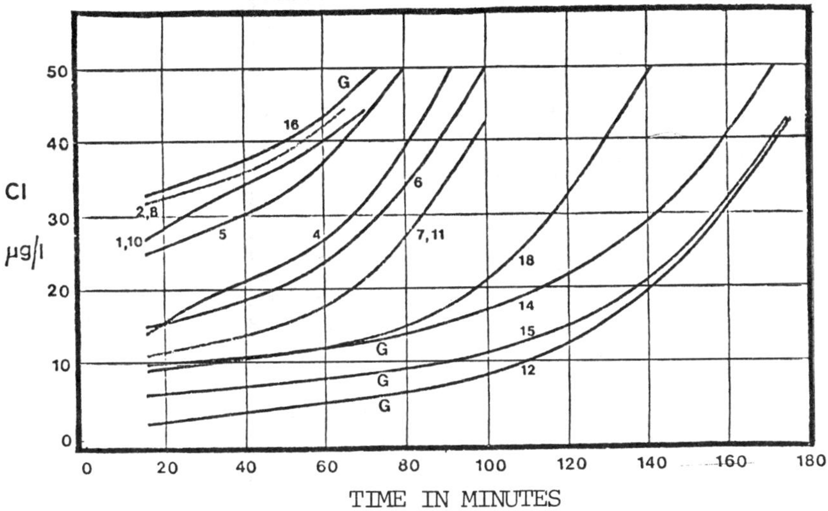

FIGURE 3

CHLORIDE LEAKAGE VS ANION RESIN H.M.S.

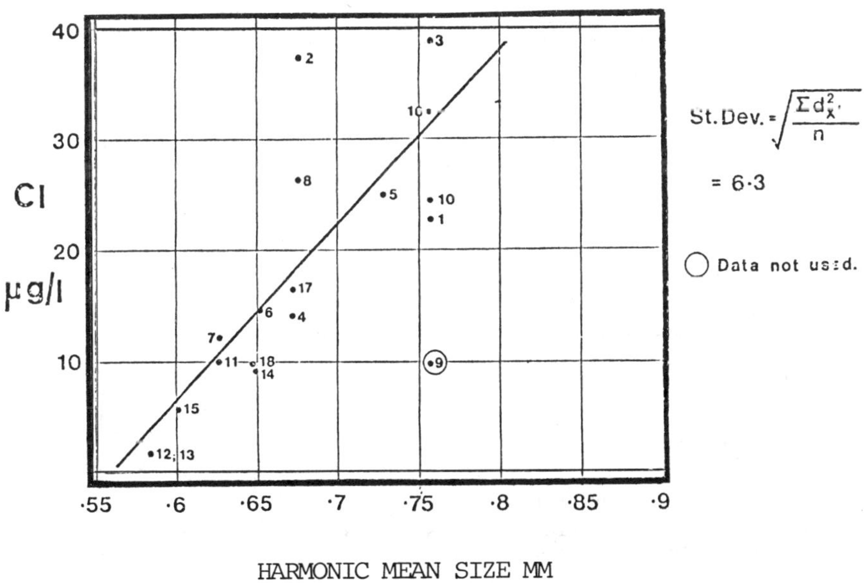

FIGURE 4

CONDUCTIVITY VS LEAKAGE

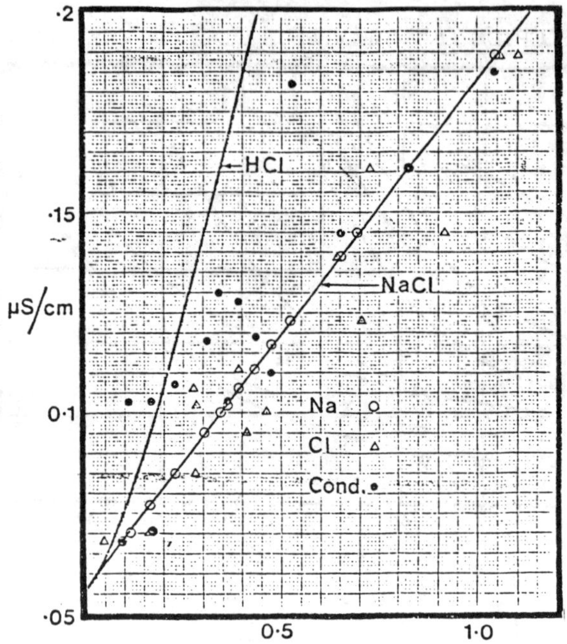

LEAKAGE MICRO EQUIV./LITRE

FIGURE 5

EFFECT OF REGEN. LEVEL ON CHLORIDE LEAKAGE

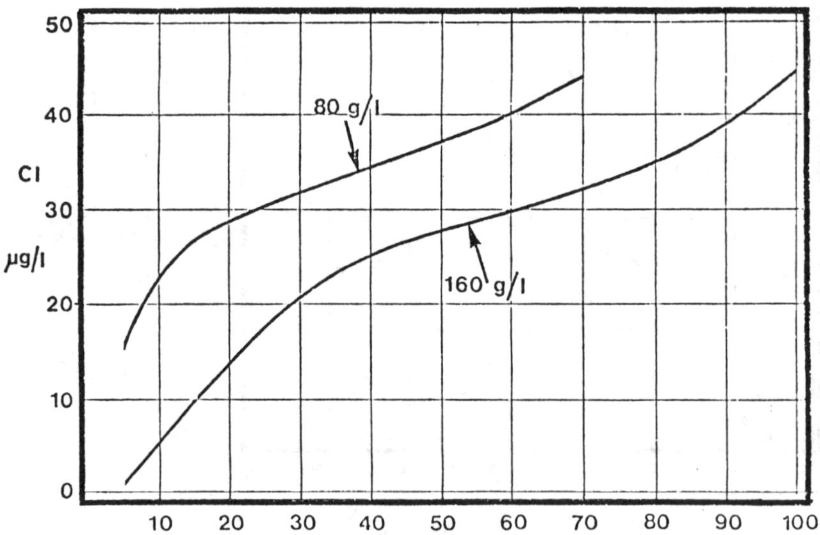

TIME IN MINUTES

FIGURE 6

PRESSURE DROP CORRECTION FACTOR FOR WALL EFFECT

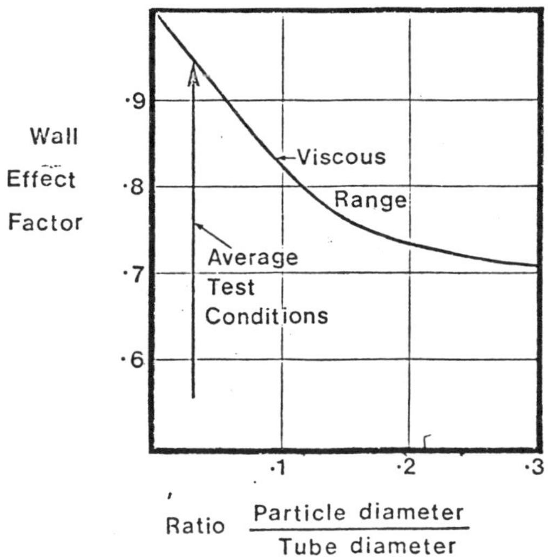

FIGURE 7

SODIUM LEAKAGE VS RESIN COMBINED H.M.S.

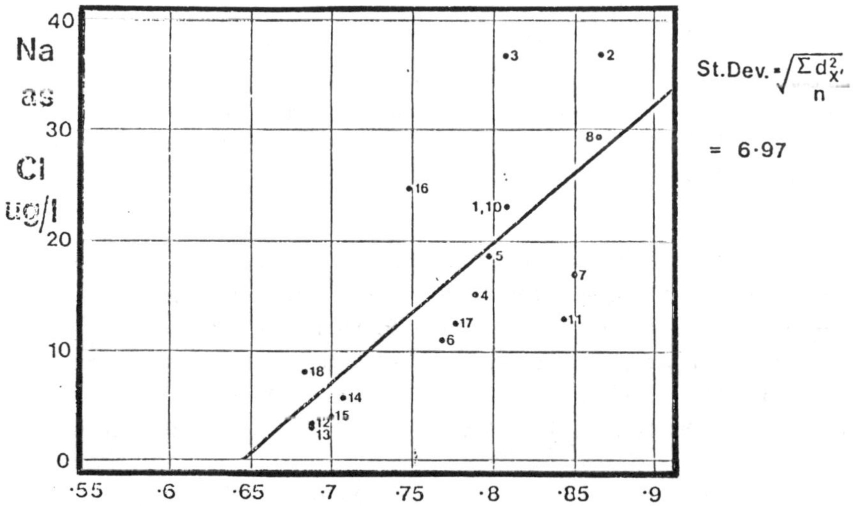

HARMONIC MEAN SIZE MM

ION EXCHANGE DESULPHATION OF FEEDWATER TO MSF EVAPORATOR. FULL SCALE EXPERIENCE.

L. Liberti*, R. Pascali**, F. Sigon**, R. Zannoni***

* I.R.S.A. - C.N.R , Via De Blasio 5, 70123 Bari, I

** E.N.E.L. - C.R.T.N., Via Rubattino 54, 20134 Milano, I

*** OFFICINE REGGIANE O.M.I., Via Vasco Agosti 27, 42100 Reggio Emilia, I

SUMMARY

To improve the performance of MSF desalination plants, it is necessary to decrease the sulphate content of seawater feeding the plant in order to avoid the scaling of the heat transfer surfaces.
The DESULF process, an ion-exchange seawater treatment, which uses a weak anionic resin (regenerated with the plant blow-down to remove sulphates) has been tested.
Data from an actual full-scale plant, treating 3360 m^3/d of seawater, are presented together with future applications.

1. INTRODUCTION

Present technology of seawater desalting by multi-stage flash (MSF) plants allows the realization and reliable operation of evaporators with a maximum brine temperature up to 90 °C, when the feedwater treatment is carried out by means of $CaCO_3$ scale inhibitors (such as polyphosphates, methacrylates, etc.), or about 110 °C when the seawater make-up is acidified and the CO_2 removed /1,2/. The concentration factor (n), is maintained around 1.8.

Higher brine temperatures are prevented by $CaSO_4$ scale formation on the heat exchange surfaces, causing the plant to worsen its operating conditions down to irreversible shut-up /3,4/.

On the other hand, since the unit water cost decreases by increasing the brine temperature /5,6/, a remarkable research and development effort has been devoted to the feedwater chemistry in order to raise the maximum operating temperature of MSF plants.

The use of strong cation exchange resins to reduce Ca++ concentration in sea water has been investigated with unsatisfactory results, owing to the unfavourable Ca++ selectivity of these resins /7,8,9/.

It was suggested to treat chemically sea water with barium chloride to remove sulphates as $BaSO_4$ precipitate /10-13/, but only bench scale data are presently available.

Pilot plant tests to inhibit $CaSO_4$ precipitation by means of particles (alumina, Ni, Cu) finely dispersed in the brine have pointed out serious operating difficulties /14-16/.

A different fate has occurred to the DESULF process, developed by IRSA (Water Research Institute of Italy) in 1969, which uses weak anion resins regenerated in Cl-form by means of the plant blow-down /17/. After being successfully tested in a laboratory and in a modular 100 m^3/d pilot plant up to 140 °C /18, 19/, the DESULF process has been employed for the first time on a large scale in the new 3360 m^3/d MSF plant built by REGGIANE OMI, on account of Cassa per il Mezzogiorno, to desalt sea water at the electricity power station of ENEL in Bari.

This paper reports the main results obtained after 5000 hours of demonstration running of Bari's plant. The latter provided basic design data for application of the DESULF process to two more MSF plants, in Kuwait (11000 m^3/d) and in Gela, Italy, (27,000 m^3/d), also described in this note.

2. THE DESULF PROCESS

At a given pH, the use of suitable, weakly basic anion resins in Cl-form permits to remove sulphates from sea water, according to the ion exchange reaction:

$$2\ RHCl + SO_4^{=} \qquad (RH)_2SO_4 + 2\ Cl^- \qquad (1)$$

The process dependence on the operating parameters has been extensively studied with laboratory experiments /20-23/, then validated on pilot plant /19,24/. In particular, the effect of feed flow-rate as well as that of brine concentration and pH values have been studied.

Owing to the high selectivity towards sulphates of the resin employed, reaction (1) is shifted to the right even at high flow rates (40 l/lrh) with related throughout values up to 20 l/lr. In such conditions, high exchange efficiencies (80-:-85 %) are obtained with a concentration factor (n) of 2.5.

Moreover, the use of weak anionic resins causes the process to be sensitive to pH changes. When operating with a weakly basic resin largely in chloride form, we have to consider the contemporary hydrolysis reaction (2):

$$RHCl + H_2O \qquad R + H_3O^+ + Cl^- \qquad (2)$$

which becomes more pronounced as the water pH value increases.

Reaction (2), competitive with the (1), occurs until the resin has reached a pH of equilibrium, this explains the buffer effect of the weak anionic resins towards slight excess or shortcomings during the acid dealkalisation pretreatment of the sea water feeding the evaporator, which otherwise would induce either corrosion on the degasifier or harmful alkaline scaling in the brine heater. On the other hand, fundamental investigation on the kinetic mechanism of reaction (1), performed during the laboratory study of the process, has disclosed that classic transport models, based on Fick and Nerst-Plank diffusion control, do not apply to this system /25-28/. As discussed in another paper presented at this ASI /29/, the possibility of a chem-control and unusual transport paths are presently being considered to explain the Cl/SO_4 exchange kinetic behaviour.

3. BARI's 3360 m^3/d DESULPHATING PLANT

To demonstrate the capability of the ion exchange desulphation pretreatment on a large scale plant, the DESULF process has been adopted for the pretreatment of the sea water feeding a new 3360 m^3/d demonstration desalting plant, financed by Cassa del Mezzogiorno and built in 1975 by REGGIANE OMI at the electricity power station of ENEL in Bari.

The plant, whose project characteristics are presented in Table 1, with a partial view and lay-out in Figs 1 and 2, operates at a maximum brine temperature of 150 °C and with a high performance ratio (18 kg distillate/kg steam). In detail, the pretreatment unit consists of two columns of 8 m^3/h containing 4.5 m^3 of resin each. The resin is a weak anionic exchanger with poly-acrylate -DVB matrix, primary and secondary amino functional groups and a crosslinking of 11 % DVB /30,31/.

The operating parameter values for the process, arising from 5,000 hours of plant operation, are presented in Table 2. Average desulphation efficiency of about 75% was obtained, varying from 70 to 85 % depending on the actual values of the plant operating conditions. Figure 3 shows the performances of the resins recorded during the last period at high temperature, (150 °C, with a make-up of 125 t/h). Analysis of the experimental results, using a half replicate factorial experimental design, permitted to correlate the desulphation efficiency to the blow-down concentration factor (n), and the cycle time (t, minutes) through the following first order equation:

$$SO_4^{=} = A + B * n + C * (t-30) \qquad (3)$$

where A, B, C are empirical constants and $SO_4^{=}$ is the sulphate concentration in the blow-down when n = 1. Equation (3) provides an acceptable fit of the data, in the following parameter range: pH 3.2 ÷ 4.0, t 30 ÷ 35 min, n 2.2 ÷ 2.52.

Direct inspection of the plant during occasional shut-off for maintenance and at the end of the 5,000 hours running period, revealed some scales on the heat exchange surfaces. Submitted to chemical analysis by X-fluorescence, the latter resulted to contain both cristalline and amorphous calcium sulphate. However, considering the duration of the running period, the temperature and the overall performance achieved, the scales may be considered negligible.

Further attention was devoted during the demonstration test to study the buffer capacity of the desulphating resin employed, according to reaction (2).

To this aim, the pH of sea water acidified before entering and that of the effluent leaving the desulphation unit, were correlated varying the amount of acid injected. The results obtained are reported in Fig. 4 as the quantity of acid (in eq/l) needed to modify the sea water pH from a certain value. As it can be seen, the system appears particularly stable in the pH range 3 to 5. Outside this range however (e.g., for an occasional acid overdosage or shortcoming up to 0.005 eq/l), the buffer capacity of the weak anion resin is so that the pH of the solution leaving it should fall between 2.5 and 6.5. The shift of experimental data of Fig. 4 towards the 0 eq/l axis is probably due to the presence of alkaline iron oxide in the plant. The buffer capacity of the resin employed has been evaluated further for the case of accidental interruption in the plant acidification system. As shown by Fig. 5, in this emergency situation the resin assures a safety period of 5-6 hours before its effluent pH arises to critical values (i.e. greater than 6).

On the basis of Bari's plant experiences it may be concluded that a performance ratio of 16 kg distillate/kg steam can be achieved with the desulphation pretreatment with an energy saving of at about 20 % compared to a traditional MSF plant (top brine temperature 115 °C and a performance ratio 6) if coupled with a thermoelectric cycle.

Furthermore the energy saving may be 60 %, if the steam is generated by a boiler used only for this task /32/.

Finally, the favourable economics of the desulphation pretreatment in single and dual purpose plant have been demonstrated, leading to a production increase up to 20 % /33/.

4. NEW DESULPHATION UNITS FOR HIGH CAPACITY MSF PLANTS

Two new desulphation units are scheduled for the next two years: the former in Kuwait and the latter in the ANIC petrochemical plant at Gela (Italy).

4.1 Kuwait desalination unit

The Reggiane O.M.I., by Reggio Emilia - Italy, are building five MSF desalination plants supplied to the Electricity and Water Kuwait Ministry; at present, four plants with poliphosphate treatment (6 MIGPD capacity each) are being built at Doha West Power Station.

A desulphation unit, designed for the treatment of a make-up flow of 480 m^3/h and a top brine temperature of 138 °C, will be linked up to the acid treatment MSF desalination plant. In Table 3 the main project data of the desalination plant are given with reference to two different operation conditions, with and without desulphation unit. In this table is evident the product increase and the make-up decrease (and consequently the acid consumption reduction) obtained by increasing the top brine temperature from 121 °C to 138 °C. The desulphation unit, designed in function of Kuwait sea water characteristics, Table 6, and the foreseen maximum top brine temperature, has a ion exchange resin volume installed of 40 cubic meters, Table 5. The start-up of the desalination plant (and desulphation unit) is planned for the end of 1983.

4.2 ANIC GELA desalination unit

The ANIC desalination unit at Gela is the biggest one in Italy and consists of four parallel lines, each one treating 27,000 m^3/d and a production of 14,400 m^3/d.

Officine Reggiane has studied and proposed, on account of Cassa per il Mezzogiorno, the realization of a desulphation unit, to be adopted on one line, with the aim of:

- restoring the project performance, presently seriously decreased by calcium sulphate scaling;
- increasing the plant capacity by raising to 127 °C the maximum brine temperature.

This desulphation unit, designed to allow the treatment of a sea water make-up flow of 1200 m^3/h and a top brine temperature of 127÷130 °C, has a ion exchange resin volume installed of 60 cubic meters. The main project characteristics of this plant are presented in Tables 3,4 and 5, while the chemical composition of the sea water is shown in Table 6.

5. CONCLUSIONS

According to the updated results, the DESULF process has shown its feasibility to solve the calcium sulphate precipitation problems when adopted for the sea water pretreatment on desalination plants.

By this we may obtain large energy savings thanks to the higher top brine temperatures, and better performances and flexibilities in traditional desalination plants.

Moreover it has been proved, on industrial scale, that the use of this kind of pretreatment allows a better pH control, increasing the safety level of plant operation, as result of the resin buffer capacity.

The pilot and bench level evaluation of the process has been successfully validated,thus providing the basis to develop further industrial applications.

BIBLIOGRAPHY

/1/ D. Borgese. Attività di ricerca e sviluppo al Centro di Ricerca Termica e Nucleare dell'ENEL nel settore della dissalazione di acque salmastre e marine. Relazione DSR/CRTN I 10-4 (1976).

/2/ N.M. Wade. A review of scale control methods. Desalination, 31 (1979), 308-320.

/3/ Multistage flash distillation desalting: state of the art. OSW R & D Progress Report 490 (1969).

/4/ Abstracts of literature on the distillation of seawater and on the use of nuclear energy for desalting. OSW R & D Progress Report 589 (1970).

/5/ Betchel. Cost studies of large multistage flash water conversion plants. OSW R&D Progress Report 116 (1964).

/6/ Betchel. Optimum brine heater outlet temperature in seawater conversion evaporator. OSW R&D Progress Report 175 (1964).

/7/ G. Klein, K.M. Makar, B.W. Tleimat, T. Vermeulen. Desing and cost of ion exchange softening. Seawater Conversion Lab. Rep. 68-2 (1968).

/8/ M.F. Mc Ilhenny. Seawater softening by ion exchange as a saline water conversion pretreatment. OSW R&D Progress Report 62 (1962).

/9/ ORNL. Continuous counter current ion exchange removal of calcium from seawater to prevent scaling of distillation equipment. OSW R&D Progress Report 309 (1967).

/10/ D.R. George, J.M. Riley. Sulphate removal from brines. U.S. Patent 3.378.336 (6-17-66/4-16-68).

/11/ G.P. Gelblum, W. Pechenick, C.E. Ennis. Evaluation of brine desulphating process as applied to desalination. OSW R&D Progress Report 289 (1967).

/12/ W.R. Grace Company. Development of precipitation processes of removal of scale formers from seawater. OSW R&D Progress Report 192 (1966).

/13/ Mason Rust. Evaluation of a 75,000 gpd seawater desulphating pilot plant. OSW R&D Progress Report 746 (1972).

/14/ Brookheaven National Laboratory. Scale control in high temperature distillation utilizing fluidized heat exchangers. OSW R&D Progress Report 571 (1970).

/15/ Baldwin, Lima, Hamilton.
Operation of the 16 stage high temperature flash pilot plant. OSW R&D Progress Report 288 (1967).

/16/ Baldwin, Lima, Hamilton. Development studies on the flash distillation process. OSW R&D Progress Report 562 (1970).

/17/ G. Boari. Brevetto italiano n. 39657 - A/69.

/18/ G. Boari, L. Liberti, C. Merli, R. Passino. Study of the SO_4/Cl exchange on a weak anion resin. Ion exchange and membranes 2 (1974).

/19/ A. Aveni, G. Boari, L. Liberti, B. Monopoli, M. Santori. Sulphate removal and dealkalization on weak anion resins of the feed water for evaporation desalting plants. Desalination (1975).

/20/ G. Boari, L. Liberti, C. Merli, R. Passino. Studio dello equilibrio dello scambio SO_4/Cl tra acqua di mare e resine scambiatrici di ioni. Quaderni dell'Istituto di Ricerca sulle Acque, 13 (1972), 7-41.

/21/ G. Boari, L. Liberti, C. Merli, R. Passino. Desolfatazione e delcalizzazione dell'acqua di mare in processi di dissalazione per evaporazione. Quaderni dell'Istituto di Ricerca sulle Acque, 13 (1972), 43-65.

/22/ V. Balice, G. Boari, L. Liberti, R. Passino, M. Santori. Impiego delle isoterme di scambio nello studio della desolfatazione dell'acqua di mare su unità di scambio ionico a letto fisso. Quaderni dell'Istituto di Ricerca sulle Acque, 13 (1972), 67-89.

/23/ V. Balice, G. Boari, L. Liberti, C. Merli, R. Passino. Studio della velocità di trasferimento degli ioni nel processo di desolfatazione dell'acquá di mare. Quaderni dell'Istituto di ricerca sulle Acque, 13 (1972), 91-109.

/24/ G. Boari, L. Liberti, R. Passino. Prevention of $CaSO_4$ scale in evaporation plants by ion exchange. J. Chromatogr. 102 (1974) 393.

/25/ L. Liberti, G. Boari, R. Passino. Chloride/Sulphate exchange on Anion Resins. Kinetic investigations. II. Particle Diffusion Rates. Desalination 25 (1978), 123-134.

/26/ L. Liberti, G. Boari, R. Passino. Chloride/Sulphate Exchange on Anion Resins. Kinetics Investigation. Chemical Control in Selective Exchangers. Desalination, 26 (1978), 181-194.

/27/ L. Liberti, G. Schmuckler. Kinetic Investigation of the

Chloride/Sulphate Exchange on Anion Exchange Resin. Desalination, 27 (1978), 253-260.

/28/ L. Liberti, I. Madi, R. Passino, L. Walis. Ion Exchange Kinetics in Selective Systems. J. Cromathogr. 201 (1980), 43-50.

/29/ L. Liberti, R. Passino, D. Petruzzelli, G. Boghetich Diffusionally vs. Chemically Controlled Kinetics in Selective Systems. Proceedings of NATO/ASI on Mass Transfer and Kinetics of Ion Exchange, Maratea May 31 - June 11 1982, Nijhoff Press, The Hague, The Netherlands, in press.

/30/ R. Pascali, V. Regis, G. Trebbi Realizzazione di un impianto dimostrativo di dissalamento da 70 t/h mediante evaporazione multistadio con processo di desolfatazione. Relazione ENEL-DSR/CRTN I - 11 (1975).

/31/ G. Boari, L. Liberti, M. Santori, L. Spinosa. Advanced evaporation plants with sulphate removal by ion-exchange. Desalination 19 (1976), 283-298.

/32/ A. Conti, R. Pascali, G. Odone. Water cost and energy saving for 70 t/h multiflash desalination plants working at 120 °C and 150 °C top brine temperature. Proceedings of the 7th International Symposium on Fresh Water from the Sea. Vol. 1, (1980), 481-496.

/33/ A. Conti, R. Pascali. Calculation of economics indexes for desalination plants with or without desulphation pretreatment. NATO/ASI on Mass Transfer and Kinetics of Ion Exchange, Maratea, May 31 - June 11, 1982. In press.

Tab. 1 - BARI DESALINATION PLANT. MAIN PROJECT CHARACTERISTICS

DISTILLATE PRODUCTION	70 t/h
RECIRCULATING BRINE	372 t/h
BLOW-DOWN	52 t/h
MAKE-UP	122 t/h
CONCENTRATION FACTOR	2.34
MAXIMUM BRINE TEMPERATURE	150 °C
PERFORMANCE RATIO	34.000 Kcal/t 16 Kg dist/Kg steam

Tab. 2 - BARI DESULPHATION UNIT. MAIN PROCESS PARAMETERS

	RANGE	OPERATING VALUES
pH MAKE-UP	3.0 5.0	4.3
pH BLOW-DOWN	7.2 8.0	7.3
F.C. BLOW-DOWN	2.0 2.7	2.6
CYCLE TIME	25 40 min	30 min
MAKE-UP	110 140 t/h	130 t/h
BLOW-DOWN	40 50 t/h	50 t/h

Tab. 3 - DOWA EAST DESALINATION UNIT. MAIN PROJECT CHARACTERISTICS.
Plant type: MSF, cross tube with brine recirculation.

	No desulphation	With desulphation
N. HEAT RECOVERY STAGES	22	22
N. HEAT REJECTION STAGES	2	2
DISTILLATE PRODUCTION	4580 t/day	5386 t/day
DISTILLATE SALINITY	30 ppm	30 ppm
PERFORMANCE RATIO	49105 Kcal/t	49702 Kcal/t
	10 Kg dist/Kg steam	10 Kg dist/Kg steam
MAXIMUM BRINE TEMPERATURE	121 °C	138 °C
CONCENTRATION FACTOR	1.66	2.04
BRINE TO HEATER	1503 t/h	1503 t/h
BLOW-DOWN	292 t/h	214 t/h
SEA WATER MAKE-UP	483 t/h	438 t/h

Tab. 4 - ANIC-GELA DESALINATION BREDA MODULE.
AMIN PROJECT CHARACTERISTICS.
Plant Type: MSF, cross tube with brine recirculation.

	No desulphation	With desulphation
N. HEAT RECOVERY STAGES	42	42
N. HEAT REJECTION STAGES	2	2
DISTILLATE PRODUCTION	14400 t/day	16030 t/day
DISTILLATE SALINITY	5 ppm	5 ppm
PERFORMANCE RATIO	40067 Kcal/t	40346 Kcal/t
	13 Kg dist/Kg steam	13 Kg dist/Kg steam
MAXIMUM BRINE TEMPERATURE	120 °C	127 °C
CONCENTRATION FACTOR	1.70	2.48
BRINE TO HEATER	4254 t/h	4500 t/h
BLOW-DOWN	833 t/h	448 t/h
SEA WATER MAKE-UP	1439 t/h	1116 t/h

Tab. 5 - DESULPHATION UNITS. MAIN PROJECT CHARACTERISTICS.

			KUWAIT	GELA
N. COLUMNS			4	4
COLUMN DIAMETER		(mm)	3000	3000
COLUMN HEIGHT: cylindrical part		(mm)	2500	2700
FILLING MATERIAL: anionic weak type resin		(l)	40000	60000
EXHAUSTION AND REGENERATION TIME		(min)	30	30
MATERIALS :	VALVES		AISI 316 L	NODULAR C.I., EPDM COAT
	PIPING (acidified sea water)		AISI 316 L	Fe 35, EBONIZED
	PIPING (brine)		A 53 GR B	Fe 35, EBONIZED
	COLUMNS		Fe 42B, EBONIZED	Fe 42B, EBONIZED

Tab. 6 - SEA WATER PHYSICAL AND CHEMICAL PARAMETERS.

		KUWAIT	GELA
pH		8.2	8.2
SULPHATES	(mg/l SO_4)	3200	3020
CHLORIDES	(mg/l Cl)	25000	21500
TOTAL HARDNESS	(mg/l $CaCO_3$)	8600	7020
CALCIUM HARDNESS	(mg/l $CaCO_3$)	1250	1030
MAGNESIUM HARDNESS	(mg/l $CaCO_3$)	7350	5990
TOTAL ALKALINITY	(mg/l $CaCO_3$)	135	130
SILICA	(mg/l SiO_2)	1.0 1.5	0.2
SALINITY	(ppm)	45000	37820
IRON	(mg/l Fe)	-	0.2
COPPER	(mg/l Cu)	-	0.02
SODIUM	(mg/l Na)	12700	11160
POTASSIUM	(mg/l K)	380	354
SEA WATER TEMPERATURE	(°C)	14 32	10 24

Fig. 1 - partial view of the plant.

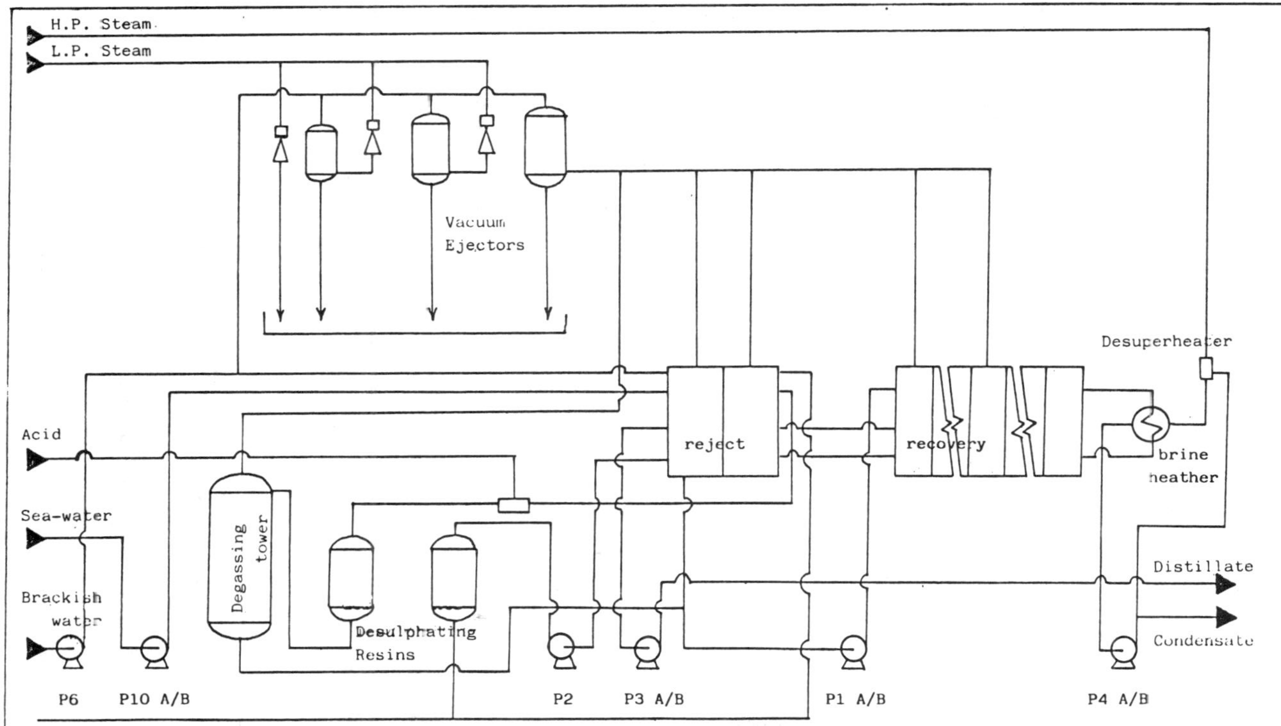

FIG. 2 MSF DESALINATION PLANT. LAY-OUT.

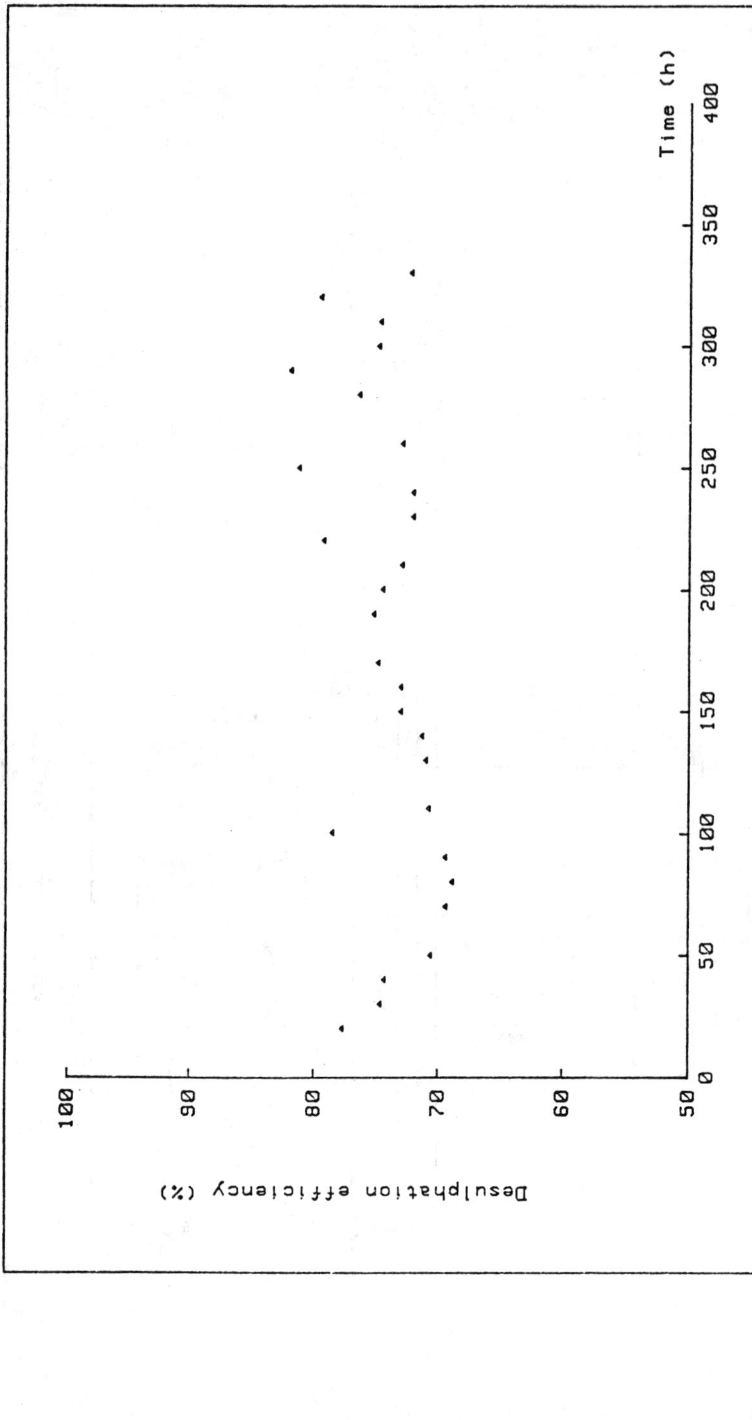

Fig. 3 BARI MSF DESALINATION PLANT. DESULPHATION EFFICIENCY.

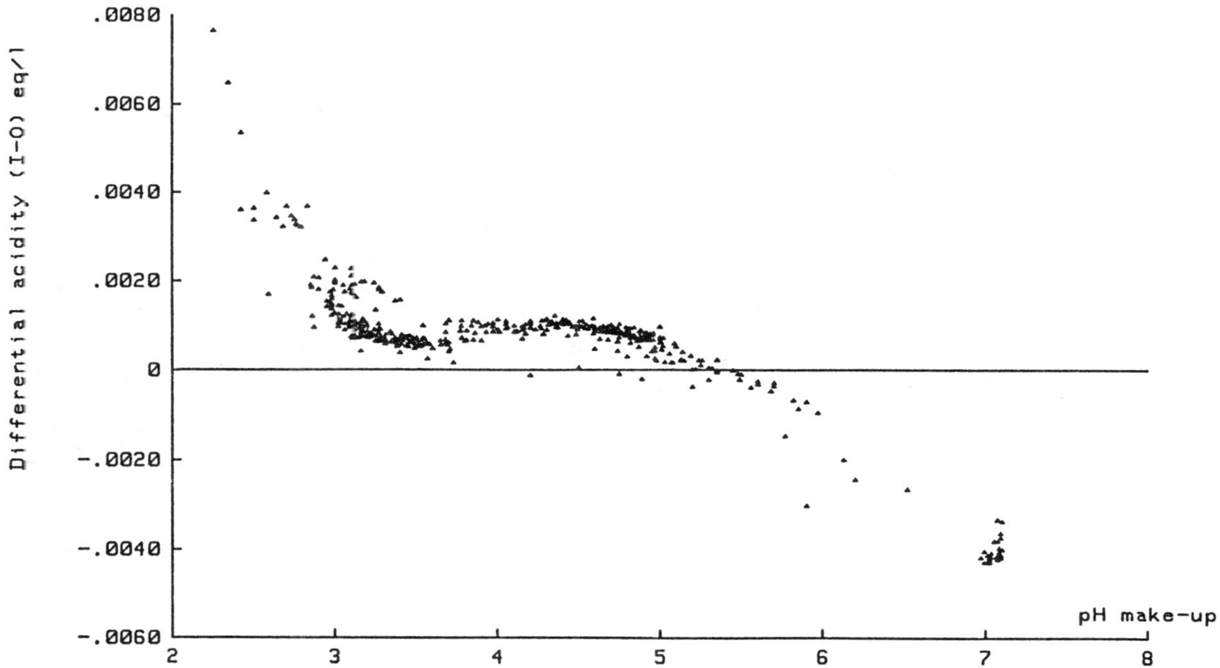

Fig. 4 BARI MSF DESALINATION PLANT. RESIN BUFFER ACTIVITY.

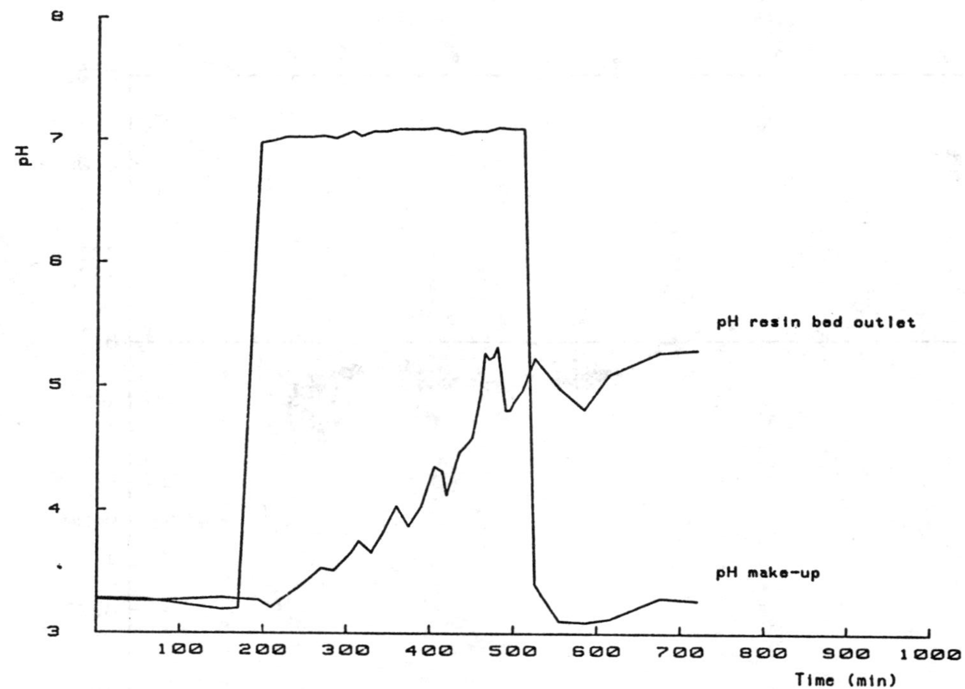

Fig. 5 BARI MSF DESALINATION PLANT. ACID FAILURE SIMULATION: RESIN EFFECT.

CALCULATION OF ECONOMIC INDEXES FOR DESALINATION PLANTS WITH AND WITHOUT DESULPHATION PRETREATMENT SYSTEM.

A. Conti, R. Pascali

ENEL - DSR/CRTN

1. CALCULATION OF ENERGY SAVING AND APPLICATION OF COST/BENEFIT ANALYSIS IN THE COMPARISON BETWEEN A TRADITIONAL 120 °C AND A NEW 150 °C (70 t/h) MULTIFLASH WITH A DESULPHATION PRETREATMENT (BARI)

In order to quantify potential savings obtained elevating top brine temperature to 150 °C (312 °F) by adopting a make-up pretreatment with a desulphation unit, both in single and in dual purpose plant, a technical and an economic analysis has been carried out using the following financial indexes:
- Payback Period (PbP)
- Rate of Return (RoR)
- Present Worth Value (PWV).

The costs have been estimated (October 1978) taking into account respectively performances verified on the demonstrative multiflash plant of Bari (performance ratio = 15 t Dist/t Steam) and those of a traditional one with a performance ratio of 6, for a production of 70 t/h. As fuel cost has increased extensively in these last years, the study has been updated at December 1981, with plant costs of 1978 increased of 20% for each year and fuel cost of 250 L/kg instead of 100 L/kg of 1978. Process characteristics of the two compared plants computed with a specific code (DISALE), are summarized on table n. 1.

For the evaluations of energy savings, both single purpose plant (auxiliary steam generator) and dual purpose plant (steam drawn from a thermoelectric cycle like Bari power plant (70 MW)), have been considered.

The values of the calculated financial indexes for 1978 and 1981 are summarized on tables 2 and 3.

Finally, a sensitivity analysis of RoR (tables 4-5-6-7) has been carried out in order to take into account the influence of:
- the uncertainty of investment
- the sectorial energy price variations
- the uncertainty of energy saved.

From the analysis of results, we can assess the following considerations:
1. Energy saving obtained using high temperature plants is higher in single purpose than in dual purpose plant (66%).
2. Advantages using high temperature plants, are higher today (December 1981) than those calculated in October 1978.
3. For single purpose plant. Rate of Return changes from 18,58 (October 1978) to 31,70 (December 1981).
4. For dual purpose plant, the rate of Return changes from 0,18 (October 1978) to 6,77 (December 1981).

2. ECONOMIC ANALYSIS OF THE ADDITION OF DESULPHATION PRETREATMENT TO A PREEXISTENT PLANT (KUWAIT)

The study has been carried out comparing the costs of a 2 MIGPD (375 t/h) multistage flash desalination plant without and with make-up desulphation pretreatment.

Top brine temperature has been raised from 112 °C to 130 °C and performance ratio from 10,0 to 10,39 t Dist/t Steam and all other consumption had been calculated with these new process values (table 8).

For the economic analysis then have been assumed basic data collected on table 9.

Final results are summarized on table 10.

Total saving adopting a pretreatment with a desulphation unit has been calculated taking into account both distillate cost reduction due to increased performance ratio, and increased production due to higher top brine temperature.

Payback period is only of 2,79 years and Rate of Return 35,42%.

Therefore all these characteristics make the desulphation process very interesting for the sea water pretreatment in high performance and advanced MSF evaporators.

	TRADITIONAL MULTIFLASH	BARI MULTIFLASH
PRODUCTION (t/hr)	70	70
TOP BRINE TEMPERATURE (°C)	120	150
STAGE NUMBER (Recovery + Resect)	19 + 3	66 + 2
PERFORMANCE RATIO (kg DIST/kg STEAM)	6	15
STEAM TO B.H. (kg/hr)	10062	1667
STEAM TO EJECTOR (kg/hr)	244	245
ELECTRIC ENERGY CONSUMPTION (kWh)	167	270
PLANT COST (S x 10^6)	1.277	2.212
HEAT TO B.H. (Kcal/hr)	6080667	2330533

Tab. 1 - PROCESS CHARACTERISTICS OF COMPARED DESALINATION PLANTS

	SINGLE PURPOSE PLANT	DUAL PURPOSE PLANT
INVESTMENT VARIATION (S)	935860	935860
ENERGY SAVING (kWh/y)	29806000	10010000
PAYBACK PERIOD (y)	4.96	14.78
RATE OF RETURN (%)	18.58	0.10
PRESENT WORTH VALUE (S)	1283680	431110

Tab. 2 - ECONOMIC INDEXES FOR SINGLE AND DUAL PURPOSE PLANT (October 1978)

	SINGLE PURPOSE PLANT	DUAL PURPOSE PLANT
INVESTMENT VARIATION (S)	1671450	1671450
ENERGY SAVING (kWh/y)	29806000	10010000
PAYBACK PERIOD (y)	3.10	9.24
RATE OF RETURN (%)	31.70	6.77
PRESENT WORTH VALUE (S)	3667653	1231739

Tab. 3 - ECONOMIC INDEXES FOR SINGLE AND DUAL PURPOSE PLANT (December 1981)

	ΔI (S)	PE (S/kWh)	ΔE (kWh/y)	RoR (%)
AVERAGE VALUES	935863	.0063	29806000	18.58
ΔI + 10%	1029449	-	-	16.44
ΔI - 10%	842276	-	-	21.11
PE + 10%	-	.0069	-	20.86
PE - 10%	-	.00569	-	16.23
ΔE + 10%	-	-	32786600	20.86
ΔE - 10%	-	-	26825400	16.23

Tab. 4 - SENSITIVITY ANALYSIS FOR SINGLE PURPOSE PLANT

	ΔI (S)	PE (S/kWh)	ΔE (kWh/y)	RoR (%)
AVERAGE VALUES	935863	.0063	10010000	.18
ΔI + 10%	1029449	-	-	-.99
ΔI - 10%	842276	-	-	1.54
PE + 10%	-	.0069	-	1.40
PE - 10%	-	.00569	-	-1.12
ΔE + 10%	-	-	11011000	1.40
ΔE - 10%	-	-	900900	-1.12

Tab. 5 - SENSITIVITY ANALYSIS FOR DUAL PURPOSE PLANT (October 1978)

	ΔI (S)	PE (S/kWh)	ΔE (kWh/y)	RoR (%)
AVERAGE VALUES	1671451	.02	29806000	31.70
ΔI + 10%	1838596	-	-	28.62
ΔI - 10%	1504305	-	-	35.42
PE + 10%	-	.0198	-	35.05
PE - 10%	-	.0163	-	28.31
ΔE + 10%	-	-	32786600	35.05
ΔE - 10%	-	-	26825400	28.31

Tab. 6 - SENSITIVITY ANALYSIS FOR SINGLE PURPOSE PLANT (December 1981)

	ΔI (S)	PE (S/kWh)	ΔE (kWh/y)	RoR (%)
AVERAGE VALUES	1671451	.02	10010000	6.77
ΔI + 10%	1838596	-	-	5.31
ΔI - 10%	1504305	-	-	8.47
PE + 10%	-	.0198	-	8.31
PE - 10%	-	.0163	-	5.16
ΔE + 10%	-	-	11011000	8.31
ΔE - 10%	-	-	9009000	5.16

Tab. 7 - SENSITIVITY ANALYSIS FOR DUAL PURPOSE PLANT (December 1981)

TRADITIONAL MULTIFLASH PLANT	
Plant cost (S)	8×10^6
Plant life (Years)	15
Interest rate (%)	3.5
Maintenance cost (2% of capital cost S/year)	160000
Plant operation cost (2.5% of capital cost S/year)	200000 (+)
Steam cost (c/kcal)	1
Electric energy cost (c/kWh)	7
Sulphuric acid cost (98% by weight c/kg)	24
DESULPHATION PLANT	
Desulphation plant cost (S)	0.7×10^6
Interest rate (%)	3.5
Resin consumption (S/year)	2000
Maintenance (S/year)	7000

(+) Personnel, administration and general expenses

Tab. 8 - BASIC ASSUMPTION FOR ECONOMIC ANALYSIS

PLANT CHARACTERISTIC	WITHOUT DESULPH.		WITH DESULPH.
Distillate produced (t/hr)	375	462	462
Top brine temperature (°C)	112	112	130
Heat to Brine Heater (kcal/hr)	19760000	23712000	22871000
Performance ratio (kg DIST/kg STEAM)	10	10	10.39
Electric power (kW)	1300	1495	1300
Sulphuric acid (kg/hr)	132	145	99
Plant availability (hr/year)	7500	7500	7500
COSTS (c/ton of distillate produced)			
Plant capital cost	24.75	23.05	20.05
Maintenance	5.63	5.31	4.33
Operation	7.04	6.63	5.39
Electric energy	24.2	22.8	19.7
Steam	52.69	51.32	49.5
Sulphuric acid	8.45	7.53	5.14
Other chemicals	2	2	2
Desulphation plant	-	-	2
Resin consumption	-	-	0.54
WATER COST (c/t)	124.83	118.64	108.65

TOTAL SAVING = 462 x 7500 x (118.64 - 108.65)/100 = 346153 (S/year)

Tab. 9 - ECONOMIC COMPARISON BETWEEN AN EXISTENT PLANT WITHOUT DESULPHATION PRETREATMENT AND THE SAME WITH DESULPHATION PRETREATMENT

	ΔI (S)	PE (S/kWh)	ΔE (kWh/y)	RoR (%)
AVERAGE VALUES	700000	.0179	14000000	35.42
ΔI + 10%	770000	-	-	32.04
ΔI - 10%	630000	-	-	39.51
PE + 10%	-	.01969	-	39.10
PE - 10%	-	.01611	-	31.70
ΔE + 10%	-	-	15400000	39.10
ΔE - 10%	-	-	12600000	31.70

Tab. 10 - SENSITIVITY ANALYSIS FOR 375 t/h MULTIFLASH WITH DESULPHATION

KINETIC STUDIES ON GEL AND MACROPOROUS ANION EXCHANGERS USING THE URANYL SULFATE/SULFATE EXCHANGE

F. X. McGarvey and E. W. Hauser

Sybron Corporation
Birmingham, New Jersey, U.S.A.

ABSTRACT

Batch contact studies have been performed with macroporous and with gel anion exchangers using the uranyl sulfate/sulfate exchange in the presence of substantial amounts of iron salts. Conditions are typical of South African test methods widely used to evaluate ion exchange resins for uranium recovery. Previous tests had shown that macroporous anion exchangers gave anomalously low values for exchange rate under these conditions.

The rate curves based on plots of uranium loading as a function of time follows the parabolic diffusion relationship with the first derivatives linear with time. Results showed lower rates for macroporous resins than for gels in the same size range. These rate values are examined as a function of residual capacity in the resin as well as from the porosity standpoint.

Salt invasion studies on the anion exchangers were to estimate porosity in aqueous systems as compared to the conventional mercury penetration methods. These results were used to develop a model for conditions in macroporous resin and gel resin in aqueous solutions.

Recent years have seen the increased use of macroporous anion exchanger for uranium recovery. This development was due to both physical and chemical observations which showed that macroporous anion exchanger had a place in this technology. One major factor was physical resistance to agitation, particularly in moving bed equipment where gel resins frequently broke down rapidly. This effect was probably due to a certain amount of flexibility in the resin structure which permitted the macroporous resin beads to show plasticity under stress.

The second effect related to the concept of an open structure which would permit rapid diffusion of the large uranyl sulfate ion through the pores in the resin structure. It was expected that, at equal crosslinking within the gel phase, a macroporous resin would load efficiently and with a well defined self-shaping front. The expectation was also that sharp elution curves would be achieved so that the concentration of the pregnant eluate would be higher than that obtained from the slower gel resin.

It is difficult to establish an accurate crosslinkage content in anion exchange resins prepared by the chloromethylation of styrene divinylbenzene copolymers due to the occurrence of secondary crosslinkage. While this is a difficult problem in gel resins, the additional variable of phase extenders makes the situation extremely complex in macroporous resin. For example, the moisture retention is used as a measure of crosslinkage for fully chloromethylated resins. When macroporous resins are involved, the moisture retention values must be divided between pore water and gel moisture. Differences have been observed between the way that gel and macroporous resins swell and shrink. These observations may be used to gain some knowledge of pore volume in aqueous solution.

Comparisons were made between macro and gel anion exchange resins for the concentration of uranium. Work was done with both the uranyl carbonate and the uranyl sulfate exchanges. The latter system is complicated by iron salts and variable EMF values in solution. Generally the macroporous anion exchangers were found to load efficiently and to strip in a fashion superior to gel resins commonly used in uranium work. Figure 1 shows the breakthrough curves for a typical loading from an acid leach liquor in a mill in Wyoming. The strip curves are shown in Figure 2 for an acid liquor while Figure 3 shows the elution curves for a carbonate system. Generally the macroporous resins are exhibiting rates equal or superior to gels although the resins are not equivalent in crosslinkage.

It came with some surprise that evaluations using batch techniques rather than column operation showed that macroporous resins did not exhibit good rates and their loadings were poor in some studies. One study made with a Canadian engineering company showed that a macroporous strong base, Ionac A-641, exhibited quite low rates when contacted with progressively dilute solutions of uranium. The rate of exchange dropped off sharply as compared with gel resins. Further studies indicated that this effect was related to porosity as measured in non-aqueous solutions. The rate was reduced about 30% when using a resin with 30% porosity. This effect carried for resin with lower porosity. Since the resins under study gave good performance in small fixed bed columns, it was evident that macroporous resins had kinetic limitations in batch type tests.

This study was undertaken to develop data under good conditions. A method currently used in South Africa was the basis for the test. This procedure(1) developed specifically for South African conditions uses an acid leach liquor with carefully controlled ferrous-ferric ratio to keep the uranium in the proper valence state. The details of the test method are given in Appendix 1. At this point, it is sufficient to say that the method is complicated by difficulty in stabilizing the solutions. Once this has been done, quite reliable results can be achieved.

The resins used in the study were conventional methods and had the characteristics listed in Table 1. Porosity values are based on the mercury penetration methods and on surface area measurements. Moisture retention was determined by the ASTM procedure and is based on chloride form resin. Capacity values were also determined on the chloride form of the resins. Volume capacity was determined using tapped down volumes and are found to be somewhat higher than values determined by standard backwashed and drained procedures.

The rate studies were performed employing the procedure used for evaluation in South Africa. The results are shown in the following figures where uranium loading is plotted as a function of time. Figure 4 summarizes the data for resins A, B and C. Figure 5 gives the results for resins D, E and F, while Figure 6 shows the values for resin G and also includes the curve for the South African standard.

The curves can be represented as parabolic functions and the slopes were measured and plotted as a function of 1/time. This resulted in a straight line with intercepts near zero for large values of time as expected from the parabolic rate curves. The slope of the curves ranged from about 3 to greater than 6. Table 2 gives the values for line slope and intercept. Generally gels give higher values than macroporous resins with substantial differences in some cases. Figure 6 shows the derivative lines for resins A, B and C. The other resins and the South African standard are shown in Figure 7.

The elution step in the uranium recovery process is as important as the loading step and is also evaluated by a batch contact as outlined in Appendix 1. Data collected on several of the resins under study are presented in Figures 9 and 10. The slopes are plotted in Figure 11 showing a linear relationship between rate of stripping and the inverse of time. All resins tested gave results slower than the South African standard but generally the macroporous resins were slower than the gel resins.

An inspection of these data show that the rate of uranium loading follows a parabolic rate curve, but the actual rate is not a function of porosity. In these particular studies, the resins having a gel structure gave somewhat higher loading and stripping rates than the macroporous resins. No relationship appeared between porosity and loading rate. Particle size variations may be involved in some cases since the size range used in the tests is quite wide. In all cases, ambient temperature (22-26°C) was employed.

These results support the Ruckenstein model(2) which considers the diffusion of the exchanging ions moving inside the porous resin with the overall rate, a complex function involving diffusion coefficients in macroporous and in microspheres containing gel structure. Patell and Turner(3) reported on tests made on the model and found that rapid diffusion in the macropores is still not sufficient to overcome the slow diffusion in the microspheres so that reaction rates are no faster than those found with gel type resins. Actually, these data support the view that they may be slower overall in cases where the gel phase is quite highly crosslinked and microsphere diffusion is slow.

The porosity of macroporous resins in aqueous systems is dependent on the degree of hydration of the gel structure and of the pores around the microspheres. Assuming that the Donnan equilibrium relationship holds, it is possible to gain some information on the porosity by the way that the resin structure responds to the presence of high concentration of electrolyte. McGarvey et al(4) has examined this property of ion exchange resins for gel structure cation exchangers. The work is being extended to macroporous cation exchangers and to anion exchangers.

Macroporous anion and gel structure anion exchangers have been subjected to the osmotic stress of concentrated salt solutions. The result is shown in Figure 12 where specific gravity of the resin structure has been plotted as a function of sodium chloride concentration. The linear change for the gel shows a typical gel response while the flat response for the macroporous shows internal changes in the macropores with a return to gel structure at the high concentrations. When gel resins are contacted with concentrated electrolyte, the resins shrink in a fashion directly related to the osmotic pressure. This results in a linear relationship between specific gravity and salt concentration. Macroporous resins shrink into their pores so that volume changes are not as pronounced. In fact, the specific gravity of macroporous resins remains quite constant as the salt concentration is increased. The tendency for these resins to float makes the measurement of specific gravity at high osmotic conditions difficult.

Measurement of specific gravity for these materials requires considerable precision and good temperature control. The constancy of the specific gravity in macroporous structures indicates that the resin bead is acting as a heterogeneous system with osmotic pressure acting mostly on the bead itself and not on microspheres. The crosslinkage of the microspheres is poorly defined in these systems due to secondary crosslinkage of the gel phase so that an exact solution to the density cannot be made at this time.

The relationship in a cation system indicates that gel density of the cation exchanger is a constant essentially independent of crosslinkage when hydrated. To calculate this value, it is necessary to know the moisture content of the gel resin and the shrinkage of the resin as a function of osmotic pressure. Such methods can be used to estimate departure from linearity and thus gain a measure of porosity in aqueous conditions.

These findings show that there are still many problems related to ion exchange processes using gel and macroporous resins. The use of batch contactors to measure resin performance for use in a columnar configuration is not recommended since the data obtained may be adversely influenced by structural matters not directly related to the final use. These data indicate that additional studies will be required before a relationship between batch contact and kinetics can be found.

REFERENCES

1. Private communication - AEB/NIM (Boydell) Oct. 1979.

2. Ruckenstein, E.; A. Vaidyanathan and G. Youngquist. Sorption by Solids with Bidisperse Pore Structure. Chem. Eng. Sci. 1971, 26, 1305.

3. Patell, S. and J. C. R. Turner. The Kinetics of Ion Exchange Using Porous Exchangers. J. Sep. Process Technol., 1980, 1(2), 31-9.

4. McGarvey, F. X.; S. M. Ziarkowski and B. Kiefer. The Influence of Salt Concentration on the Hydraulic Properties of Ion Exchange Resins. AIChE National Meeting, Orlando, Florida, February 28-March 4, 1982.

APPENDIX 1

The basic test procedures used in determining the loading and elution rates of various types of anion ion exchange resins were taken from the standard South African test methods[1].

Each type of resin was pretreated before use in the loading and elution tests. The resins were converted to the sulfate form using 10% w/w sulfuric acid until no chloride sites were left on the resin. Then each resin type was loaded using the standard pregnant solution (800 mls of solution for 20 mls of resin). The resin was stripped of the uranium using 10% sulfuric acid, converted to the hydroxide form using 4% NaOH, and then reconverted to the sulfate form using 10% w/w sulfuric acid.

The loading rate test involved the use of 7.5 ml of free wet settled resin (-20+35 mesh) placed in two liters of standard pregnant solution. The batch style equilibrium loading was determined by withdrawing, at predetermined intervals, 25 milliliter aliquots of solution and analyzing the six aliquot solutions for uranium as U_3O_8. The resulting data were used to plot the loading curve.

The elution rate test consisted of loading 12 milliliters of free wet settled resin (-20+35 mesh) in 3.0 liters of standard pregnant solution, batch style, over a 24 hour period. Ten of the 12 milliliters of resin were then placed in the two liters of 10% w/w sulfuric acid. The batch style equilibrium elution rate was then determined by withdrawing, at predetermined intervals, 25 milliliter aliquots of solution and analyzing the six aliquot solutions for uranium as U_3O_8. Using the resulting data in combination with the amount of uranium left on the resin, the elution curves were constructed.

The analytical method for determining the uranium concentration was a standard spectrophotometric determination based on a solvent extraction (isooctane-tributyl phosphate) of the uranium from an aluminum nitrate solution and color development using dibenzoyl-methane-alc solution. The resulting colored solution was read at 420 mμ, and the uranium concentration determined based on standard calibration curves. A standard concentration solution and reagent blank were run each time a set of samples was analyzed.

The composition of the standard pregnant uranium solution was found to be a critical factor in the loading of the resins.

---

1) Ref. AEB/NIM (Boydell) Oct. 79

Variations in the EMF of the solution, the iron content, the pH, the sulfate concentrations as well as an aging effect of the solution, affected the loading of the uranium on the resin. The following was the composition of the pregnant solution:

(1)	U_3O_8	0.18±.005 g/l
(2)	Fe^{+3}	2.5±0.1 g/l
(3)	SO_4^{-2}	22.0±1.0 g/l
(4)	pH	1.70±0.01

Table 1

Characteristics of Resin Used in Study

Sample	Type	Kind	Total Capacity		Strong Base Capacity		Water Retention, %	Porosity %
			meq/gm	meq/ml	meq/gm	meq/ml		
A	I	Gel	4.4	1.40	3.50	1.10	55	-
B	I	Gel	4.2	1.10	4.00	1.00	52	-
C	I	Macro	3.9	1.15	3.90	1.10	56	10
D	I	Macro	3.9	1.3	3.9	1.25	53	30
E	II	Macro	3.6	1.3	3.5	1.25	51	20
F	II	Macro	3.5	1.1	3.3	1.05	55	10
G	II	Macro	3.8	1.1	3.6	1.05	58	35

Table 2

Slope and Intercept Values

Sample	Type	Kind	Slope	Intercept
A	I	Gel	6.54	0.005
B	I	Gel	5.13	0.00
C	I	Macro	2.87	0.00
D	I	Macro	4.41	0.016
E	II	Macro	4.98	0.00
F	II	Macro	4.35	0.00
G	II	Macro	5.04	0.013
H*	-	-	4.44	0.014

*South African Standard

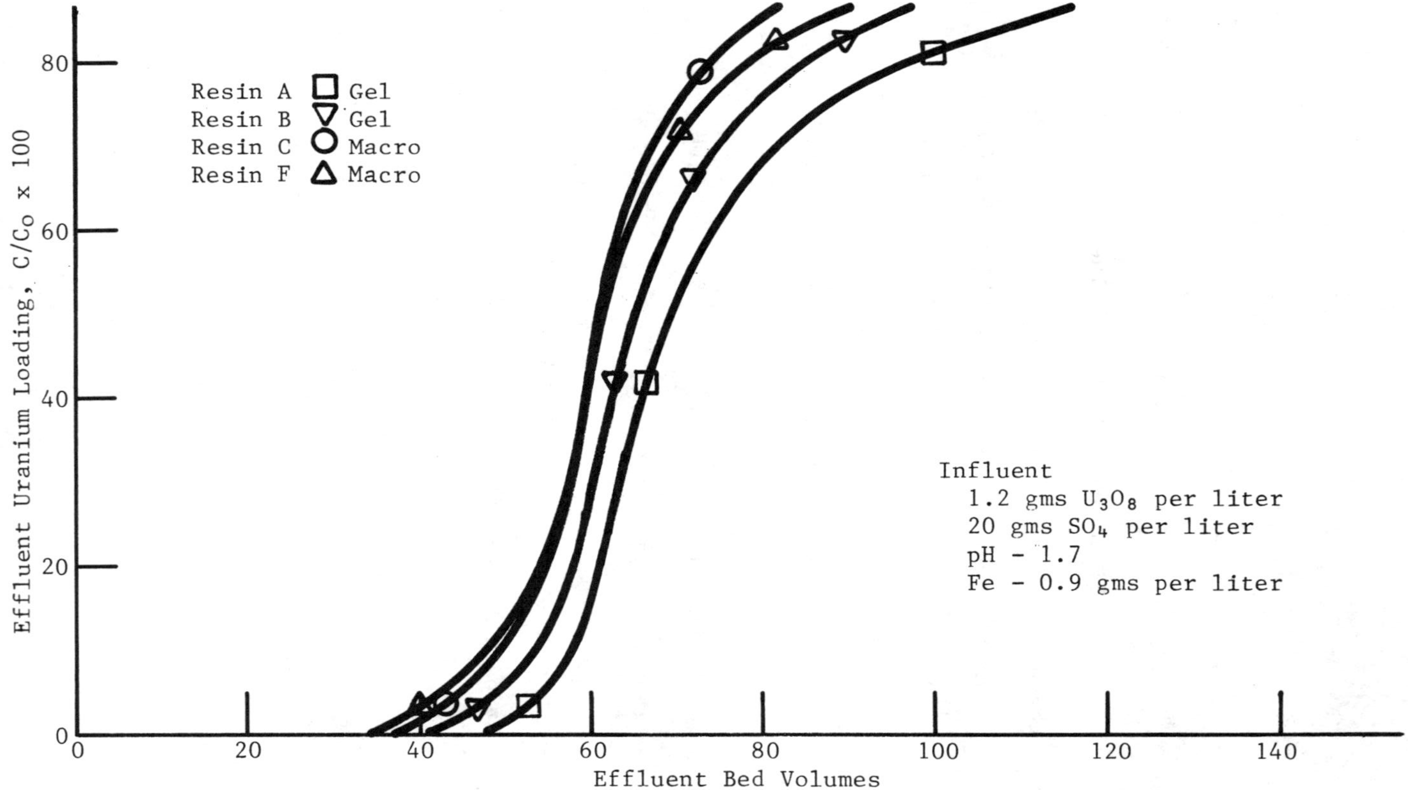

Figure 1: Loading Uranyl Sulfate on Strong Base Anion Exchangers

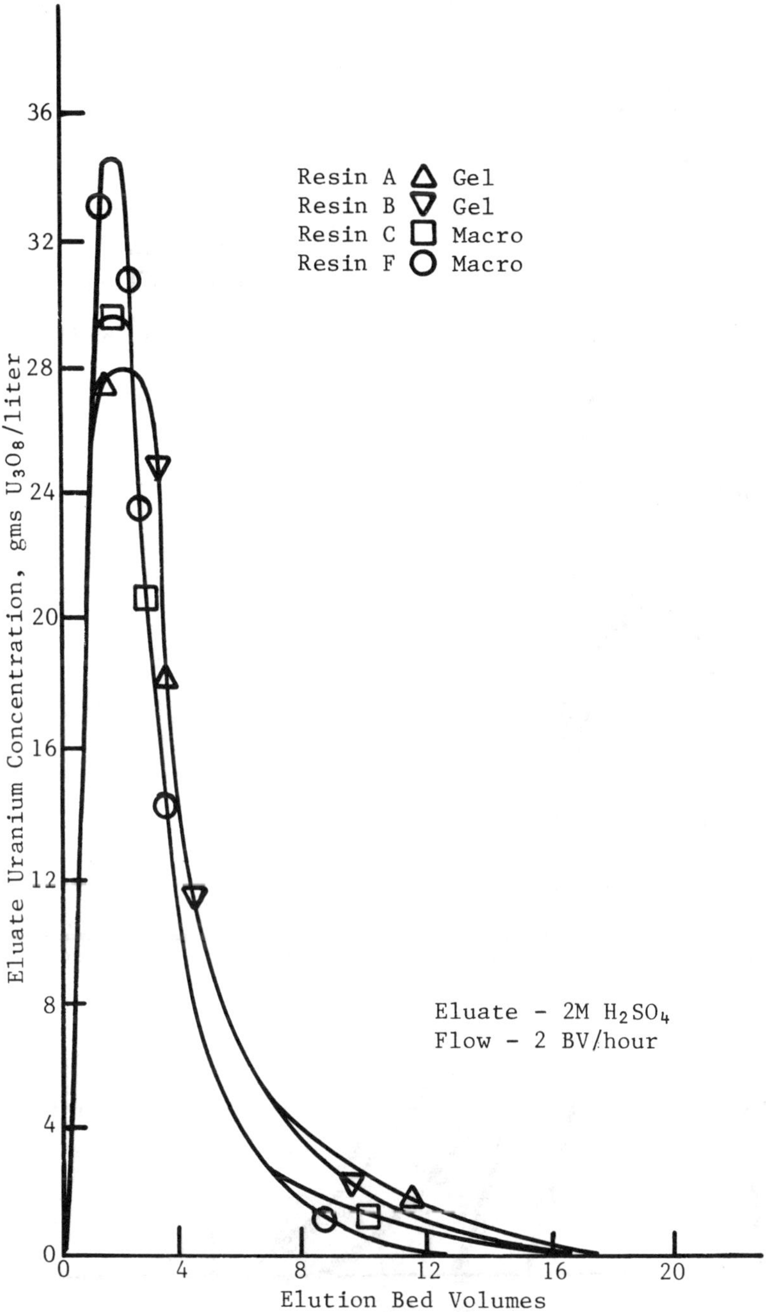

Figure 2: Elution Curves Using 2M Sulfuric Acid

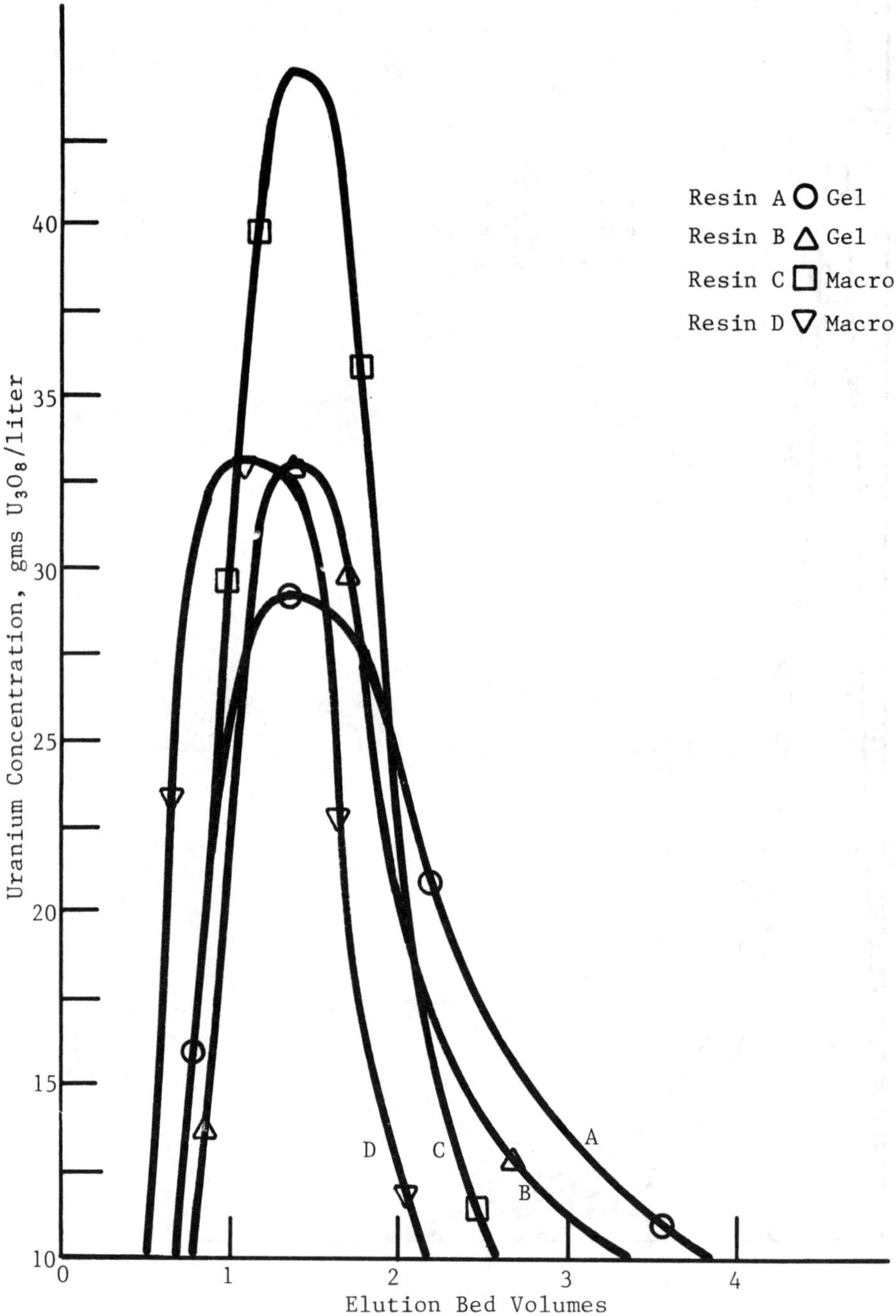

Figure 3: Elution Curves Using 1M Ammonium Bicarbonate

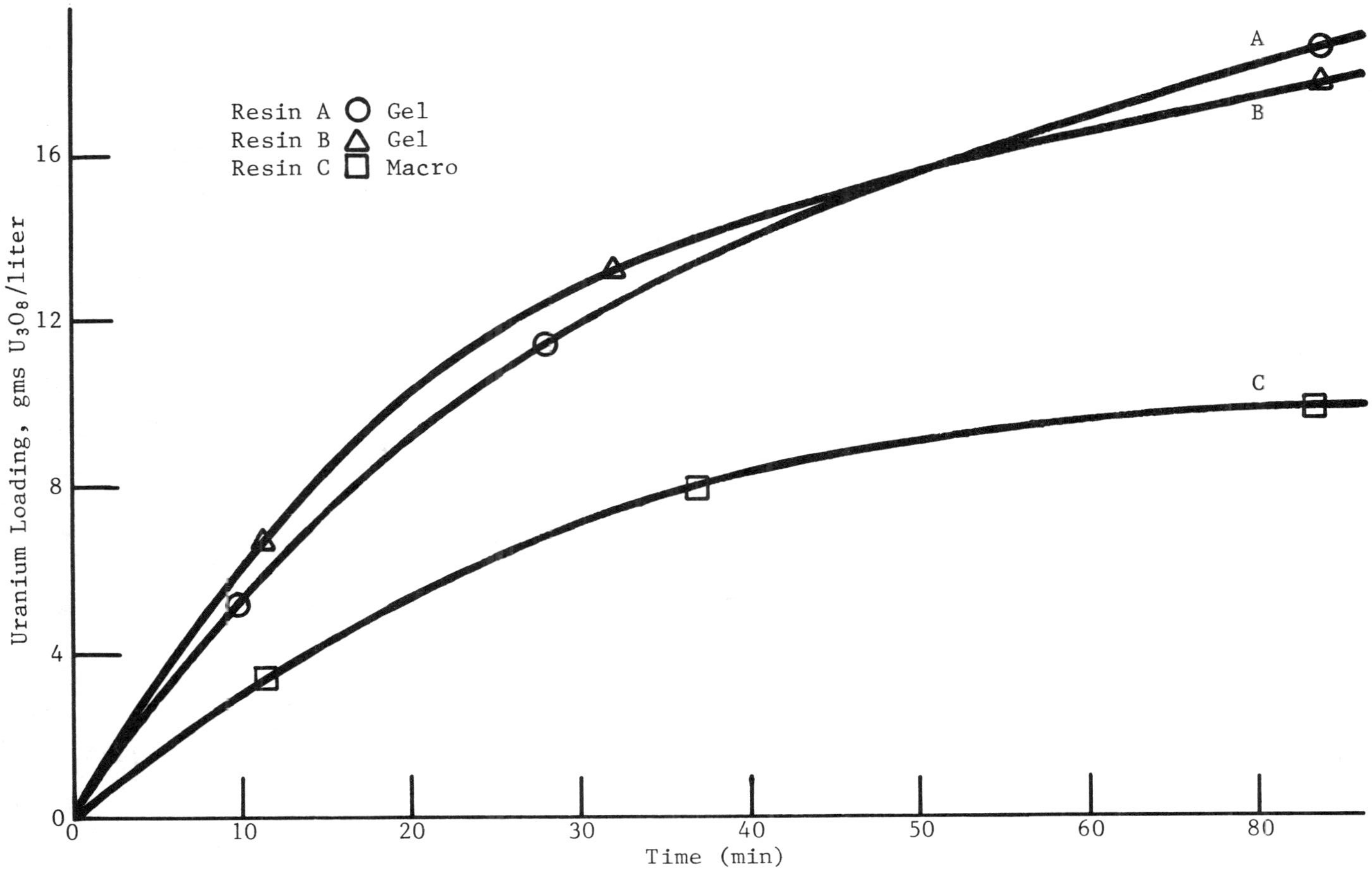

Figure 4: Loading of Uranium as Func-ion of Time, Resins A, B and C

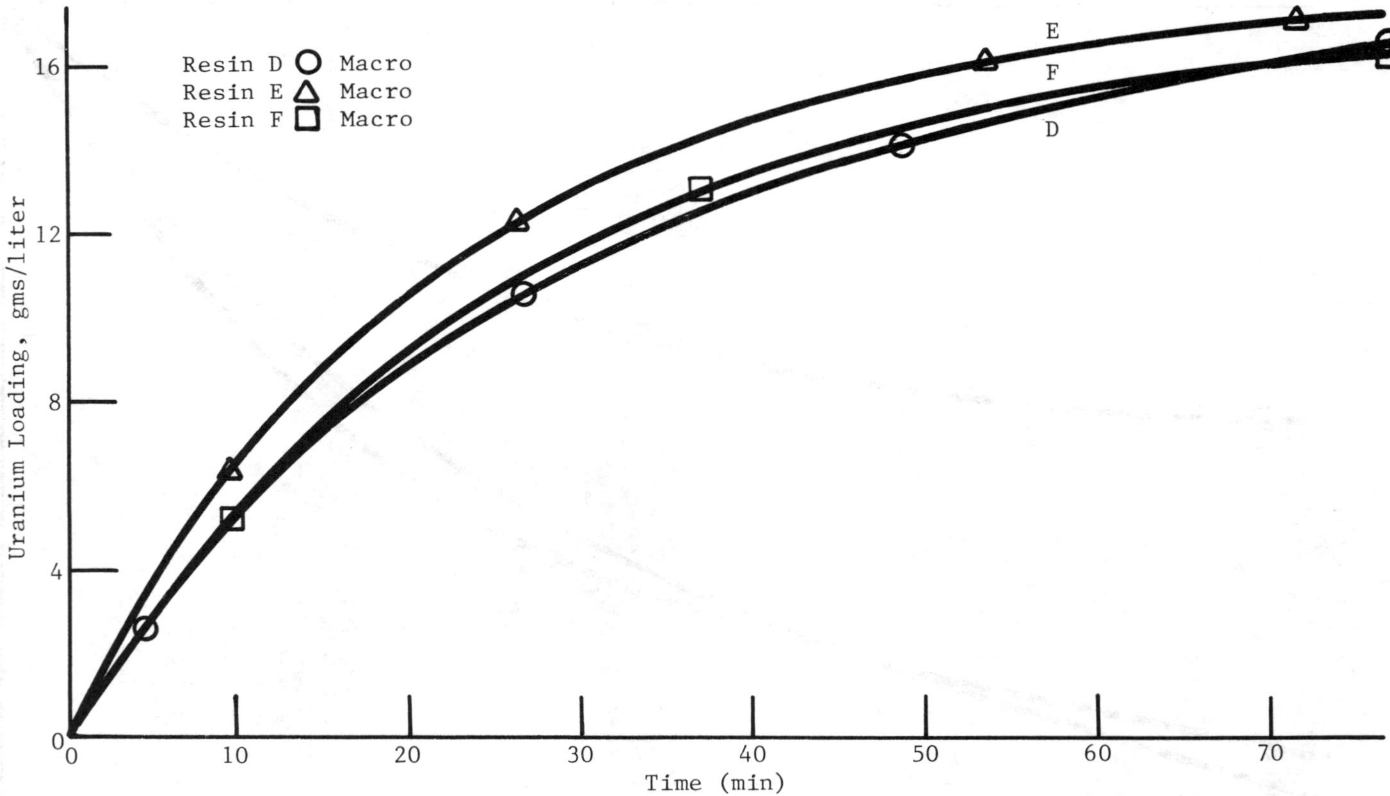

Figure 5: Loading of Uranium as Function of Time, Resins D, E and F

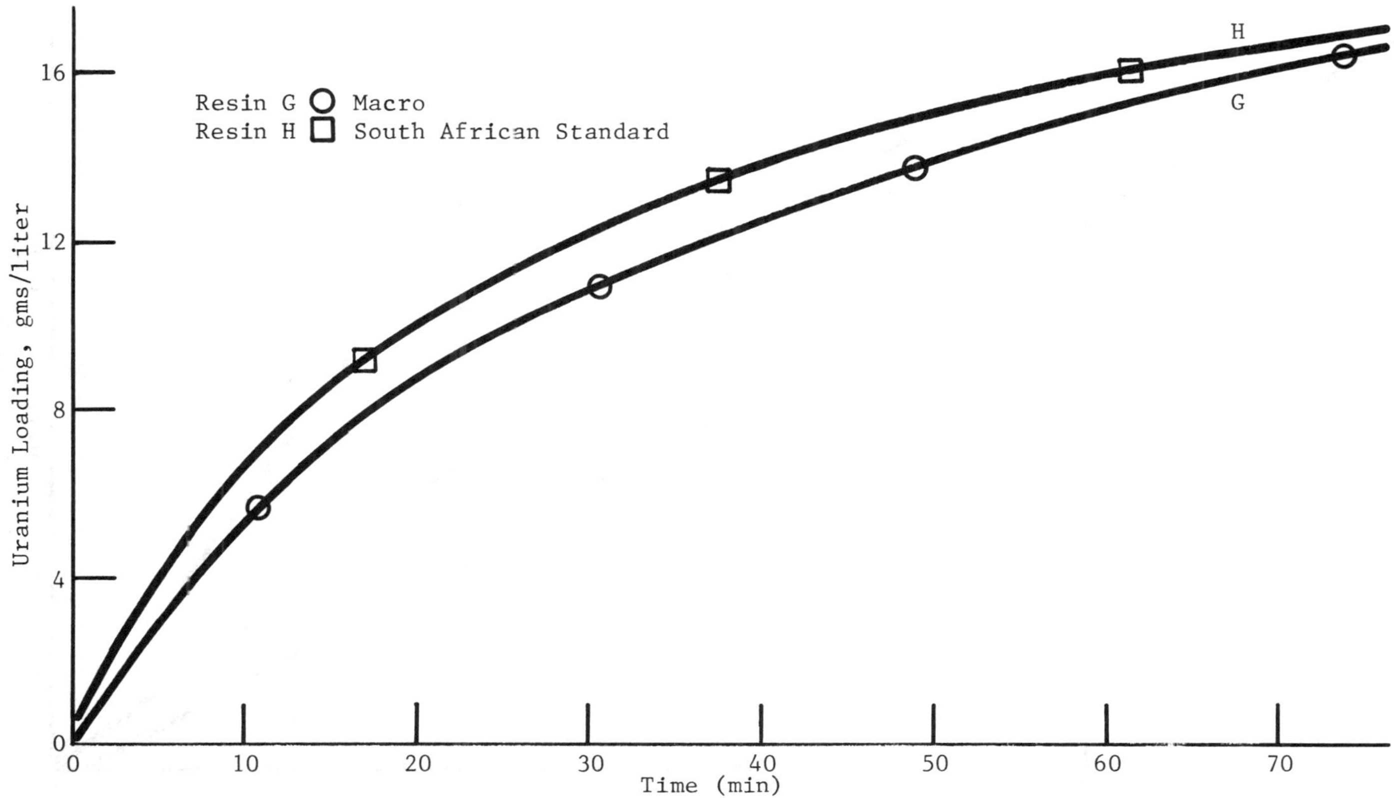

Figure 6: Loading of Uranium as a Function of Time, Resin G and South African Standard

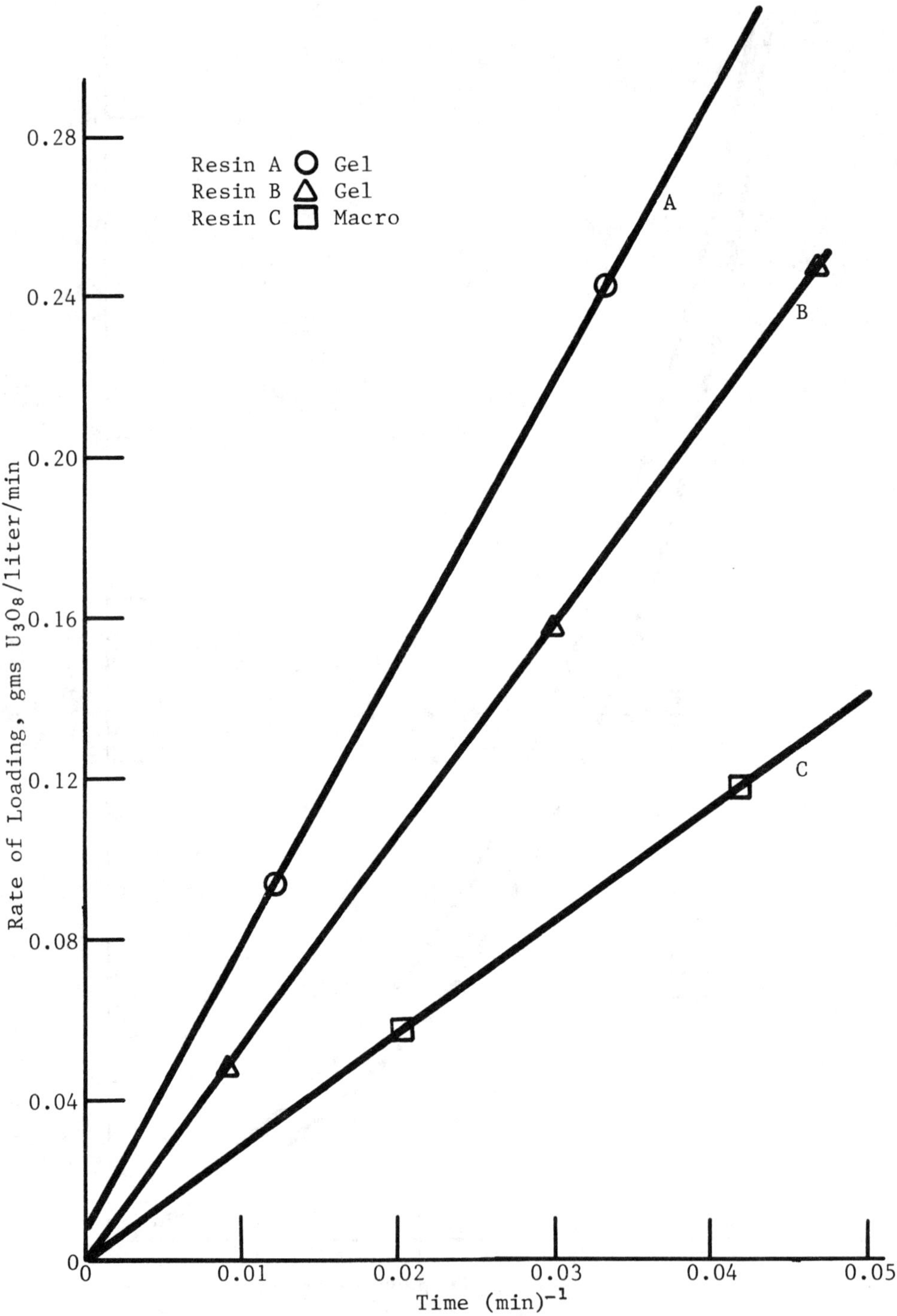

Figure 7: Rate of Uranium Loading as an Inverse Function of Time

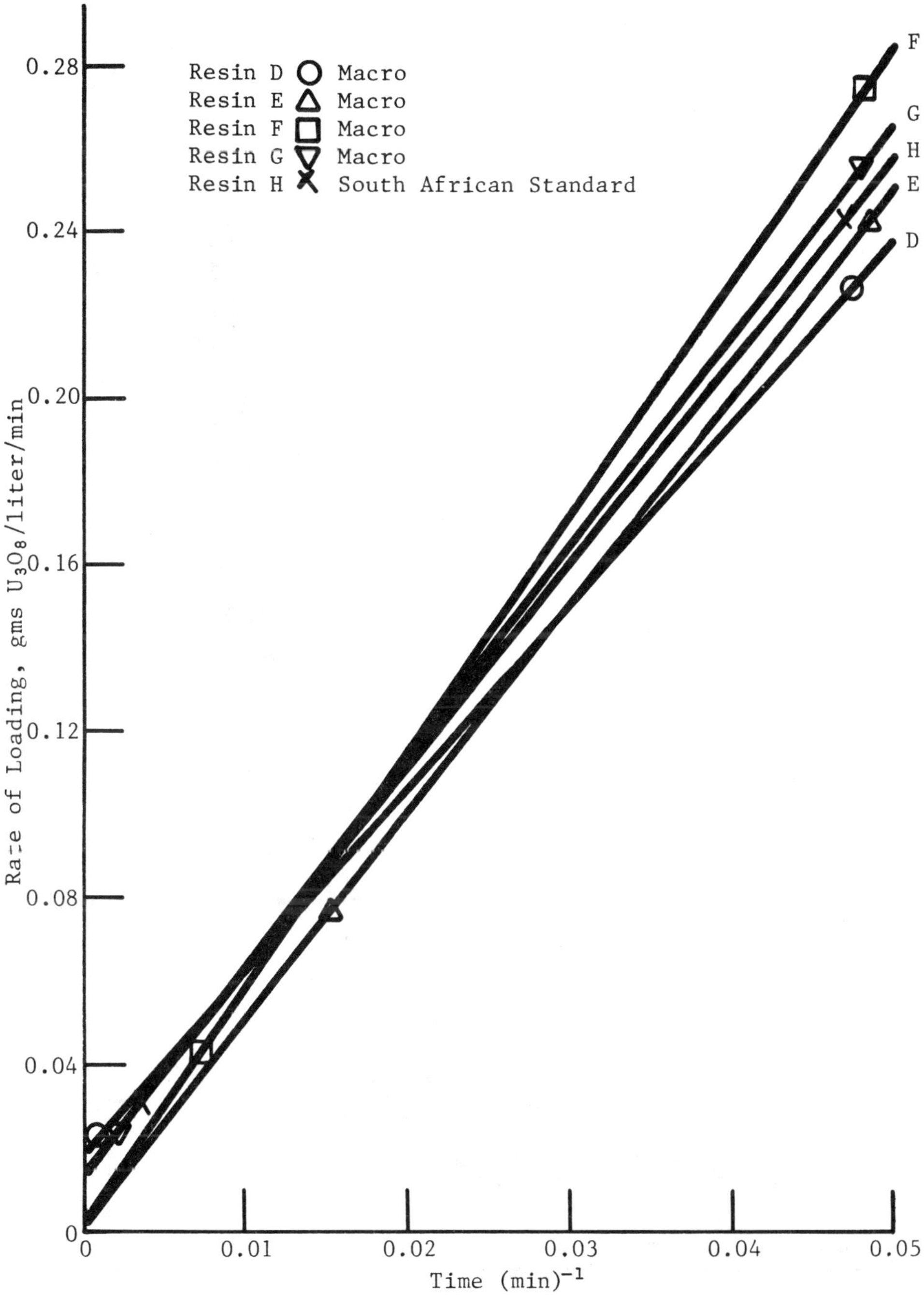

Figure 8: Rate of Uranium Loading as a Inverse Function of Time

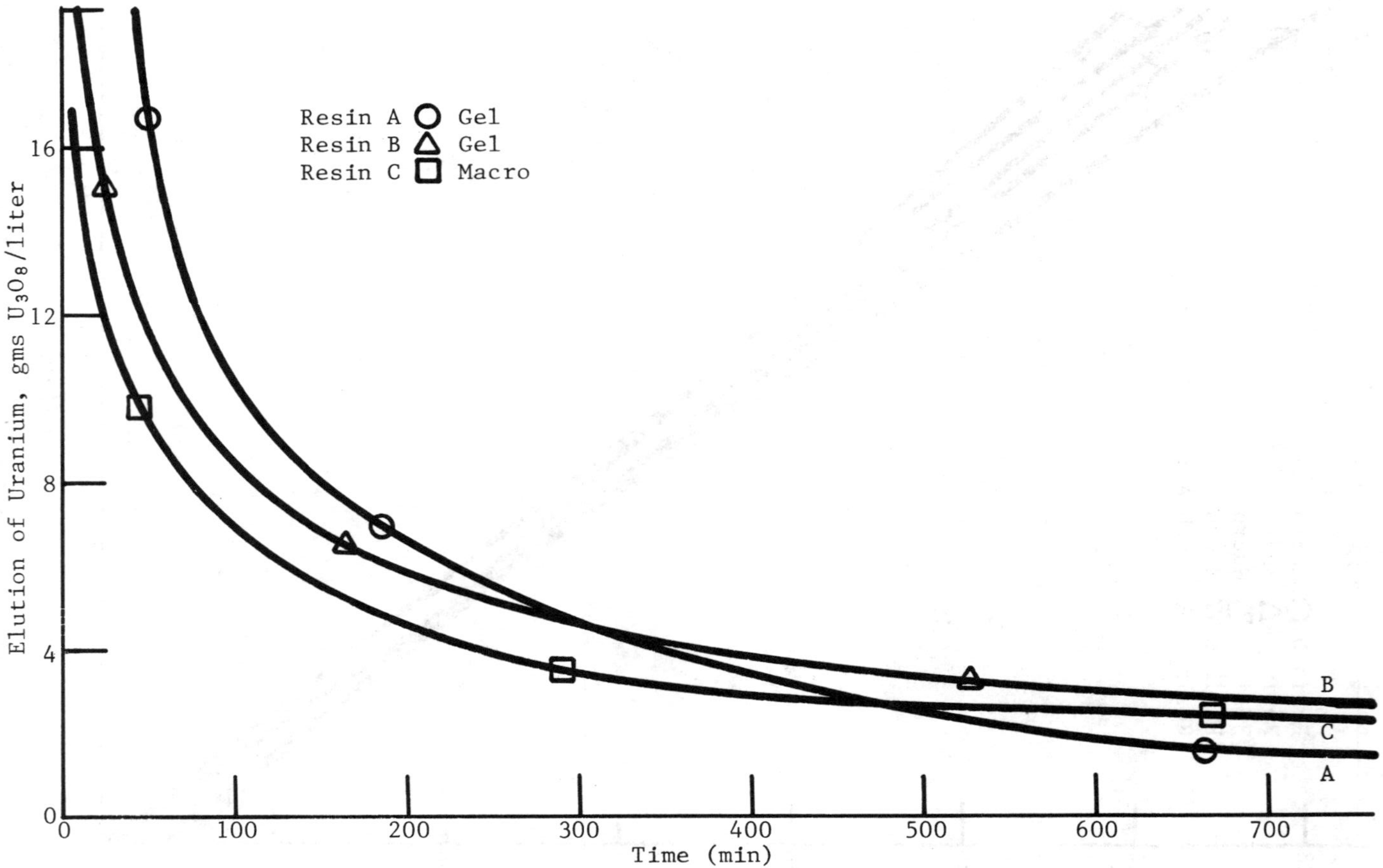

Figure 9: Stripping of Uranium by Sulfuric Acid

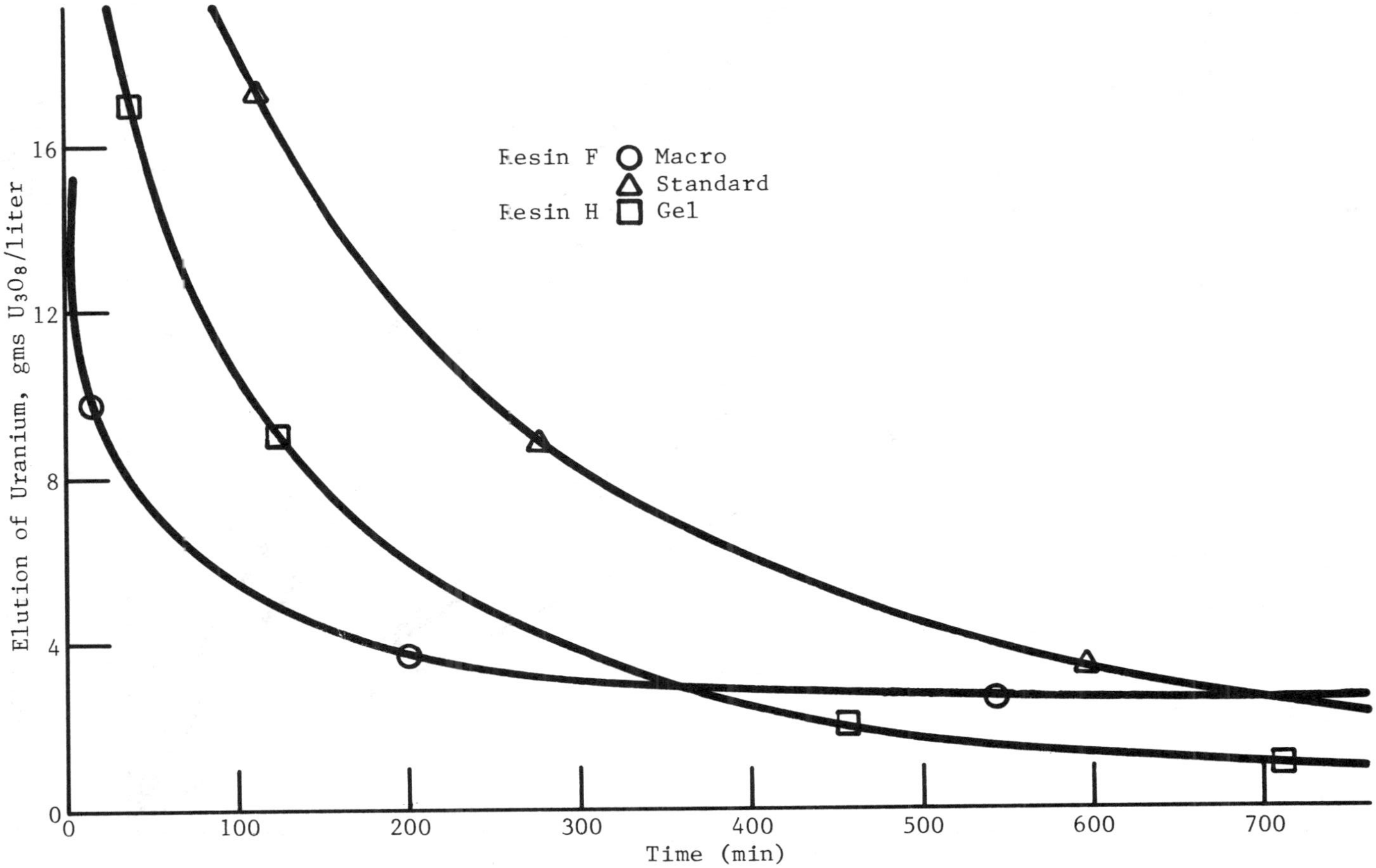

Figure 10: Stripping of Uranium by Sulfuric Acid

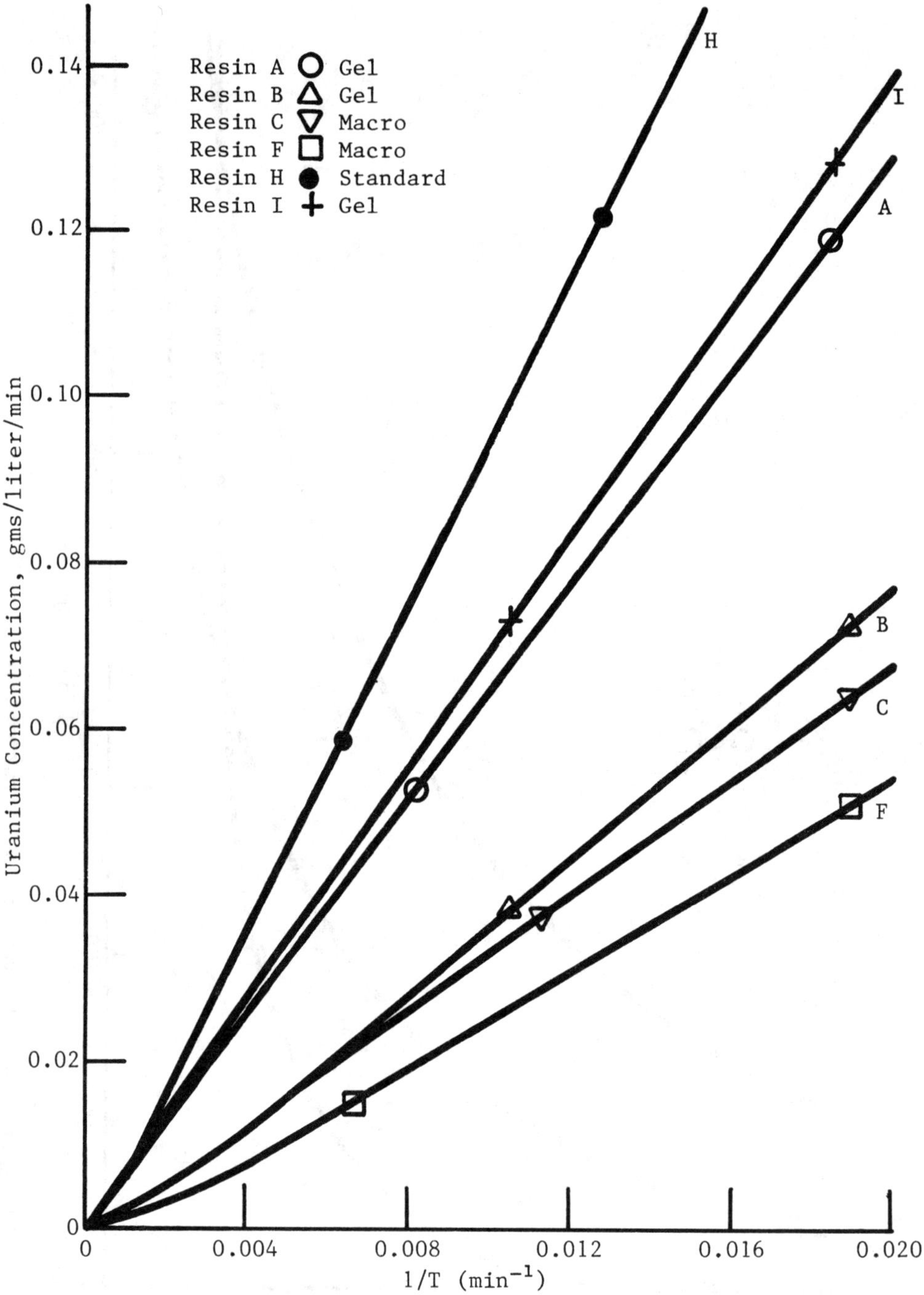

Figure 11: Rate of Elution as Function of Inverse Time

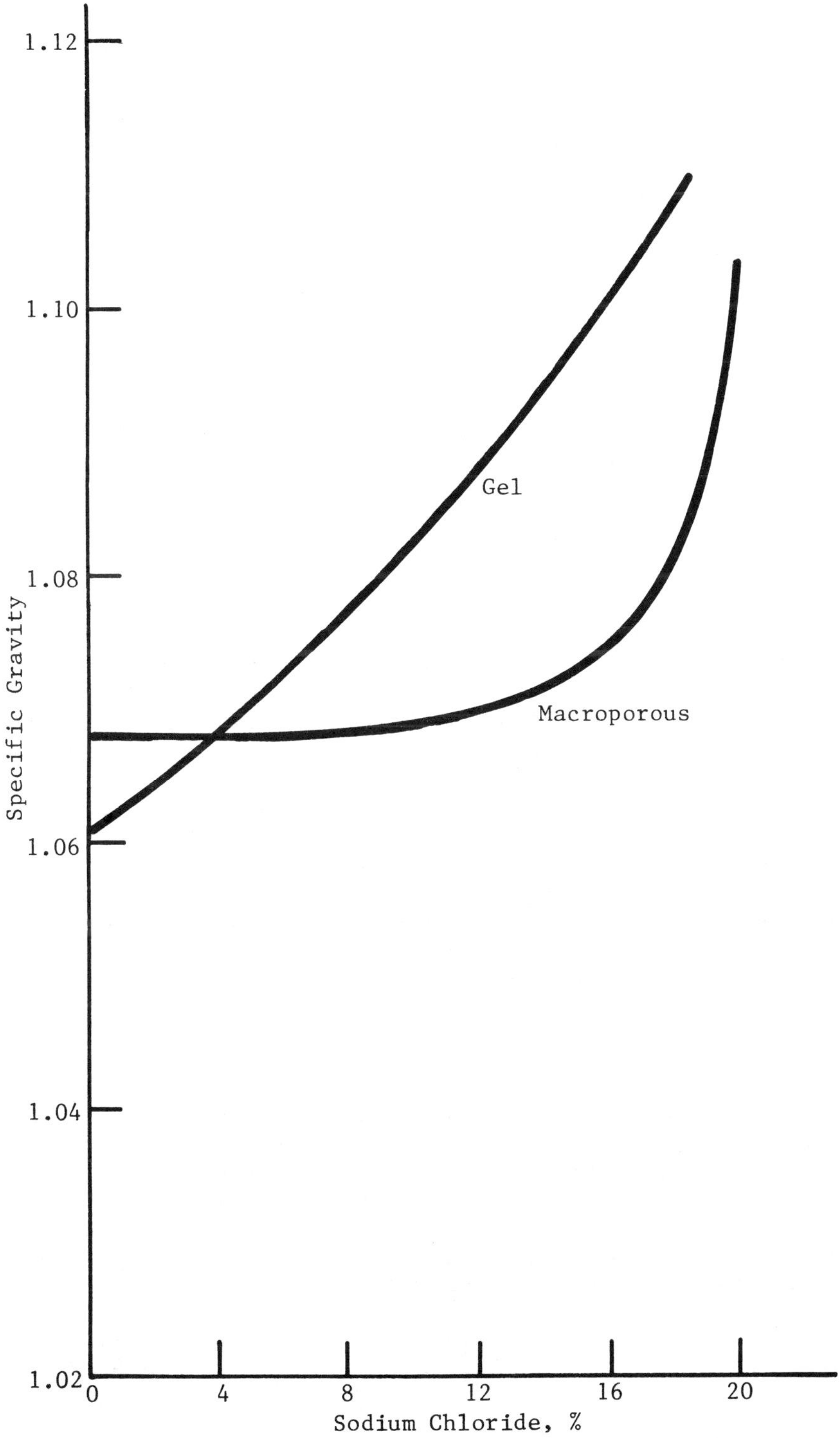

Figure 12: Specific Gravity of Resin Beads as a Function of Salt Concentration

SALT CONSERVATION, SELECTIVITY REVERSAL AND BREAKTHROUGH DETECTION IN ION EXCHANGE FOR NITRATE REMOVAL

Dennis Clifford, Ph.D., P.Eng.
Liou-Liang Horng and Chieh-Chien Lin
Environmental Engineering Program
University of Houston, University Park
Houston, Texas 77004

ABSTRACT

A single-bed ion-exchange process for nitrate removal from water supplies has been studied extensively in the laboratory and in the field. In the conventional mode, nitrate removal is achieved by passing a portion of the feedwater through a fixed bed of chloride-form, strong-base anion resin, and blending the nitrate-free effluent with raw water to produce a blended product water with less than 10 mg/L NO_3-N.

The theoretical and actual effects of sulfate concentration, sulfate/nitrate selectivity, resin type and empty bed contact time were studied during the exhaustion cycle. During pilot-scale field tests in Glendale, Arizona it was found that the end of a run could be reliably determined by bed volumes throughput or by the pH increase (0.6-1.0 pH) which accompanied nitrate breakthrough.

Regeneration was accomplished using NaCl but was not particularly efficient. In fact, for complete regeneration (3.5 eq Cl^-/eq resin), 10 eq of Cl^- were typically required for each eq of nitrate removed from the feedwater. It was discovered that substantial savings in NaCl could be achieved by using dilute (0.25-0.50 N NaCl) regenerants which minimize the effect of electroselectivity reversal on sulfate elution. Also, partial regeneration (1.0 eq Cl^-/eq resin) followed by 7.0 mg/L NO_3-N leakage during exhaustion (using a 20 mg/L NO_3-N feedwater) was found to be feasible and less costly than complete regeneration with zero leakage and bypass blending.

A 4000 m^3/day (blended product water) treatment plant to reduce nitrate from 21 mg/L to less than 10 mg/l in a 600 mg/L TDS background containing 100 mg/L SO_4^{2-} would cost about $300,000. Operating costs including capital recovery would be approximately 0.07 $/$m^3$ of product water.

INTRODUCTION

Nitrate-N levels exceeding either the WHO European "not recommended" limit of 22.6 mg/L [1] or the USEPA maximum contaminant limit (MCL) of 10.0 mg/L [2] are found in groundwaters or surface waters throughout the world. At present, single-bed chloride-form anion exchange using strong-base resins is, for economic reasons, the treatment process of choice. Fixed [3] and continuous [4] beds are being used in a very limited number of full-scale applications.

While alternative non-polluting processes [5, 6] are being developed, efforts are also being made to minimize the salt usage in the chloride exchange process. These salt conservation efforts are the primary focus of this paper. Nitrate breakthrough detection is also described. The importance of breakthrough detection cannot be overstated because nitrate is concentrated in the ion exchange process and can typically be eluted from the column at a concentration between 1.5 and 3.0 times its feed concentration.

Some of the research results presented here were obtained from bench scale laboratory studies conducted at the University of Houston. Most of our results, however, were obtained using the UH/EPA Mobile Drinking Water Treatment Facility [7], shown in Figure 1.

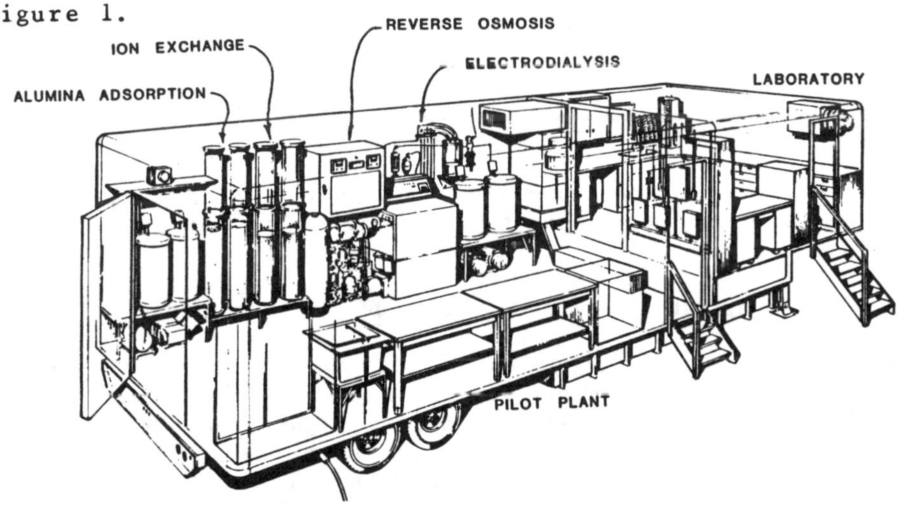

FIGURE 1 Interior drawing of the mobile Drinking Water Treatment Research Facility built and operated by the University of Houston for the U.S. Environmental Protection Agency.

For the nitrate removal studies, it was located in Glendale, Arizona at a well which had been removed from the water distribution system due to high nitrate levels. The raw water contained 19-25 mg/L NO_3-N, 123 mg/L Cl^-, 42.5 mg/L SO_4^{2-}, 124 mg/L HCO_3^-, 532 mg/L TDS, and 23.1 mg/L SiO_2; it had a pH of 8.0. The nitrate removal field studies in Glendale took place during a 15-month period ending in September of 1982.

Considerable reference is made in this paper to the salt conservation research results presented by Guter [8]. He operated a chloride ion-exchange pilot plant for several years in McFarland, CA which is now constructing a full-scale plant to produce 3800 m^3 of drinking water containing approximately 7.0 mg/L NO_3-N.

CHLORIDE-FORM ANION EXCHANGE PROCESS

Selectivity Sequences

The nitrate removal concept is simple enough. Feedwater anions including nitrate are exchanged for chloride on a strong-base anion resin. The nitrate-free effluent is blended with bypass raw water to produce a product water of acceptable nitrate concentration. Just prior to nitrate breakthrough, if this point can be detected, column exhaustion is terminated and the resin is either completely ($>$ 95% NO_3-N removal) or partially regenerated with NaCl. The exhaustion and regeneration steps are shown in Figure 2.

The keys to understanding and operating the process are (1) the selectivity sequence during exhaustion and (2) the selectivity reversal during regeneration. In dilute feedwaters ($<$ 0.010 N), all commercially available resins exhibit the following preference for the common anions:

$$\text{sulfate} > \text{nitrate} > \text{chloride} > \text{bicarbonate}$$

In more concentrated solutions ($>$ 0.010 N), as found in saline groundwaters, the positions of sulfate and nitrate are reversed and the sequence is

$$\text{nitrate} > \text{sulfate} > \text{chloride} > \text{bicarbonate}$$

In very concentrated solutions ($>$ 0.25 N) like those used during regeneration, divalent sulfate becomes the least preferred ion, much less preferred than even chloride. Figure 3 depicts a pair of nitrate/sulfate binary ion exchange isotherms at two different levels of total concentration. Sulfate is preferred about 3 to 1

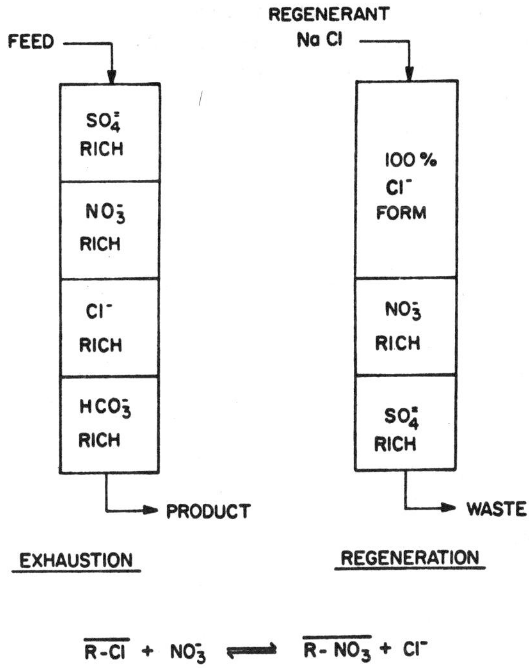

FIGURE 2 Exhaustion and regeneration steps in the chloride-form ion exchange process for nitrate removal. Sulfate bypasses nitrate during cocurrent regeneration.

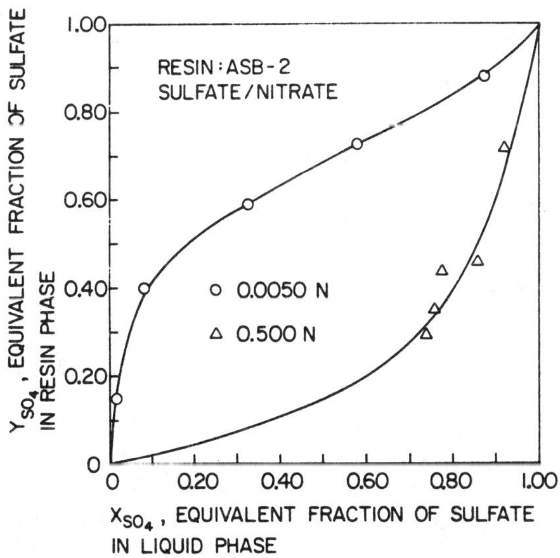

FIGURE 3 Binary sulfate/nitrate ion exchange isotherms showing selectivity reversal between dilute and concentrated solutions for a quaternary, type 2 resin.

in dilute (0.0050 N) solution while nitrate is preferred by about 4 to 1 in concentrated solution (0.500 N). Such a selectivity reversal does not occur in monovalent/monovalant exchange as indicated by the single nitrate/chloride isotherm, independent of concentration, shown in Figure 4.

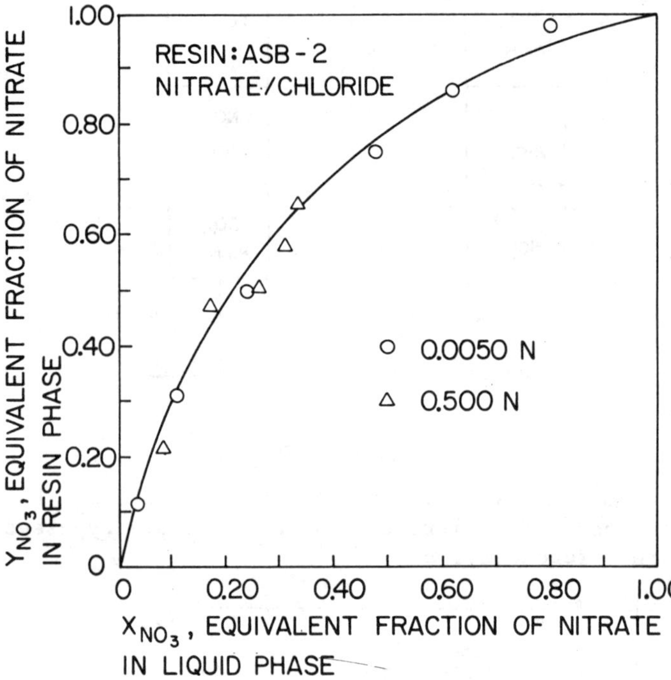

FIGURE 4 Binary ion exchange isotherm for monovalent/monovalent exchange in dulute and concentrated solutions for a quaternary, type 2 resin.

This selectivity reversal during cocurrent regeneration causes sulfate to bypass nitrate during regeneration as shown in Figure 2. Consequently, the least preferred sulfate is eluted almost immediately by NaCl during regeneration as depicted in Figure 5--a typical concentration history of sulfate and nitrate eluted from an exhausted nitrate removal column. In studying Figure 5, it is evident that partial regeneration is feasible because some nitrate is eluted with the sulfate. In the case shown, 4 BV of regenerant would elute nearly all the sulfate and about one half of the nitrate. If this partially regenerated resin were mixed, it could be put back into nitrate removal service as described originally by Gauntlett [9] and later by Guter [8]. High nitrate leakage results from such partially regenerated columns and bypass blending must be eliminated if the NO_3-N level is to be kept below the standard.

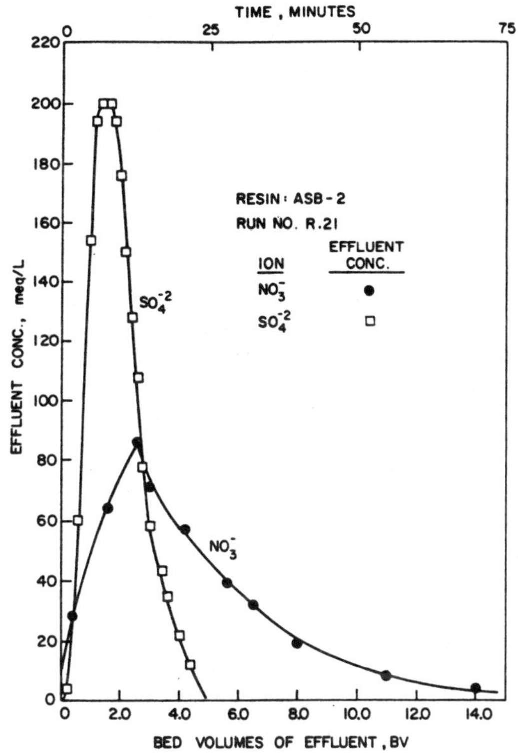

FIGURE 5 Typical elution curves for sulfate and nitrate during regeneration with 0.5 N NaCl of a nitrate removal resin column exhausted with simulated Glendale Water.

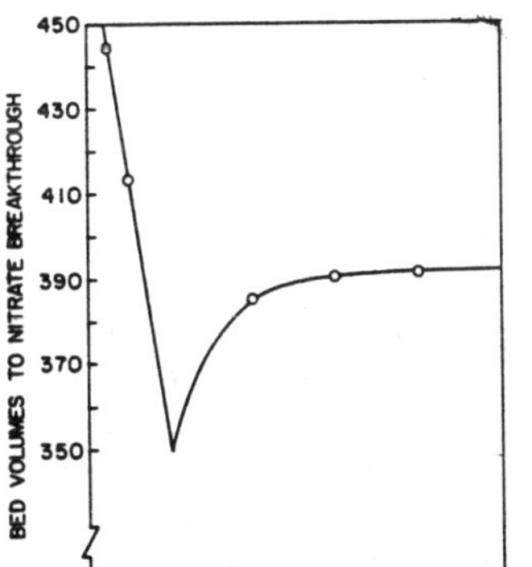

FIGURE 6 The effect of sulfate/nitrate separation factor, $\alpha_{S/N}$, on the amount of water (BV) which can be treated before nitrate breakthrough with Glendale, AZ raw water using quaternary type 2 resin in the chloride form.

The ionic strength of the aqueous phase and the resin properties, e.g., matrix and functionality, are the factors which determine when a given resin will start to prefer divalent ions like sulfate over monovalent ions like nitrate and chloride. The determinants of divalent/monovalent selectivities of resins as related to nitrate removal have been studied by Clifford [6] and more recently by Guter [8].

Determinants of Selectivity

The distance of charge separation in the ion exchange polymer was proposed by Clifford [6, 10] as being the primary determinant of divalent ion selectivity. Divalent ions, it was reasoned, prefer two closely spaced charges as are found in anion reins with nitrogen in the matrix (acrylic, epoxies) as opposed to pendant from it as in the polystyrene resins. Using experimental polystyrene resins with large quaternary groups like $R\text{-}N(CH_2CH_3)_3^+$ and $R\text{-}N(CH_2CH_2CH_3)_3^+$, Guter showed that sulfate was less preferred than nitrate even in very dilute (< 0.010 N) solution. It appears that the larger the quaternary groups, the farther apart they must be and the less likely it is for them to satisfy divalent ions.

Of course, distance of charge separation cannot fully explain a selectivity sequence. Many physicochemical factors are involved. Another important one is the polarity of both the matrix and functional groups. Nitrate much prefers hydrophobic phases, so making the quaternary groups larger and larger by adding aliphatic $-CH_2$ groups helps in two ways to improve nitrate/sulfate selectivity. The resin becomes a more hydrophobic environment for nitrate and the charged sites are forced further apart, making it more difficult for sulfate to be accommodated [6, p. 117].

Nitrate Breakthrough Prediction

The BV's treated before nitrate breakthrough is not a simple function of the sulfate/nitrate separation factor, $\alpha_{S/N}$, as can be seen in Figure 6. If the resin prefers sulfate ($\alpha_{S/N} > 1.0$), an increase in $\alpha_{S/N}$ actually improves the column run length. For the nitrate selective resins, run length is continually improved as $\alpha_{S/N}$ decreases. The shortest runs are predicted for the case of $\alpha_{S/N} = 1.0$. The plot in Figure 6 is based on multicomponent chromatography theory as proposed by Helfferich [11] and practiced by Clifford [12]. It has been found to be a powerful tool in predicting the influences of the many possible process variables on nitrate breakthrough.

EXHAUSTION OF CHLORIDE-FORM RESIN

Effects of EBCT and Resin Type

Chloride ion exchange using strong-base anion resins was the only ion exchange process studied at Glendale, AZ. All but two of the 30 ion-exchange column runs were done in a 20.3 cm (8 in) diameter column with a bed depth of 75 cm (29.5 in) containing 24.3 L (0.86 ft^3) of resin.

Neither the particular strong-base resin used nor the empty bed contact time (EBCT) had significant effects on the nitrate breakthrough point which typically occurred near 400 BV. The resin and EBCT variations are shown in Table 1 and Table 2 while Figure 7 illustrates a typical set of breakthrough curves for all the major ions in the feedwater. The variable nature of the column effluent is clearly seen in this figure.

TABLE 1. Effects of Anion Resin on Bed Volumes to Nitrate Breakthrough

Resin,	BV, L	Resin Capacity meq/mL	BV to NO_3^- Breakthrough	THROUGHPUT $\frac{\text{equiv. ions fed}}{\text{equiv. resin capacity}}$
ASB-1	47.0	1.39	414	2.39
ASB-2	47.0	1.33	414	2.43
ASB-2	24.3	1.33	404	2.43
DOWEX-11	24.3	1.17	346	2.27

Note: Throughput compensates for variations in resin capacity. ASB-1 is a quaternary, gel type 1; ASB-2 is a quaternary, gel, type 2 resin.

TABLE 2. Effect of Empty Bed Contact Time on Bed Volumes to Nitrate Breakthrough for ASB-2 Resin

Loading Rate gal/minft^2	EBCT min.	BV to NO_3^- Breakthrough
1.95	9.44	406
4.54	4.05	404

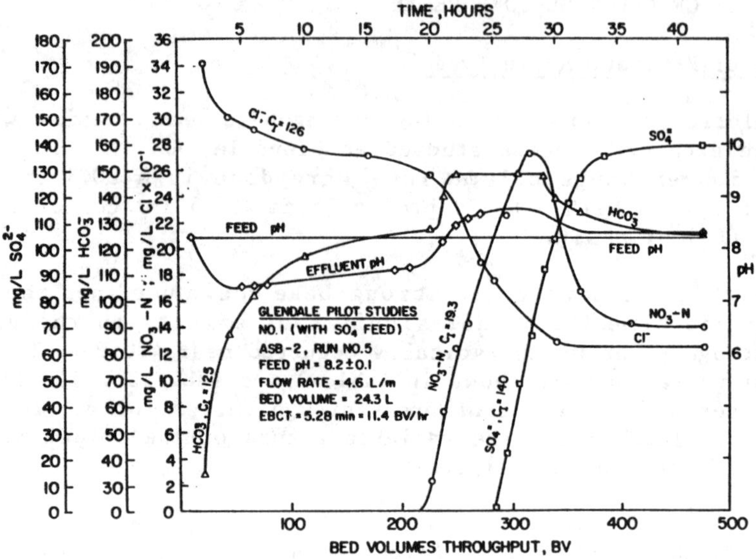

FIGURE 7 Breakthrough curves for the ions emanating from a completely regenerated chloride-form anion exchanger in Glendale, AZ. Note the pH change accompanying nitrate breakthrough.

Breakthrough Prediction by pH Change

It was noted in Figure 7 and verified in other pilot-scale and laboratory runs that a measurable pH change, in this case, about 1.0 pH unit, occurred at nearly the same time as nitrate breakthrough. The research engineer noticed this after the first few runs and thereafter could always shut the column down at nitrate breakthrough by monitoring the column effluent pH. The empirical rule is that when column effluent pH and influent pH are equal, the run must be terminated due to nitrate breakthrough.

We have found that the pH increase which occurs near nitrate breakthrough is due to a carbonate wave leaving the column at the same time as nitrate. The magnitude and sharpness of the nitrate-coincident pH change depend on the resin used, and the bicarbonate and sulfate concentrations. We believe that this pH change is reproducible and that it will usually be possible to use ΔpH to anticipate nitrate breakthrough in municipal nitrate removal applications. This conclusion is supported by much lab and pilot scale data from both nitrate and arsenate removal studies.

The Effect of Sulfate on Nitrate Breakthrough

As previously stated, sulfate is preferred to nitrate by all available strong-base anion resins. For this reason an increase in sulfate concentration decreases the bed volumes (BV) to nitrate breakthrough even if the total ionic concentration (meq/L) remains constant. This effect was demonstrated in the pilot studies by spiking the feedwater with sodium sulfate. The results are shown in Table 3 where it is seen that adding 100 ppm sulfate decreases the BV to nitrate breakthrough by 40 percent.

TABLE 3. Effects of Added Sulfate on Bed Volumes to Nitrate Breakthrough for ASB-2 Resin

SO_4^{2-} Conc. mg/L	Feed TDS meq/L	BV to NO_3 Breakthrough	BV to SO_4^{2-} Breakthrough
42.5	7.80	400	750
140	9.83	240	290
310	13.4	180	160

One beneficial effect of an increasing sulfate concentration is that it sharpens and enlarges the pH change which occurs at nitrate breakthrough. The pH change is very evident in Figure 7 at 140 mg/L sulfate; it is not early so sharp at 42.5 mg/L sulfate--the unspiked concentration of Glendale water.

REGENERATION WITH NaCl

Complete Regeneration--Effect of NaCl Concentration

Following an exhaustion run, the resin may be partially or completely regenerated. The usual ion exchange practice for demineralization and softening is complete regeneration to minimize contaminant ion leakage on subsequent runs. This typically involves using a two- to five-fold stoichiometric excess of regenerant. Our studies in Glendale indicated that the regenerant concentration (0.25 to 3.0 N NaCl) significantly affected the amount of excess regenerant required to remove at least 95 percent of the nitrate from the exhausted resin. Table 4 indicates that there is a direct relationship between the amount of NaCl required and NaCl concentration. The number of chloride ions required per resin site, i.e., equiv. Cl^-/equiv. resin, is cut in half as the NaCl concentration is reduced from 2.0 N (12 percent NaCl) to 0.5 N (3 percent NaCl). For a resin with a 1.2 equiv./L exchange capacity, this quartering of regenerant concentration will only double the required regenerant volume from 4.2 to 8.4 BV. The increased regenerant volume is insignificant because the regenerant can be made up using raw water. It is permissible to use raw water because a small amount of nitrate leakage (< 1.0 ppm) is tolerable during exhaustion and bypass blending.

The reason suggested for the increasing regeneration efficiency with decreasing NaCl concentration is based on a lessening of the selectivity reversal in sulfate/chloride exchange with the more dilute regenerants. Divalent sulfate becomes increasingly less preferred as the regenerant ionic strength increases and it is of less help in driving off the nitrate during cocurrent regeneration.

TABLE 4. Effect of NaCl Concentration on Regeneration Efficiency

NaCl N	NaCl Conc. %	equiv. Cl / equiv. resin	equiv. Cl- / equiv. NO_3^- removed
0.25	1.5	2.8	7.0
0.50	3	3.5	8.8
1.00	6	5.2	13.0
2.00	12	7.0	17.5
3.00	18	8.7	21.8

Partial Regeneration--Incomplete Nitrate Removal

Guter [8] has recommended partial regeneration, e.g., 1.0 equiv. Cl^-/equiv. resin to reduce NaCl consumption at the expense of high (6-10 ppm) nitrate leakage and elimination of the raw water bypass. Using both lab and pilot scale columns, we have verified that this is feasible but only if the partially regenerated resin is well mixed after regeneration. Extensive back washing as recommended by Guter didn't provide the required mixing and led to nitrate leakage in excess of the MCL in our Glendale pilot studies. We have eliminated this excessive leakage in laboratory columns by removing, physically mixing, then repacking the resins following partial regeneration. This elimination of excessive leakage can be seen in Figure 8 where the performance of an unmixed, heterogeneous bed is compared to that of a well-mixed, homogeneous bed. Both beds had been previously exhausted to nitrate breakthrough using resins in the pure chloride form and then regenerated with 1 equiv. Cl^-/ equiv. resin. This is partial regeneration since it takes 3.5 equiv. Cl^-/equiv. resin to strip at least 95% of this nitrate from the spent resin using 0.5 N NaCl.

A mathematical model [11, 17], which assumes equilibrium and constant separation factors, closely predicts the observed leakage from beds containing 10-30 percent nitrate and 70-90 percent chloride if they are homogeneous. A comparison between actual and predicted values is made in Figure 9. Here it can be seen that nitrate leakage is well predicted by constant separation factor multicomponent, chromatography theory assuming a constant feed of Cl^-, NO_3^-, HCO_3^-, and SO_4^{2-} to a homogeneous, two-component, presaturated bed containing 80% chloride and 20% nitrate.

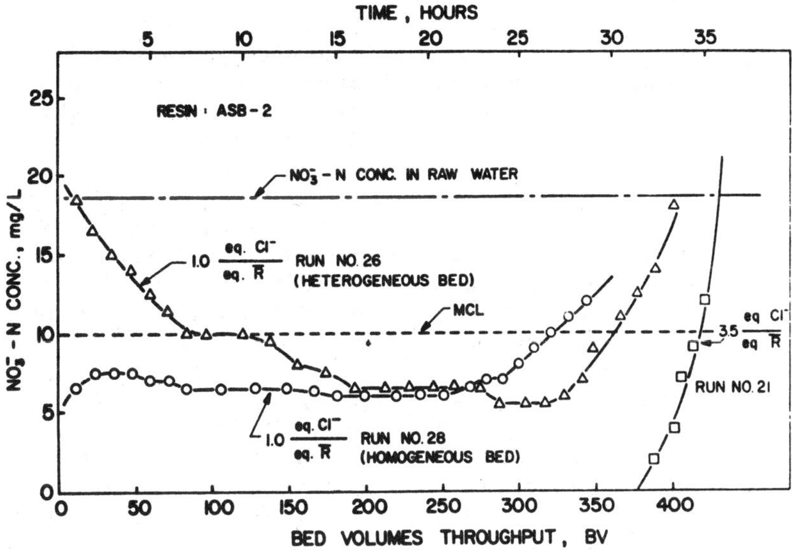

FIGURE 8 Nitrate breakthrough curves following complete regeneration (3.5 eqiv. Cl^-/equiv. resin) and partial regeneration (1.0 equiv. Cl^-/equiv. resin) with and without resin mixing. Column feed is simulated Glendale water.

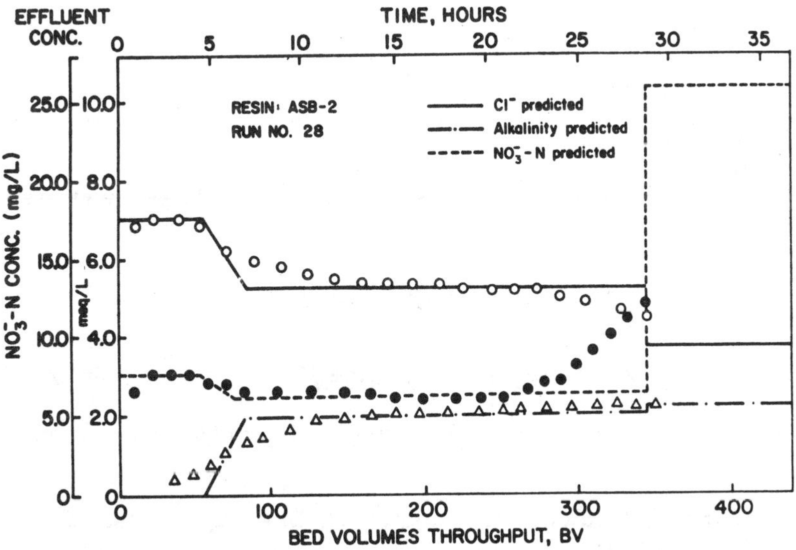

FIGURE 9 A comparison of predicted and actual concentration histories for ions emanating from a partially regenerated bed receiving simulated Glendale water feed.

APPROXIMATE COSTS FOR NITRATE REMOVAL

Rough cost estimates were made for nitrate removal by chloride ion exchange assuming that our pilot-scale findings and those of Guter [8] could be put into commercial practice. A rough estimate is 0.071 $\$/m^3$ (27 cents/1000 US gallons) of product water for 4000 m^3/day (10^6 gallon/day). This estimate includes salt consumption, capital recovery (12% interest), labor, energy, resin replacement every five years, and miscellaneous operating and maintenance expenses. Further assumptions in the cost estimate were nitrate-N reduction from 21 to 7.0 mg/L, sulfate less than 100 mg/L, complete regeneration with 0.5 N NaCl, and a bypass flow = 30% of the blended product water flow.

Partial regeneration with 1.0 equiv. Cl^-/equiv. resin was compared to complete regeneration with dilute (0.25 N) NaCl on the basis of grams of NaCl required per liter of product water containing 7.0 ppm NO_3-N. After accounting for longer run length and raw water bypass in the complete regeneration process, it was found that partial regeneration used 15 percent less salt in treating the Glendale water. This modest advantage of partial regeneration tends to be offset by the smaller pH at nitrate breakthrough, the need to regenerate more often, and the requirement for air or mechanical mixing of the resin bed after regeneration. Guter [8] claims that the advantage of partial regeneration becomes greater as sulfate concentration increases.

REFERENCES

1. World Health Organization, "European Standards for Drinking Water," 2nd Ed., Geneva, 1970.

2. U.S. Environmental Protection Agency, "National Interim Primary Drinking Water Regulations," EPA-570/9-76-003, 1972.

3. Green, L.A. Water Pollution Control, V. 77, No. 4, p. 478, 1978.

4. Holzmacher, R.G., Water and Sewage Works, p. 210, July 1971.

5. Holl, W. and B. Kiehling, Water Research, V. 15, p. 1027, 1981.

6. Clifford, D.A. and W.J. Weber, Jr., EPA Report No. EPA-600/2-78-052, June 1978.

7. Clifford, D.A. and M.R. Bilimoria, Final EPA Report on Grant No. CR-806073, also University of Houston CE report UH CE 83XX.

8. Guter, G.A., USEPA Interim Report EPA-600/S2-81-029; also final EPA Report on Grant R-805900, 1982.

9. Gauntlett, R.B. Water Treatment and Examination, V. 24, Part III, p. 172, 1975.

10. Clifford, D.A. and W.J. Weber, Jr., Reactive Polymers, V. 1, p. 77, 1983.

11. Helfferich, F.G., Multicomponent Chromatography, Xerox Univ. Microfilms, Ann Arbor, originally Marcel Dekker, 1970.

12. Clifford, D.A., Ind. & Eng. Chem. Fundamentals, V. 21, p. 141, May 1982.

ACKNOWLEDGEMENT

This work was supported by a grant from the U.S. Environmental Protection Agency--Grant Number CR 807939, Mr. Sorg, Project Officer.

EQUILIBRIUM AND DIFFUSION RATE EFFECTS OF UNIVALENT AND DIVALENT IONS IN A BIFUNCTIONAL RESIN

Asaye Mamo, Thomas Heeb, Kent S. Knaebel

Department of Chemical Engineering, The Ohio State University
USA

INTRODUCTION

Although considerable effort has been invested in research of equilibria and rates of exchange for monofunctional ion exchange, little has been published on these topics for bifunctional resins. Nevertheless, bifunctional resins have been the central issue in research on fixed bed, thermal cycling processes, such as the Sirotherm process and cycling zone adsorption. A major portion of the potential applications of the thermal-cycling processes involves purification of dilute, aqueous, mixed salt solutions. Consequently, the properties of such resins, while in contact with mixed salt solutions, are relevant.

The preponderance of work on bifunctional resins and thermally cycled adsorption processes has been oriented toward single dissolved salts, the majority of which have been sodium chloride. In spite of this, many potential applications involve mixtures of dissolved salts, which have differing valences and the associated range of resin affinities. The high affinity of a divalent component would imply a diminished exchange capacity for a univalent component in a mixture. This would result in a reduced extent of purification of the mixture compared with an analogous solution comprised of just univalent dissolved salts.

These expectations are supported by observations of ion exchange with univalent and divalent anions of Liberti and co-workers (1,2,3). There is also direct analogy to competitive adsorption of mixed organic species on activated carbon by Fritz and Schlunder (4,5,6) and others(7,8).

Nevertheless, it is envisaged that the differences in properties (other than affinity) for different solute may lead to greater separations by selection of the cycling domain than would be available through ordinary adsorption with analogous conditions and properties. For example, relatively large ions of high selectivity may be separated from their smaller counterparts by exploiting the difference in diffusivities through high-frequency cycling of temperature in a fixed bed process. But, before estimates of the potential improvement of performance may be made, it has been necessary to measure the relevant equilibrium and diffusion rate parameters.

Thus, there are three major topics of this report: ion-exchange equilibria and temperature effect, ion-exchange rates, and estimated column dynamics, all based on univalent and divalent salts (NaCl, Na_2SO_4, $CaCl_2$) in contact with a bifunctional resin (Amberlite XD-5, which is a product of Rohm and Haas Company).

EQUILIBRIUM BEHAVIOR

In order to obtain a basis for evaluating the potential quality of separation in a thermally cycled process, equilibrium isotherms were obtained by batchwise experiments. The resin was prepared by thoroughly regenerating it in a fixed bed with deionized water at 95°C. The pH of the resin was adjusted to 5.7, which has been shown elsewhere (9) to be the value at which maximum exchange capacity is observed. Two samples of Amberlite XD-5 were examined, from lot numbers SW-79/0145 and LM-812. Initial equilibrium experiments with sodium chloride showed that the latter had significantly higher capacity than the former. This may have resulted, however, from the extreme nature of the regeneration, and it could be diminished in normal regeneration by hot feed solution.

Since the sample from lot number LM-812 exhibited superior capacity for NaCl, it was selected as the basis for comparison of capacities of univalent and divalent species.

The equilibrium experiments were performed in 25 m glass vials, each of which was filled with a mass of dry, regenerated resin and a volume of standard solution. For the experiments at 25°C, twenty-four hours elapsed before the equilibrium state was measured, while at 95°C the interval was twelve hours. The equilibrium composition was determined by extracting a 1 ml sample by a syringe and diluting that to make 50 ml. The equilibrium concentration was determined by measuring the conductance of the diluted sample at 25°C, then multiplying the resulting concentration by the dilution ratio. This technique was found to be accurate within 3%, by trials with standard solutions.

The results of the equilibrium experiments of XD-5 resin (lot number LM-812) are shown in Figures 1,2, and 3, respectively for NaCl, $CaCl_2$ and Na_2SO_4. The relative affinities may be interpreted from the distances between the respective isotherms for the components. The affinities are ordered as

$$Ca^{+2} >> Na^{+}$$
$$SO_4^{-2} > Cl^{-}$$

These are similar to those observed in common ion exchange materials for similar concentrations.

An unusual feature of these isotherm shapes is that, two distinct mechanisms appear to have influence on the capacity of the resin for these solutes. The mechanisms, as inferred from the equilibrium curves, amount to ordinary saturation of the functional groups and a separate effect that is dependent upon the solution composition, and which may be related to the effect of diminished selectivity for higher valence species as ionic strength increases, which is observed in common, monofunctional ion exchangers. If physical adsorption were present, rather than chemical exchange, the composition-dependent effect could be ascribed to the immobilization of the solute in the voids, rather than at specific sites. But, in that case the quantity of solute immobilized in the voids would be positively correlated with the concentration of the surrounding fluid (10).

Although the theory to account for the second mechanism is still being developed, it is apparent that the isotherms may be quantitatively expressed as

$$q^* = \frac{AC}{1 + BC} + DC$$

where A and B are parameters that are strongly dependent on temperature, but D is less so.

As mentioned previously, two samples of XD-5 resin were examined in terms of capacity for NaCl. It was found that the sample from lot number LM-812 exhibited much greater capacity at 25°C than did the sample from lot number SW-79/0145. These results are shown in Figure 4. The results at 95°C were indistinguishable, however.

Previous work with Amberlite XD-5 has recently been published by Hsu and Pigford (11). In fact, they present equilibrium isotherm data for NaCl, $CaCl_2$, and $MgCl_2$. They also observed regions of linearity and non-linearity of the isotherms, although they chose to split the solution concentration domain according to the behavior, and expressed the capacity in coupled equations. They also discussed several physical properties of the XD-5 resin.

In comparing their isotherms for NaCl with those in Figure 4, it appears that their sample was similar to that of lot number SW-79/0145. This is also implied by the difference in ultimate capacities for $CaCl_2$ (i.e., at high concentration). They observed an ultimate capacity of about 1.0 equiv/kg while the value shown here in Figure 2 is about 1.6 equiv/kg.

Finally, as will be discussed in the section on column dynamics, it is essential to have equilibrium data over the full range of conditions in order to develop an accurate mathematical analysis or design of a system. Accordingly, an effort has been made here to examine a broad range of compositions so that for most applications it will be unnecessary to extrapolate the equilibrium data.

RATE PHENOMENA

Two constraints on fixed bed, cyclic exchange operations are film diffusion and intraparticle diffusion. The latter may be viewed by a variety of mechanisms, however, which depend on the species, conditions, and structure of the solid phase. In contrast, film diffusion depends strongly on the fluid dynamics of the packed bed. As a result, for a given quantity of ion-exchange material (which is dictated by equilibrium constraints for a specific separation), the effect of film diffusion may be essentially eliminated by selection of appropriate column geometry.

Furthermore, in thermally cycled processes, it may be possible to improve the extent of isolation of ions which diffuse slowly by operating at a high cycling frequency. For example, if the time constant for diffusion were larger than the cycle period, the concentration profile within the ion exchange particles will not become uniform (12). Thus, separation may be enhanced by this rate limiting step.

In order to estimate the feasibility of this concept for specific components, it is necessary to measure their mass transfer coefficients that may be applied to the theory. Accordingly, several batchwise rate experiments were conducted with the XD-5 (lot number SW-79/0145) resin. The procedure involved exposure of resin to solutions of compositions that were different from that of prior equilibration. The apparatus was a 1.5 ℓ reaction flask with a spinning basket containing resin. Concentration was monitored via solution conductance, which was recorded by a microcomputer-based acquisition system.

The parameters that were examined include temperature and stirring speed. Particle size, initial solution concentration and initial resin loading were fixed. The concentration versus time data were analyzed by assuming that the rate limiting step was diffusion through the pores. This rate process is described by Fick's second law, in which the diffusivity is assumed to be constant. The equation is

$$\frac{\partial}{\partial t}(\varepsilon_p c + \rho\, q^*) = \varepsilon_p D\, \nabla^2 c$$

where c and q* represent the fluid and solid phase concentrations of solute in the pores, based on the surface area of the particle that is occupied by pores.

The solution is presented elsewhere (13), but the result is

$$F = \frac{C_o - C}{C_o - C_\infty} = 1 - \Sigma\, \frac{6\alpha(1+\alpha)\exp(-D^*\, t\, x_i/R^2)}{9 + 9\alpha + \alpha^2\, x_i^2}$$

where C,Co and C_∞ are the solution concentrations at any time, initially and at equilibrium, respectively. F is the fractional uptake of solute by the resin. The effective diffusivity is related to the pores fluid diffusivity is related to the pore fluid diffusivity by

$$D = (1 + \frac{\rho}{\varepsilon_p}\frac{\partial q^*}{\partial c})\, D^*$$

The results of the rate experiments are summarized in Table 1.

COLUMN DYNAMICS

In the preceding sections, the major constraints of separation performance have been discussed. It is apparent that the limits that are imposed by the equilibrium condition are more greatly dependent upon the species present than are the rate-limiting properties. Nevertheless, application of these constraints has two bearings. First, at low fluid velocities and low cycling frequencies, the fluid and ion exchange material may be essentially at equilibrium, at a given axial position. At the other extreme, that is high velocity and frequency, dissipation due to intraparticle diffusion may dominate the rate and therefore the extent of exchange.

It has been shown elsewhere that these two extremes may be dealt with by two analytical techniques. When the local equilibrium assumption is valid, nonlinearities in the isotherm may have a strong influence on the separation performance, and these may be taken into account by applying the method of characteristics to solve the continuity equation. The technique that is applicable to the case when dissipation is important requires linearization of the isotherm, as well as other effects that depend on composition or temperature. Additional details are discussed in (12).

One particularly useful, yet simple, result of the method of characteristics solution is the expression for determining the maximum concentration shift that may be encountered upon a shift in temperature. This is expressed as

$$K C_1 + q_1^* = K C_o + q_o^* = F$$

where the subscripts o and 1 refer to alternate half cycles, while F and K are constants for a particular set of operating conditions. By inserting the form of the isotherm that was suggested in a previous section, i.e., having additive Langmuir and linear concentration dependence, yields a simple quadratic equation that may be solved directly. The result is nearly identical to that obtained from just the Langmuir form. When multiple solutes are present, coupled nonlinear equations must be solved, which involves an iterative solution.

CONCLUSIONS AND DISCUSSION

Previous efforts to analyze thermally cycled ion-exchange processes were limited by insufficient equilibrium and rate data. In particular, the area that appeared least understood was that involving both univalent and divalent species. Since accruate techniques as well as adequate data are required for analysis or design, additional data were gathered. The equilibrium data for sodium chloride and sodium sulfate exhibit maximum values of capacity at moderate concentration levels. In contrast, calcium chloride exhibits typical saturation behavior. The bifunctional resin was Amberlite XD-5, which is a product of Rohm and Haas Company. The ion exchange rate experiments indicated that there are only small differences in the effective intraparticle diffusion constant, which would not be reasonable to employ for enhancing separations.

Finally, the suggested form of the equilibrium isotherm, which provides for a combination of linear and Langmuir forms, was applied to the result of the method of characteristics solution for determining the concentration shift that is associated with a temperature shift. The result was a simple quadratic equation that applies for a single solute. Multiple solutes would require solution of coupled quadratic equations.

ACKNOWLEDGEMENT

Financial assistance from the Separation Processes Program Area of the National Science Foundation is gratefully acknowledged, as is the contribution of resin for this study by the Rohm and Haas Company.

REFERENCES

(1) Liberti, L.; Boari, G.; Passino, R.; Desalination, 25, 2, 123 (1978).

(2) Liberti, L.; Boari, G.; Passino, R.; Desalination, 26, 2, 181 (1978).

(3) Liberti, L; Schmuckler, G.: Desalination, 27, 3, 253 (1978).

(4) Fritz, W.; Schlunder, E.U.; Chem. Eng. Sci., 29, 1279 (1974).

(5) Fritz, W.; Schlunder, E.U.; Chem. Eng. Sci., 36, 721 (1981).

(6) Fritz, W.; Merk, W.; Schlunder, E.U.; Chem. Eng. Sci., 36, 731 (1981).

(7) Oscik, J.; Jaroniec, M.; Derylo, A.; AIChE J., 27, 3, 524 (1981)

(8) Janoniec, M.; Derylo, A.; Chem. Eng. Sci., 36, 1017 (1981).

(9) Knaebel, K.S., Ph.D. Dissertation, University of Delaware, Newark, Del., 1980.

(10) Vieth, W.R.; Howell, J.M.; Hsieh, J.H., J. Membr. Sci., 1, 177 (1976).

(11) Hsu, T-B; Pigford, R.L., Ind. Eng. Chem. Prod. Res. Dev., 20, 361 (1981).

(12) Knaebel, K.S.; Pigford, R.L., Ind. Eng. Chem. Fundam., (in-press, 1983).

(13) Knaebel, K.S.; Cobb, D.D.; Shih, T.T.; Pigford, R.L., Ind. Eng. Chem. Fundam., 18, 175 (1979).

Table 1.

Diffusion Rate Experimental Results

Experiment	Component	Speed rpm	V/M l/kg	T C	D* cm^2/s $x10^7$	D cm^2/s $x10^5$
1	NaCl	100	250	25	.27	.40
2	NaCl	350	250	25	1.09	1.62
3	$CaCl_2$	350	250	25	0.50	1.38
4	$CaCl_2$	350	250	35	1.57	4.89
5	Na_2SO_4	350	250	25	1.30	2.46
6	Na_2SO_4	350	250	35	1.33	2.21

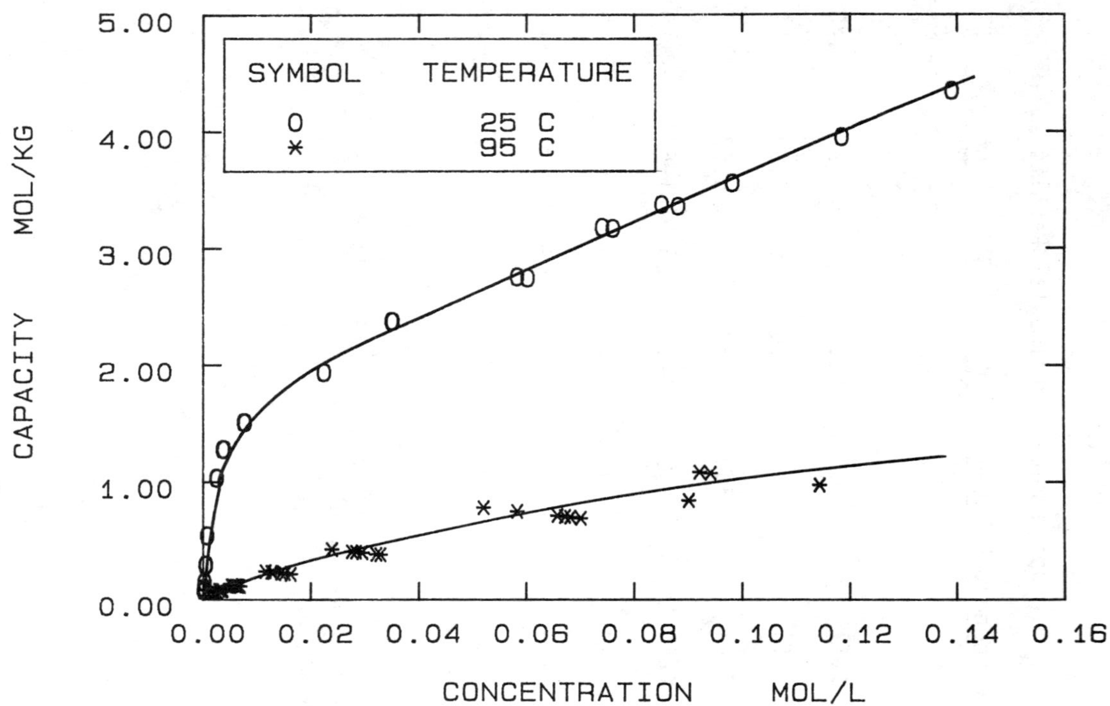

FIGURE 1. Equilibrium isotherms for Sodium Chloride at 25 C and 95 C, with pH at 5.7. XD-5 Resin Lot Number: LM-812.

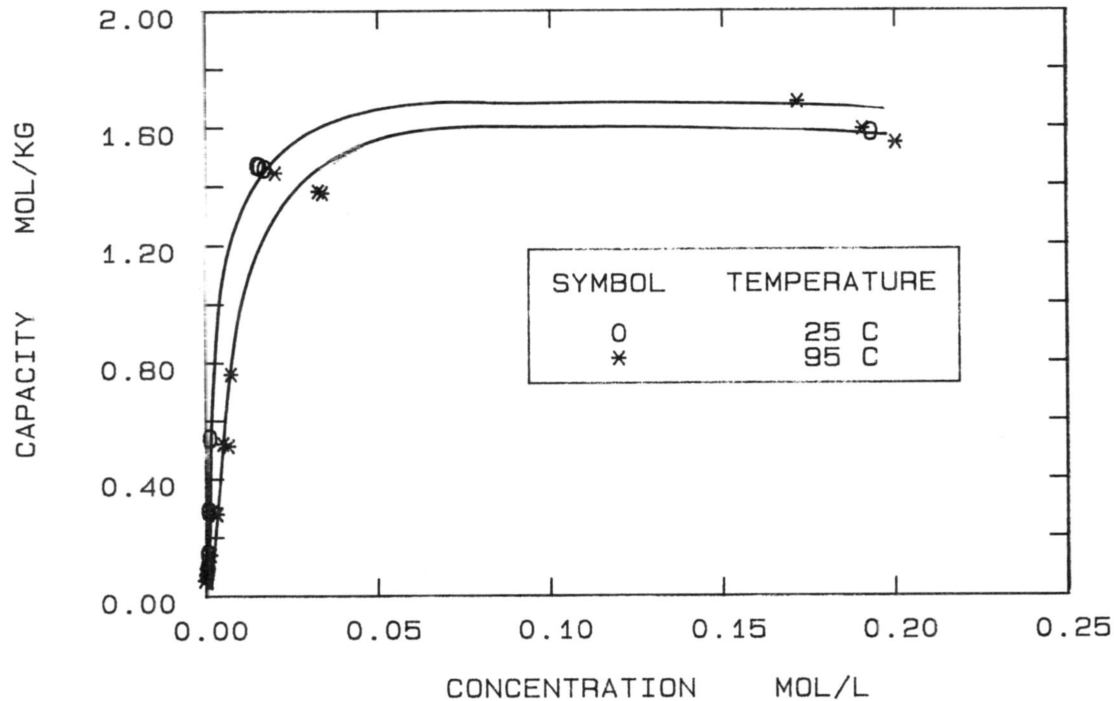

FIGURE 2. Equilibrium isotherms for Calcium Chloride at 25 C and 95 C, with pH at 5.7. XD-5 Resin Lot Number: SW-79/0145.

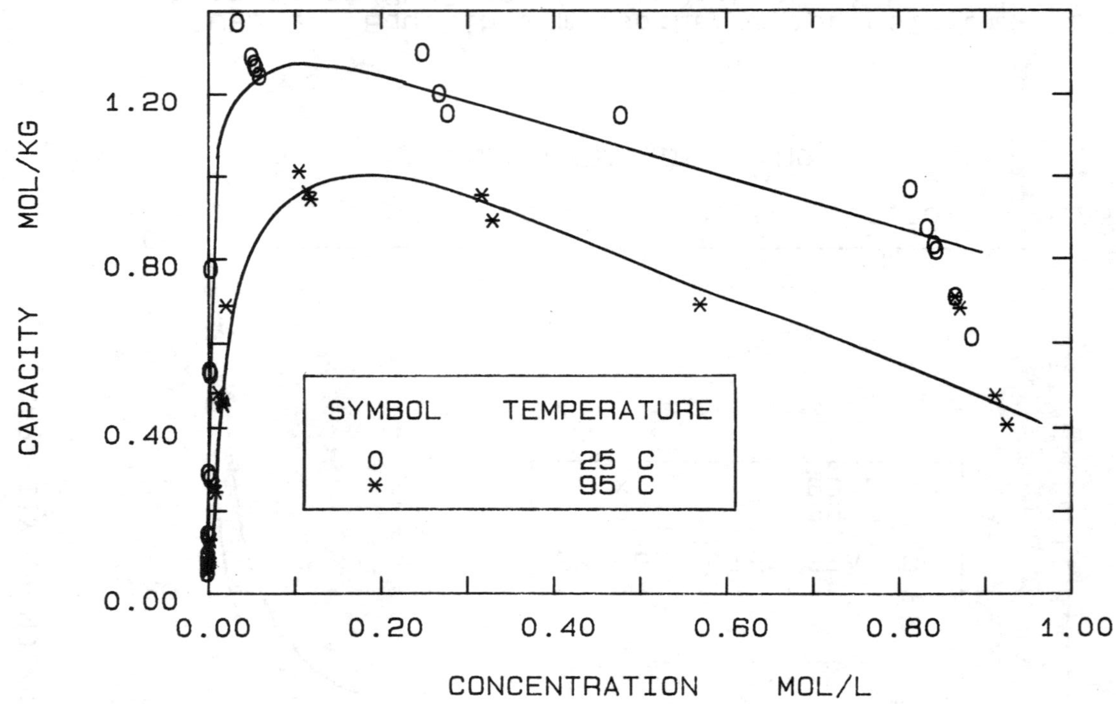

FIGURE 3. Equilibrium isotherms for Sodium Sulfate at 25 C and 95 C, with pH at 5.7. XD-5 Resin Lot Number: SW-79/0145.

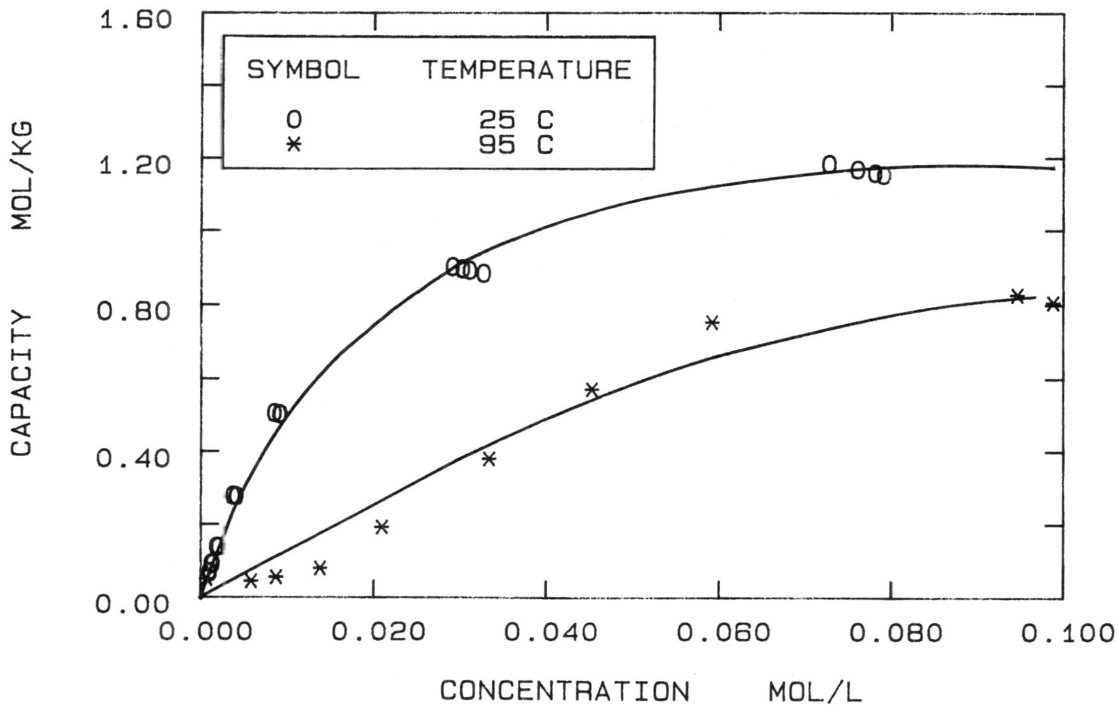

FIGURE 4. Equilibrium isotherms for Sodium Chloride at 25 C and 95 C, with pH at 5.7. XD-5 Resin Lot Number: SW-79/0145.

ANALYSIS OF ION EXCHANGE PHENOMENA OCCURRING IN SILICATE GLASSES IMMERSED IN A MOLTEN SALT BATH

G. Battaglin, M. Berti, G. Della Mea, P. Mazzoldi, M. Guglielmi[*]

Istituto di Fisica, Unità GNSM-CNR, Via Marzolo n.8, 35100 Padova, Italy

[*] Istituto di Chimica Industriale, Via Marzolo n.9, 35100 Padova, Italy

It has long been known that when a glass is immersed in a molten salt bath, ions move in and out of the glass via an ion exchange mechanism occurring at the glass/salt interface (1).

If the exchange is carried out at a temperature well below the glass transition one, the network of the glass cannot adjust to the different sizes of the exchanging ions and the result is the appearance of stresses on the surface layer interested by the phenomenon (2). If the salt bath ions are larger than the host in the glass, a compression layer is formed at the surface of the glass highly increasing its technological strength (3). Stresses and compositional changes results also in an increasing of the refractive index of the layer where exchange took place: very low loss optical waveguides can be obtained with ion exchange methods (4). Up to now, the early stages of the phenomenon, in which the major changes of the physical properties of the surfaces take place, have not been satisfactory investigated because of the lack of an experimental technique which associates the necessary depth resolution in determining the depth concentration profiles of the exchanged ions and the need that the analysing probe would not modify the actual profiles.

In the present work we investigated the $Na^+ \rightleftarrows K^+$ and $Li^+ \rightleftarrows K^+$ exchange processes in silicate glasses immersed in a molten KNO_3 bath at the temperature of 380°C. Exchange times varied between 15 s and

8 hours.

Samples have been analysed with a non destructive nuclear technique, the Rutherford backscattering (5). Such a technique consists in the energy analysis of light ions, generally $^4He^+$ ions, produced and accelerated in a Van de Graaf accelerator, a nuclear physics machine, which struck the sample surface and are elastically scattered in the backward direction by nuclear interactions with the elements present in a surface layer of the target. Such a technique is completely non destructive and allows the determination of concentration profiles in the first 2 microns below the surface with a depth resolution which usually is about 300 Å, but, in some cases, it can be lowered to 100 Å or less. The adopted technique is then particularly suitable for the analysis of the early stages of the investigated reactions. The molar composition of the glasses utilized in the present experiment are reported in Table I.

	Glass "a"	Glass "b"
SiO_2	80.7 %	67.9 %
Al_2O_3	4.2 %	9.2 %
Li_2O	------	22.9 %
Na_2O	15.1 %	------

Table I: Molar composition of glass "a", utilized in the $Na^+ \rightleftarrows K^+$ exchange, and of glass "b", utilized in the $Li^+ \rightleftarrows K^+$ exchange experiments.

The Rutherford backscattering (R.B.S.) spectra obtained from the two glasses before and after the 30 seconds exchange are reported in Figure 1.

Using standard RBS analysis procedure (5) we obtained from the experimental spectra the K depth concentration profiles. Some of them are reported in Figure 2.

It is clear the marked difference between the potassium profiles in the $Na^+ \rightleftarrows K^+$ and in the $Li^+ \rightleftarrows K^+$ exchanges.

The former case could be analysed in terms of a simple error function impurity profile, whereas in the latter case there is the formation of a plateau region followed by a steep tail. In both cases the surface potassium concentration seems to be time dependent, but this could be an artifact in the present experimental conditions: in fact, after the extraction from the salt bath, samples cannot be immediately cooled, otherwise they would break in very small fragments, but they remain hot for several seconds and the diffusion in this cooling stage can flatten the starting profiles. The smoothing of the profiles masks some features which can be connected to the concentration dependence of the interdiffusion coefficient, so we supposed it constant in calculations.

In the case of the $Na^+ \rightleftarrows K^+$ exchange we solved the diffusion equation

$$\frac{\partial C(x,t)}{\partial t} = \tilde{D}\,\frac{\partial^2 C(x,t)}{\partial x^2} - S\,\frac{\partial C(x,t)}{\partial x} \tag{1}$$

with the boundary conditions appropriate to the whole analysed process (diffusion during exchange and during cooling).

In eq.1 $C(x,t)$ is the K concentration in the glass at time t and depth x, $\tilde{D}$ is the interdiffusion coefficient (which is assumed concentration independent but temperature dependent) and S is a parameter connected to mobilities and to the partial stress tensor. The best fit of the experimental profiles was obtained with the values of

$$\tilde{D}\ (380^\circ C) = 5 \times 10^{-11}\ cm^2/s$$

$$S\ (380^\circ C) = 10^{-7}\ cm/s\ .$$

The obtained $\tilde{D}$ value is in close agreement with those reported in literature for similar exchange conditions (6).

We did not succeed in describing in a similar simple manner the potassium profiles resulting after the $Li^+ \rightleftarrows K^+$ exchange process. In fact, the sharp advancing front of the penetrant in the glass can explained only in terms of a discontinuity in the diffusion coefficient: above a critical value of the penetrant ion concentration stresses are so high that they cause a change in the average size of the sites available for cation jumps, leading to an increase in the diffusion coefficient.

Order of magnitude calculations of the interdiffusion coefficient in the present experimental conditions give a value of $\tilde{D} \leq 10^{-12}\ cm^2/s$ for the low concentration interdiffusion stage and a value of $\tilde{D} \geq 10^{-9}\ cm^2/s$ for the interdiffusion in the highly stressed region.

It is interesting to note that there should be a critical ionic size ratio between the exchanging ions for observing the formation of a flat-topped profile. In fact, as can be seen in Table II, where we grouped our results together with some other results present in the current literature, when the ratio of the ionic radius of the cation of the salt to the ionic radius of the mobile cation of the host matrix is lower than about 1.5 an error function-like diffusion profile is observed, but when such a ratio is higher than about 1.5 the formation of a plateau in the diffusion profile is detected.

Exchange	Ionic radius ratio	Flat profile	Reference
$Ag^+ \rightleftarrows Na^+$	1.33	No	(7)
$K^+ \rightleftarrows Na^+$	1.40	No	present work
$Rb^+ \rightleftarrows Na^+$	1.56	Yes	(4)
$Cs^+ \rightleftarrows Na^+$	1.78	Yes	(7)
$K^+ \rightleftarrows Li^+$	2.22	Yes	present work

Table II: Ionic radius ratios and appearance of a flat topped profile.

Finally we also observed that the surface K concentrations after very long times exchanges (concentrations which are not affected by the diffusion in the cooling stage after the exchange) are constant and lower than the initial concentration of Na or Li ions in the untreated glasses. These data are reported in Table III:

Atomic concentrations (10^{21}atoms/cm^3)

	Li	Na	K
glass "a" untreated	---	7.0	---
glass "a" 8h exchange	---	1.2	5.8
glass "b" untreated	11.6	---	---
glass "b" 8h exchange	2.3	---	9.3

Table III: Atomic concentrations of the exchanging ions in glasses "a" and "b" before and after the 8 h exchange.

Our work is in progress to clarify the formation of the highly stressed region and its evolution as a function of temperature and time. Naturally we will work to obtain as lower as pos sible cooling times after the high temperature exchange in order to follow more in details the exchange process, in particular to obtain reliable information on the possible time dependence of the K surface concentration and on the concentration dependence of the interdiffusion coefficients.

REFERENCES

1. Shulze, G. Ann. Physik 40 (1913) 335.
2. Doremus, R.H. J. Am. Ceram. Soc. 57 (1974) 478.
3. Hale, D.K. Nature 217 (1968) 1115.
4. Neuman, V., Parriaux and L.M. Walpita. Electr. Lett. 15 (1979) 704.
5. Chu, W.K., Mayer, J.W. and M.A. Nicolet. Backscattering Spectrometry (Academic Press, New York, 1978).
6. Frischat, G.H. Ionic Diffusion in Oxyde Glasses. Diffusion Monograph Series (Trans. Tech. Publ., Bay Village, Ohio, 1975).
7. O'Conner, D.J., Palmer, D.W., Townsend, P.D., Morris, J., Pitt, C.W., Neuman, Z. and L.M. Walpita. Nucl. Instr. and Meth. 182/183 (1981) 797.

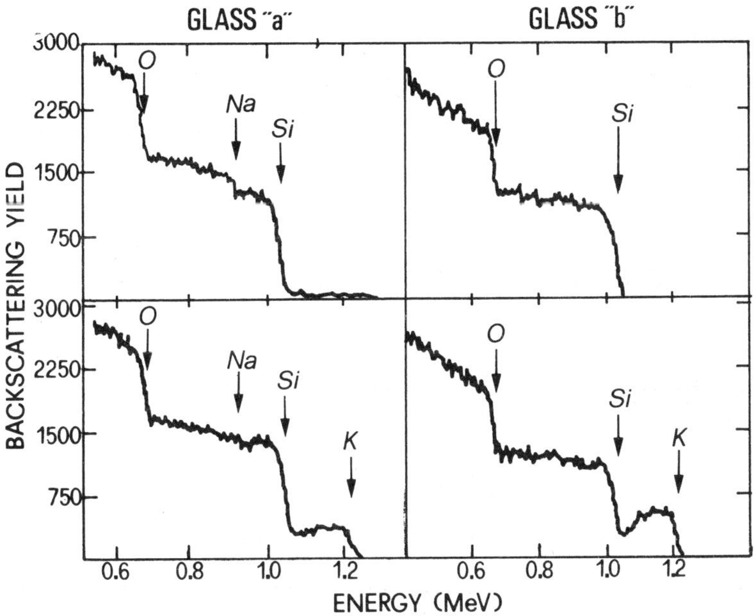

Figure 1: RBS spectra obtained from glasses "a" and "b" before (upper part) and after the 30 seconds exchange (lower part of the figure). Arrows indicate the signals corresponding to the different elements contained in the samples.

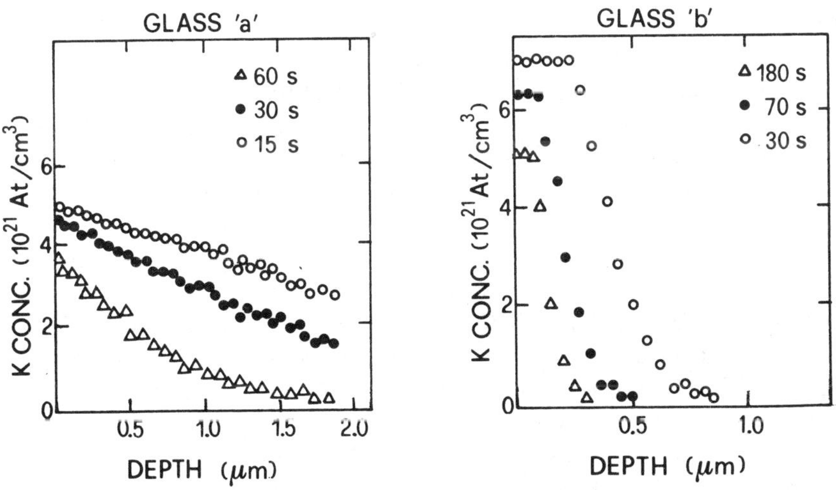

Figure 2: Typical K depth concentration profile obtained, with the RBS technique, from glasses "a" and "b" after different exchanging times.

THE RIM-NUT PROCESS FOR RECOVERY OF N/P FERTILIZER FROM SEWAGE.START-UP OF BARI's PLANT.

L.Liberti,G.Boari,N.Limoni,C.Longobardi,R.Passino

Istituto di Ricerca Sulle Acque-Consiglio Nazionale delle Ricerche,5 via De Blasio,70123 Bari,Italy

Nutrients such as ammonia and phosphates,coming mainly from secondary urban effluents from aerobic treatment, are generally recognized as principal source of eutrophication of surface and coastal waters.As more stringent regulations on nutrient discharge are being implemented, specific treatment to abate nutrients from sewage are applied worldwide,such as precipitation of phosphates and biological nitrification-denitrification of ammonia.

These destructive treatments,however,have serious drawbacks:the former produces considerable amounts of chemical sludges to be disposed off;the latter is prone to bulking phenomena,while its efficiency drops noticeably at low temperature.Furthermore,none of them allows to recover the potential fertilizer value of nutrients.

Selective ion exchange is being considered as a promising technique to remove nutrients from sewage and to recover them as valuable products.After several years of research at IRSA-CNR,the RIM-NUT process has been developed to remove and recover ammoniumphosphate from sewage by ion exchange.[1,2] The process (see Fig.1) has the following characteristics:two ion exchange resins,cationic and anionic in series,treat the wastewater removing selectively both nutrients;furthermore,residual suspended matter is removed by filtration,while bioresistant organics and viable organisms are also absorbed.

Neutral 0.6M NaCl is used for resin regeneration(which may easily substituted by sea water in coastal plants). Cationic regeneration is performed with three separate fractions (8 BV each) of solution,taken respectively from reservoirs SC1,SC2 and SC3 in the order.While the last two fractions are recycled directly into SC1 and SC2,the head fraction (which contains approx.70% of ammonia exchanged) is collected into SC4 for the precipitation treatment. Similarly,anionic regeneration is done with two aliquots (2 BV each) of neutral 0.6M NaCl taken from SA1 and SP. The latter is recycled directly into SA1,while the former is collected into SC5 to be precipitated.A two-stage precipitation procedure is applied continuously on regeneration head fractions in reservoirs SC4 and SC5.Na_2CO_3 is added first to precipitate Ca and heavy metals in SC4;then Mg and PO_4 salts are added so that magnesiumammoniumphosphate,a valuable slow release fertilizer,is precipitated in SC5 in proper operating conditions.The fertilizer is finally transferred in the thickner SC6,from which the tail-fraction of regenerant is transferred to SC3.

As described in detail elsewhere[3], a fully continuous automatic laboratory plant,operated for approx.one year (1500 cycles,or 120,000 BV treated) with a secondary effluent coming from a sanitary station near Bari,showed the reliability of the RIM-NUT process.Average removal of both nutrients in the range of 85-95% was observed.140 g/d of 99% pure fertilizer per equivalent inhabitant were recovered.Heavy metal content is very low (see Tab.I).

Based on these results,the construction of a 200 m^3/d demonstration plant has been completed in West Bari Sanitation Station in 1982;a typical secondary effluent analysis is shown in Tab.II(conventional activated sludge plus nitrification).Phosphate and ammonia concentrations are quite high due to local low water allowance(200 l/d,person). Tab.III shows operating conditions for the demonstration plant,start-up of which occurred manually in May 1982. Results collected during start-up operation are shown in Figs.2 and 3 and in Tab.IV.140 BV of secondary effluent are treated with removal efficiencies of 80 and 95% for cationic and anionic resins respectively,with final concentration of nutrients well within italian limits.

Performance of the precipitation section confirmed the results obtained in laboratory trials.

Substantial improvement of effluent quality was observed du to 55% residual TOC and 99.8% total coliforms removals.

Economic comparison of RIM-NUT and conventional treatments(i.e.,precipitation +(de)nitrification) showed that capital costs are in the same order of magnitude,while running costs may be very low with this process,depending on revenue from sale of fertilizer recovered.

The demonstration plant will permit confirmation of these results by long term tests to verify the potential of the RIM-NUT process as tertiary treatment for water reuse and nutrient recovery.In particular,optimization of biological (de)nitrification will be considered,so to have an N:P=1:1 ratio in the influent to the ion exchange plant. In this case,fertilizer production will be cut to 30-40 g/d per person,but running costs will be much lower,due to the reduced addition of phosphate missing to stoichiometry.

References

1.L.Liberti,G.Boari,R.Passino,Phosphate and ammonia recovery from secondary effluents by selective ion exchange with production of a slow release fertilizer,Water Res.13(1979)65

2.L.Liberti,G.Boari,R.Passino,It.Pat.No.47912-A/81,27 Feb. 1981

3.L.Liberti,N.Limoni,R.Passino,D.Petruzzelli,Ammonium phosphate recovery from urban sewage by selective ion exchange,in L.Pawlowski ed.,Physicochemical Methods for Water and Wastewater Treatment,Pergamon Press,Oxford, 1980,73.

Tab.I.Fertilizer production with the 340 l/d laboratory RIM-NUT plant (period June 6-July 27 1981)

A) Mass balance

- effluent treated	2.13 m^3
-ave.NH_4^+ in secondary effluent	3.44 mol/m^3
-ave.NH_4^+ in tertiary effluent	0.60 mol/m^3
-ave.removal efficiency	83%
-nutrients removed	6.04 mol(*)

B) Chemical analysis

- fertilizer recovered 1431 g
- stoichiometry $Ca:Mg:NH_4:PO_4:H_2O$ =0.02:1:1.02:1.08:6
- chemical formula $MgNH_4PO_4 \cdot 6H_2O$

C) Concentration of metals

	(mg/kg fertilizer)	(mg/kg $CaCO_3$)
Fe	27	80
Cr	14	15
Pb	10	25
Cu	6	43
Cd	traces	8
Ni	traces	110
Al	traces	58

(*) corresponding to 1480 g of $MgNH_4PO_4 \cdot 6H_2O$

Tab.II. Average composition of West Bari' secondary effluent (mg/l)

Chlorides (as Cl)	161
Alkalinity (as $CaCO_3$)	415
Sulphates (as SO_4)	38
Phosphates (as P)	11
Nitrates (as NO_3)	0
Nitrites (as NO_2)	0
Ammonium (as NH_4)	60
Potassium (as K)	30
Sodium (as Na)	130
Calcium (as Ca)	30
Magnesium (as Mg)	6
BOD_5 (as O_2)	32
COD (as O_2)	107
pH	7.4
Fecal coliforms	$17x10^4$ (MPN/100 cc)

Tab.III.Operating conditions of West Bari' 200 m^3/d RIM-NUT demonstration plant

	Cationic Section	Anionic Section
resin	natural zeolite	strong base
No.of column	1 + 1	1 + 1
resin volume(m^3/col)	0.45	0.33
bed depth (m)	1.2	1.2
column diameter (m)	0.7	0.6
a) <u>exhaustion</u>		
flow rate(BV/h,↓)	20	27
duration (h)	7	7
b) <u>regeneration</u>		
regenerant	0.6M NaCl	0.6M NaCl
dosage(BV/col)	3x8	2x2
flow rate(BV/h,↓)	12	3
duration (h)	2	2

Tab.IV.Typical results from West Bari' RIM-NUT plant (May 1981)

date	run	exchange capacity (mol/m_r^3)	influent (mg/l)	effluent (mg/l)	removal (%)	ave.TOC removal (%)
cationic section						
21/5	□	390(NH_3)	60(NH_3)	10(NH_3)	84	
25/5	△	355 "	58 "	12 "	79	
anionic section						
25/5	△	52(PO_4)	12(PO_4)	0.5(PO_4)	96	55
27/5	○	47 "	11 "	0.5 "	95	55

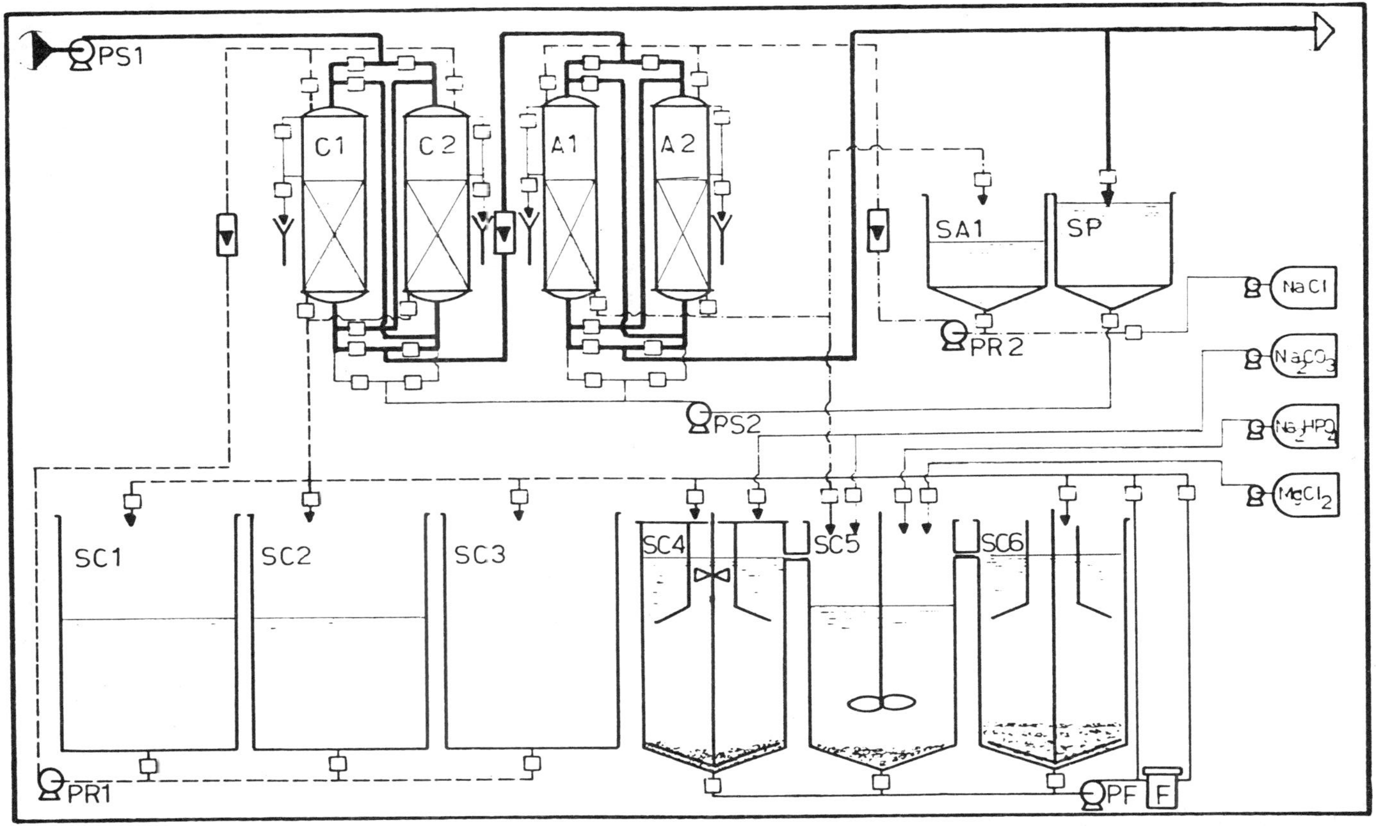

Fig.1. Flow-sheet of the RIM-NUT process.

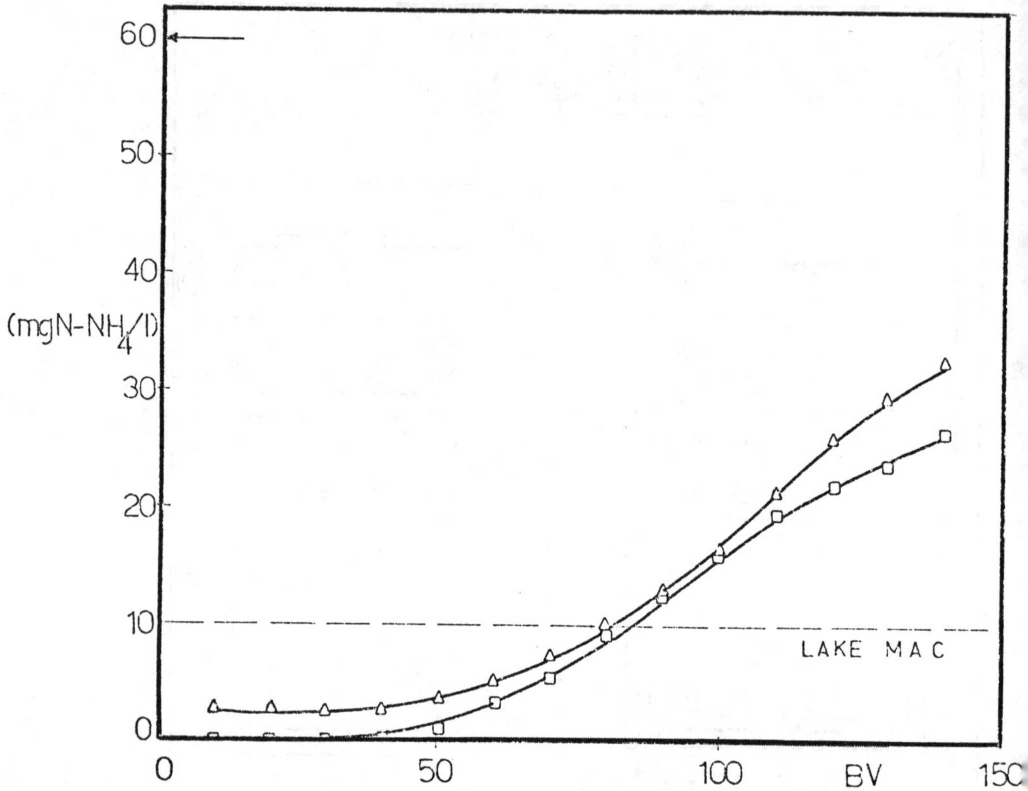

Fig.2. Typical breakthrough histories from cationic section of the West Bari's RIM-NUT demonstration plant (□ May 21,1982; △ May 25,1982)

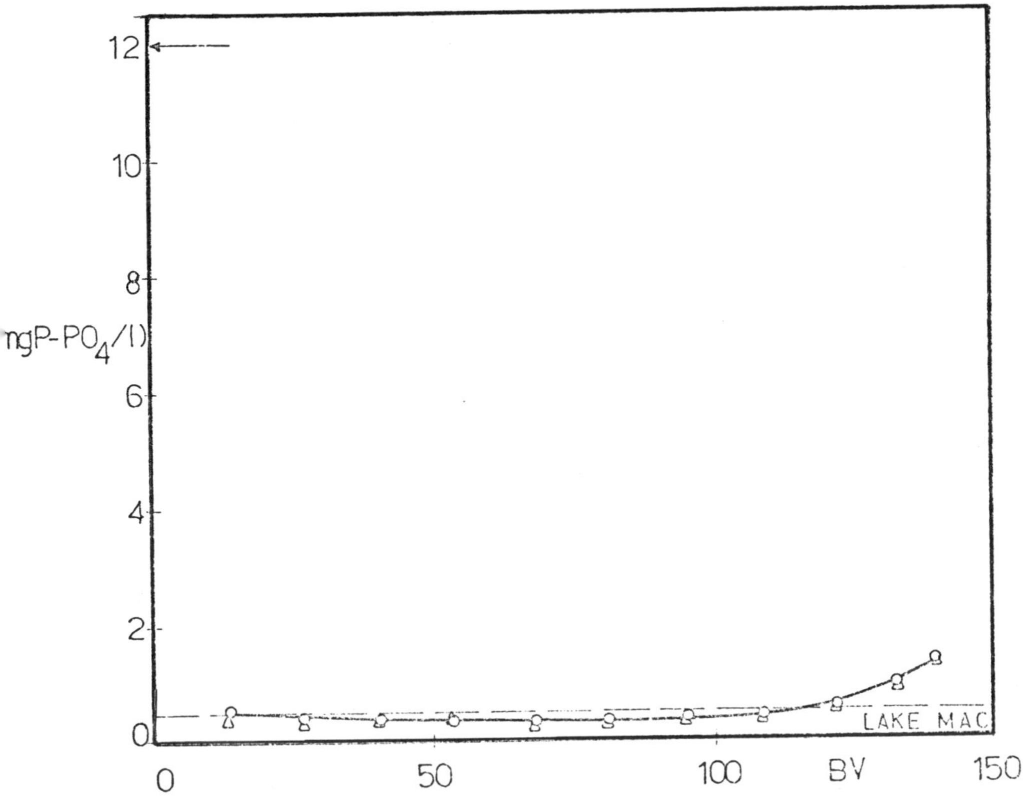

Fig.3.Typical breakthough histories from the anionic section of the West Bari's RIM-NUT demonstration plant (Δ May 25,1982; o May 27,1982)

ION EXCHANGE FOR THE RECYCLING OF WASTEWATER CONSTITUENTS

Brian BOLTO and Lucjan PAWLOWSKI
(CSIRO Division of Chemical Technology,
P.O. Box 310, South Melbourne 3205, Australia).

ABSTRACT

Recycling of the constituents of wastewater requires efficient and cheap separation methods. Pollutants must be removed in a concentrated form to facilitate their recycling. Similarly, the raw water must be purified to a level necessary for water reuse. One very efficient method for the separation of ions from liquids is ion exchange, which has a greater utility for the treatment of wastewater than is currently realised.

The application of ion exchange to the recovery of metals, nutrients and rare chemicals (as well as water) from industrial effluents and sewage is summarized. Attention is given to ways of improving the economics of recovery. One method of making the recovery processes simpler and less costly is to employ magnetic resins. Because of their excellent kinetic properties the size of the equipment needed is minimized. The present and future possible applications of magnetic resins are reviewed.

INTRODUCTION

The American philosopher, B. Fuller, remarked that "a pollutant is a wasted resource". Looking through Man's industrial activities, one finds many examples which confirm Fuller's opinion. Water is possibly the most important substance in Nature, and its value as a resource diminishes with pollution.

Generally, the direct consumption of water in industrial processes is rather low. Water is mainly used as a reaction medium, for washing purposes and for the transportation of materials and energy. During its use water becomes a medium for some chemicals which are often toxic to the environment.

In recent years, more and more attention has been given to multiple reuse of water because the world's fresh water resource is shrinking. There has been increasing development of technologies which allow for the purification of wastewater to the extent that the recovered water can be recycled to production.

However, less attention has been paid to recycling the chemicals contained in wastewater, partly because their value in such a dilute state is usually lower than that of water, and therefore the economic motives for their separation are not strong enough. Nevertheless, there are circumstances in which it may be economic to recover certain components from wastewater, particularly when the component occurs in large quantities. As non-renewable resources become exhausted their market price increases to the point where their recovery becomes an economic proposition. Alternatively, punitive social costs are imposed on the polluter who passes them onto the consumer.

An example is the processing of nitrogen compounds. About 0.3% of ammonia produced is discharged to the environment as a pollutant, and losses of nitric acid are even higher, reaching 1.3%. Clearly, the recovery of such amounts of nitrogen compounds should not be neglected. The losses of heavy metals in metal finishing operations are even higher.

By implementing recycling one may change Fuller's statement to "a pollutant is a recoverable resource". However, to develop this type of approach for the pollutants in wastewater one has to have suitable separation and concentration techniques. One of the most promising methods is ion exchange. The intention of this paper is to present ion exchange as a useful tool for solving some of today's environmental and resource problems.

GENERAL DESCRIPTION OF ION-EXCHANGE PROCESSES

Ion exchange allows the separation of ionic solutes from liquids, and of one solute from another. The selection of the method of separation depends above all on the composition of the solution to be treated and to what level the impurities have to be removed. One may distinguish two kinds of operation: selective removal of ionic impurities and full deionization. There are some types of highly concentrated ionic wastes consisting of exhausted processing solutions, one example being spent plating baths. During processing the bath picks up impurities and the bath liquor deteriorates. These wastes cause a serious threat to the environment as they contain a high concentration of pollutants. Ion exchange provides an answer to this type of waste problem because it makes possible the removal of the contaminating ions, so that the spent bath becomes useful again. The conditioning of the spent solutions may be done by selective removal of undesirable ions and by separation of acid from salt, using the method of acid retardation (60).

In many cases industrial wastewaters such as rinse waters and condensates contain only undesirable ions, so that full deionization is required. In other cases deionization is necessary to implement a closed-loop system in plating operations, in which case both chemicals and water are recycled.

Examples of both selective removal and full deionization are mentioned below. A comprehensive review of these areas has been prepared for publication elsewhere (59).

Selective removal of ionic impurities

It is often necessary to remove one kind of ionic impurity from a mixture of impurities contained in a wastewater. Generally, the application of ion exchange for this purpose permits the replacement of undesirable ions by others which are not harmful to the environment.

A typical example of this concept is in the recovery of precious metals, where the treatment process aims for complete retention. Water reuse is of far less economic significance here and, often, the volume of effluent is relatively small. Some plating effluents contain gold in the form of an anionic complex $[Au(CN)_2]$. A standard strongly basic anion exchange or a quaternized polyvinylpyridine anion exchanger can be used for removal of the anion.

In one application where the plating rinse water contained 4-5 mg Au/L, a two-column system was used for its adsorption (2). A weakly acidic cation exchanger in the H-form (Diaion WK20) was used, presumably to remove other heavy metal cations and to lower the pH, followed by a weakly basic anion exchanger (Lewatit MP62). The anionic gold was adsorbed on the anion exchanger.

Another plating effluent, containing anionic gold together with organic acids and their salts, was treated with a weakly basic anion exchanger at pH ∿3, on a three-column system arranged in series (3). The gold was retained on the first bed and the other solutes on the second and third beds. For the removal of anionic gold from highly acidic solutions, a polyisothiouronium resin (Srafion NMRR) can be used. Total capacities in 80g of metal/L of resin (lb) for various precious metals are as follows: Rh: 80, Au: 340, Pt: 200, Pd: 120. Nevertheless for gold the operating capacity of this resin is low (55-110g Au/L resin) (lb). Because the value of the gold retained in the resin greatly exceeds that of the spent resin, it is common practice to incinerate the resin at 1000°C to recover the metal. However, a solution of 5% NaOH + 5% NaCN is also used for regeneration of the resin.

Another precious metal widely used in the plating and photographic industry is silver. In plating, silver is used as a cyanide complex with various additives, and the resultant effluents can contain $[Ag(CN)_2]^-$, KCN and K_2CO_3. For the adsorption of the silver complex, a strongly basic anion exchanger is employed, such as Amberlite IRA-400 (4). In another application two ion-exchange columns in series were used; one containing a weakly acidic cation exchanger and the other a weakly basic anion exchanger (5).

In a further example, a plating effluent containing Ag: 14.2, Cu: 3.8, CN: 17 and COD: 5.4 mg/L was passed through a series of beds - the first a weakly basic resin, the second a resin of medium basicity, and the third a strongly basic anion exchanger (the last one to prevent leakage of CN^-). Demineralized water was obtained (containing very small amounts of Ag^+, Cu^{+2} and CN^-, each <0.1 mg/L). The silver was recovered from the first bed (6).

The recovery of silver is very important in the photographic industry, the effluents from which contain the metal in an anionic form, $[Ag(S_2O_3)_2]^{-3}$. A strongly basic anion exchanger is usually used for its recovery (7). For example, with Duolite A101D, a decrease in the silver content from 7 to 0.05 mg/L is obtained. The weakly basic resin Amberlite IRA 68 has also been used for the recovery of silver from photographic effluents. The weakly basic resin has a lower capacity, but is easy to regenerate with an alkaline solution which also allows a higher concentration of silver to be obtained in the regeneration effluent. Electrolysis is used for the final recovery of the silver (8,9). Because of the lower value of silver, incineration of the exhausted resin is not practised and regeneration is achieved with 5% NaOH and 5% KCN(5) or solutions of NH_4Cl(7).

The recovery of selenium, the salts of which are used in the electrolytic colouring of metals, is also practised by ion exchange. The rinse water as well as the exhausted bath liquor contains selenium oxy-acids. A two-column system is used to recover the selenium: the first column, containing a chelate resin, removes heavy metal ions while the second, containing an anion exchanger, removes the selenium oxy-acids. The anion exchanger is regenerated with 0.5N NaOH and the regeneration effluent contains 5.5g Se/L (10-12).

Tin is also used in electrolytic coating operations. The rinse water can be treated with a two-column system: the first containing a strongly acidic cation exchanger and the second a weakly basic resin. The tin is recovered from the latter by elution with caustic soda (13).

A strongly basic anion exchanger is used for the recovery of molybdenum (14) and arsenic ions (15). An ion-exchange method is also used for the recovery of lithium (16). An interesting method was developed for recovery of boric acid from radioactive wastewater. The effluent is cooled to 4-10°C, then passed through an anion exchanger bed which retains the boric acid. The resin is regenerated with warm (60-70°C) water (17). An ion-exchange method has been used for the recovery of vanadium from the effluent of zirconium - vanadium pigment manufacture (18).

Mercury is one of the worst pollutants, having caused serious disasters in Japan and Iran. Most of the mercury arises from chlorine and caustic manufacturing plants. Reuse of mercury from wastewater treatment reduces the cost of wastewater treatment by about 25% (1a). The Imac TMR resin is a polymeric mercaptan in which the thio (-SH) groups have a high affinity for Hg (19,20). The operational ion-exchange capacity depends on the mercury concentration in the effluent and ranges from 48g Hg/L for a mercury concentration of 0.6 mg Hg/L to 100g Hg/L for a mercury concentration of 10 mg Hg/L. Concentrated (35%) HCl was used to regenerate the exhausted resin (16).

A resin known as Srafion NMRR, which contains isothiouronium groups, is also used for mercury removal. Its capacity for chloride ions is 3.1 meq/g resin and for mercury 0.545g Hg/g of resin. The resin can be regenerated with a 5% solution of thiourea containing 2.2g HCl/L. Other mercury selective anion exchangers are known.

It is also possible to use conventional resins for the removal of mercury (21). After oxidation of Hg^0 to Hg^{+2} with NaOCl at pH 2.5-3.5, the mercury can be adsorbed as the complex $(Hg\ Cl_4)^{-2}$ onto an ion exchanger such as Wofatit SBW or Varion AD. When 12N HCl is employed for regeneration, 85% of the mercury can be eluted from the ion-exchange bed.

A very important application of ion exchange is in the recovery of copper from leaching operations in copper production. Because of the low pH (<2) normal chelate resins do not adsorb the copper ions. Two specific resins have been developed, which have a high affinity for copper over other metal ions (22-25). They are based on N(hydroxy-alkyl)picolylamines. One of the resins (Dowex XFS-4195) can even take up copper ions from 3N H_2SO_4, but a more concentrated acid (10N) is required for regeneration. The other resin (Dowex XFS-4196) can take up copper ions from less acidic solutions having pH > 1.5, and can be regenerated adequately with sulfuric acid of concentration 100g H_2SO_4/L. The regeneration effluent contains 33g Cu^{+2}/L and 40g H_2SO_4/L. It can be used for production of copper in the electro-winning process. The consumption of H_2SO_4 amounts to about 1.7g/g of Cu, i.e. only 10% greater than the stoichiometric requirement (1.54 g/l) (24). Recently a new resin (Dowex XFS 43084) has been developed which is similar to Dowex XFS 4196, but which has a higher preference for copper over iron ions (24).

There are some wastewaters which contain up to several grams per litre of ammonium sulfate and low concentrations of copper ions. It is possible to use such effluents for irrigation, but the copper ions have to be removed. An ion-exchange method has been used for this purpose to produce concentrated copper sulfate, which may be recycled (26).

Another interesting application of ion exchange is in the treatment of wastewater from gas desulfurization. The effluent contains, among other things, nickel and vanadium, which can be recovered. The effluent is oxidized and its pH is adjusted to 5-10 before it is passed through a chelate resin bed where nickel ions are adsorbed. The pH of the effluent from this operation falls to below 5. The solution is then passed through the next chelate resin bed where vanadium ions are adsorbed (27). On regeneration, the concentrated solutions obtained yield useful by-products.

Zinc may be removed from wastewaters as the cation. However, its removal as the anion $(ZnCl_4)^{-2}$ is also possible. In the presence of HCl zinc forms the anionic complex which can be easily retained on an anion exchanger bed. By washing the exhausted anion exchanger with water one may obtain pure $ZnCl_2$(28). This concept allows separation of zinc ions from a regeneration effluent containing a mixture of metals; for example, a strongly acidic cation exchanger is used for the removal of metals from mixed plating effluents. By using an excess of HCl for the regeneration of the exhausted cation exchanger one obtains a mixture of the chlorides. Further treatment of this regeneration effluent with an anion exchange resin results in the selective adsorption of $(ZnCl_4)^{-2}$.

Zinc is used in the koalin industry. A strongly acid cation exchanger in the Na^+ form can be used to remove zinc ions from effluents. With the use of 300g NaCl/L resin for regeneration it is possible to obtain a capacity of 32-40g Zn^{+2}/L resin and a zinc leakage below 0.4 mg Zn^{+2}/L (29). There is also a problem with the removal of zinc from the blow-down of water from cooling towers, and ion exchange should be considered a useful solution (30).

Let us now summarize some of the methods for the selective removal of heavy metals. Fortunately, the selectivity of even strongly acidic cation exchangers favours the uptake of polyvalent cations from dilute solutions. Weakly acidic resins (carboxylic acid type) also have a high affinity for heavy metals. Thus, even with the use of conventional resins, it is possible to obtain selective removal of heavy metals from effluents. In the case of the weakly acidic resins, regeneration is very easily accomplished with acidic solutions, as these resins have a high affinity for hydrogen ions. However, because of this they are sensitive to the pH of the solution to be treated and are not able to take up metal ions in an acidic environment.

Specific resins for the selective removal of ions are also known. They are documented in the literature (1c, 31); some commercially available types are listed in Table 1. Often, other ions are also taken up by the ion-exchange resin. Direct elution will lead to a regeneration effluent which contains high levels of possibly undesirable ions. In these cases, pretreatment of the exhausted resin with some of the desirable ions is recommended. These ions will replace the undesirable ions in the ion-exchange bed, so that the final regeneration effluent will contain mainly the desirable ions.

The choice of a chelate or a carboxylic resin is dependent on the particular case. For example, the effectiveness of the weakly acidic resin Amberlite DP-1 is equivalent to that of the chelate resin Amberlite IRC 718 for the removal of zinc ions. However, when a 10% HCl regenerant is used at a flow rate of 8 BV/hr, the concentration of recovered zinc ions is higher for the carboxylic acid resin. Similar results were obtained with effluents containing Pb^{+2} or Cd^{+2} ions. However, with an effluent containing copper ions together with ammonium sulfate, the chelate resin is superior. Both resins are sensitive to flow-rate levels, the carboxylic acid resin being the more affected of the two, within capacity decreasing as the flow rate increases (32).

In the application of a strongly acidic cation exchanger to the recovery of metal ions from solutions which contain a high level of alkali metal ions, the ratio of ions taken up by the cation exchanger makes it impossible to recycle the regeneration effluents. This occurs, for example, in the recovery of zinc from the effluents of synthetic fibre plants (where the concentrations are ∿540 mg Zn^{+2}/L and 10 300 mg Na^{+}/L. For recycling to be practical, the high content of Na_2SO_4 should be lowered to less than 20%. This lowering could be achieved by a two-step regeneration process: elution with 1% H_2SO_4 for partial removal of Na^{+}, followed by regeneration with 10% H_2SO_4 for elution of Zn^{+2} from the bed (33). A similar effect can be obtained by pretreatment of the ion-exchange bed with a solution containing Zn^{+2}, which displaces the sodium ions from the cation exchanger and, thus, improves the Zn^{+2}/Na^{+} ratio in the bed.

The removal of microbial and plant nutrients from wastewater assists in the prevention of eutrophication of receiving waters. Sewage contributes the largest amount of nutrients with agricultural run-off and the fertilizer industry also making significant contributions. Ion-exchange methods can be applied to the removal of nutrients from various wastewaters.

An interesting attempt to use ion exchange for the recovery of ammonium and phosphate ions has been made by Liberti (34-36). The method is based on the selective adsorption of ammonium ions on clinoptilolite and of phosphate on an anion exchanger, and on the precipitation of $MgNH_4PO_4$ from the regeneration effluents.

Full deionization

In many industrial plants, one can separate-out the various components from wastewaters, chiefly various rinse waters and condensates. Undesirable ions must be removed to detoxify the water; they may have value as reclaimed chemicals. Full deionization is the best purification procedure for such wastewaters as it produces demineralized water which simplifies recycling. The general flowsheet of such a process is pictured in Fig. 1.

One of the important applications of this approach is the closed-loop systems of plating operations, in which both reclaimed chemicals and water are returned to production. Once-through rinsing operations use tap water. However, its use significantly complicates the reclamation of chemicals because the water contains ions which will contaminate the recovered solutions. Therefore, only demineralized water should be used for rinsing.

The treatment units consist of:

- separation of suspended matter
- removal of organic matter
- decationization of rinse water
- deanionization of rinse water

For the separation of suspended and organic matter, one of the established conventional methods can be used without any difficulty.

The decationization step is similar to that performed in the decationization of boiler feed water. A strongly acidic cation exchanger must be used because of the low pH (<2) of the rinse water. Some attention must also be given to the oxidizing properties of the rinse water. The regeneration of the exhausted cation exchange bed is performed in a conventional way, preferably with the cheaper sulfuric acid but hydrochloric acid is also recommended (37,38). Regeneration with hydrochloric acid eliminates the difficulty of $CaSO_4$ precipitates forming when the water balance is maintained by topping up the system with fresh but hard water (38). Because of the strong bonding of polyvalent heavy metal cations to the resin, a high dose of acid is required for regeneration. Amounts ranging from 320g H_2SO_4/L of resin at 10% concentration (39) to 490g at 7%

concentration (40) have been recommended. A decrease in the consumption of acid to a level below 200g/L of resin is possible when the heavy metals are precipitated prior to decationization. The regeneration effluent (which contains a mixture of $Fe_2(SO_4)_3$, $Cr_2(SO_4)_3$, Na_2SO_4 and H_2SO_4, or the corresponding chlorides and hydrochloric acid) does not have any value and is therefore discharged after neutralization with lime, followed by clarification. The amount of generated wastewater does not exceed 2-5% of the total volume of rinse water.

The anion-exchanger unit is the more important for the recovery of chemicals. Deanionization can be performed with or without recovery of chromic acid. With small amounts of rinse water, it may be simpler not to recycle the chromic acid in which case regeneration of the anion exchanger is performed with a 4% solution of NaOH. Chromates present in the regeneration effluent are reduced to trivalent chromic ions, and, after neutralization with lime, the clarified wastewater is discharged to the sewer. For the conversion of chromates to chromic acid, the regeneration effluent from the anion exchange (a mixture of Na_2CrO_4 and NaOH) has to be decationized with a strongly acidic cation exchanger.

For the adsorption of chromic acid onto an anion-exchange bed, it is possible to use strongly basic (37) or weakly basic anion exchange resins (41). Strongly basic resins permit low leakage of chromate (<1 mg/L), but they require greater amounts of NaOH for regeneration since the efficiency of regeneration is lower; whereas, a weakly basic resin has a higher level of leakage but consumes less NaOH. The use of a weakly basic resin has some advantages over a strongly basic one because the closed-loop system does not require a very low content of Cr(VI) in contrast to the stringent limits imposed for environmental reasons when discharging the wastewater. In rinse tank systems, parts of the metals are immersed in the rinse water, where the concentration of Cr(VI) is kept at a level of 50-200 mg/L by the continuously flowing water which carries out some of the Cr (VI) (equivalent to the'drag out' from plating baths). When make up water containing some Cr(VI) is used, at a concentration less than that in the rinse tank, any increase in the concentration of Cr(VI) in the rinse-tank is prevented, provided faster rinse flow rates are used.

Clearly then, the level of leakage of Cr(VI) is not the most important factor in a closed-loop rinsing system. The critical factor is the decrease in the consumption of NaOH which has a very strong influence on the economics of rinse water pretreatment. Another extremely important parameter is the concentration of the regeneration effluent. For regeneration of the strongly basic anion exchanger in the conventional manner, a dose of 80-140g NaOH/L of resin at 4% concentration is recommended (42). However, as was shown by the use of weakly basic resins regenerated with highly concentrated caustic soda (41), it is possible to get regeneration effluents containing up to 160g Na_2CrO_4/L. In cases where a hot plating bath is used, it is then possible to recycle the regeneration effluent without the need for its concentration by evaporation. The recovery of nickel in metal finishing plants is only possible if the nickel-bearing effluent is separated from the other process streams. Because the nickel is usually used as the sulfate, its recovery in the sulfate form means that sulfuric acid must be used for regeneration. The nickel is taken-up on a strongly acidic cation exchanger in the decationizer part of the ion-exchange plant. Because the regeneration effluent contains too much sulfuric acid, a two-step procedure is recommended. Firstly, that part of the regeneration effluent which is rich in nickel sulfate is neutralized with $NiCO_3$ and is reused to reconstitute the solutions in the bath. The remainder is then collected and used for the first part of regeneration. The decationized effluent is deanionized on an anion-exchange unit, and the demineralized water is recycled. The regeneration effluent from the anion-exchange unit is discharged as waste. About 96-160g H_2SO_4/L of resin is required for regeneration of the cation exchanger and the capacity of the resin is 19-29 Ni^{+2}/L (37). Other metal-bearing rinse waters, including copper-, zinc-, cadmium-, cobalt-ones, can be similarly recycled.

Another area where full deionization can be used is in the purification of condensates. The concentration of solutions by evaporation produces condensates which may contain several to a few thousand milligrams of solutes per litre. One of the best known examples is the recycling of condensate constituents (water, ammonia and nitrates) from the fertilizer industry. Full demineralization of the condensates by ion exchange leads to the production of water and concentrated ammonium salts. The water is used as boiler feed water. Decationized condensate from the processing of ammonium nitrate can be used instead of demineralized water to absorb nitrogen oxides in the processing of nitric acid.

The ion-exchange processes used to recycle the condensate constituents of the nitrogen industry can be conveniently and arbitrarily divided into two parts: the recovery unit and the water polishing unit. This facilitates arrangement of the published data.

The main aim of the recovery unit is the rough separation of water and solutes and to increase the concentration of the latter to the highest possible level; i.e. to divide the condensate into roughly demineralized water and concentrated salts. Such roughly demineralized water can be used for some purposes without further treatment. However, for many potential uses (e.g. as boiler feed water) it is too dirty, and a polishing unit is often required. A mixed bed can be used to polish the roughly demineralized condensate from the processing of ammonium nitrate or sulfate. However, polishing of the analogous roughly demineralized condensate from the processing of ammonia depends on its chemical composition which varies from plant to plant. In some cases the condensate contains only ammonium carbonate, but often significant amounts of chloride and sulfate are present also. Efficient polishing methods are based on processes for the demineralization of low salinity waters. In many cases, the placement of a weakly basic anion-exchanger bed before the mixed bed leads to savings in caustic soda and to prolonged working cycles for the mixed bed. The regeneration of the polishing units should be carried out in the same manner as in conventional ion-exchange demineralization plants. Furthermore, roughly demineralized water can be mixed with raw water in a conventional ion-exchange plant; all nitrogen plants possess such units for the processing of boiler feed water as well as for some process requirements.

The use of conventional ion-exchange equipment simplifies the plant used for the recycling of the condensate constituents. The most important factor affecting the economics of the ion-exchange method is the concentration of the regeneration effluent. Special continuous counter-current columns are used to increase its concentration. There are several specially designed systems which are suitable for treatment of nitrogen fertilizer effluents. The Chem-Seps system is based on the Higgins contactor. The regenerants are a 22% nitric acid solution for the cation exchanger, and a 7% ammonia solution for the weakly basic anion exchanger. The strength of the regeneration effluent varies from 16-24% ammonium nitrate (43,44), and normally over 97% removal of nitrogen is anticipated. Another continuous counter-current system has been developed by Arion (45,46).

It is based on a specially designed column which dewaters the resin at the end of each cycle. The cation exchanger is regenerated with 45-60% nitric solution at 20°C and the weakly basic ion exchanger with 20-25% ammonia solution. The regeneration effluents are recycled for the manufacture of ammonium nitrate.

The use of concentrated solutions of nitric acid (30-60%) for the regeneration of fixed beds has been investigated also (99-103). The main problem is to ensure uniform flow of the regeneration effluent through the ion-exchange bed. Because the regenerant occupies only a small part of the bed, it moves through the bed as a chromatographic band. The method permits the production of regeneration effluents up to 26% ammonium nitrate.

If regeneration effluents containing ammonium sulfate are introduced into the processing of ammonium nitrate, the physical durability of the ammonium nitrate granules is increased, which improves their quality as a fertilizer. Attention has therefore been given to the use of sulfuric acid for regenerating the cation exchanger. This approach has the added advantage that sulfuric acid is a much weaker oxidizing agent than nitric acid. With concentrated sulfuric acid, a regeneration effluent up to 24% of $(NH_4)_2SO_4$ can be obtained (47). Hence, sulfuric acid can also be used to regenerate the cation exchanger for recycling the regeneration effluent.

In all cases the chemicals (nitric acid, sulfuric acid and ammonia) are returned to the manufacturing system. On their way through the ion-exchange unit they pick up some water. As additional energy must be expected to evaporate-off unwanted (excess) water, the efficiency of regeneration should be evaluated on the basis of the amount of unwanted "dilution water" recycled by the regenerating agents (nitric acid, sulfuric acid, ammonia) per amount of recovered chemicals or amount of treated effluent (47).

NEW DIRECTIONS IN THE DEVELOPMENT OF ION EXCHANGE METHODS

Whilst the separation of a particular wastewater into its components is not chemically difficult, the only methods of real practical significance are those which assure economical performance of the process. The same applies to the application of ion exchange to the recycling of wastewater constituents. Here, the efficient separation of wastewater into concentrated salts and water is often too expensive, especially for large amounts of wastewater. Therefore, in addition to the development of new technological flowsheets for the process, equally important are improvements in the unit operations and ion-exchange materials.

In this regard, CSIRO researchers have devised magnetic resins, which have a considerable number of advantages (48,49). Because small (ca. 200 μm across) resin particles are employed, rates of adsorption are significantly enhanced. Because the intensity of processing is increased, the same result can be achieved with less resin and a smaller plant. The magnetized resins are easy to separate from treated waters, and can be readily transferred, making them ideal for use in continuous contactors, several forms of which have been studied (49,50). All involve truly continuous transfer of resin, rather than intermittent transfer as is the case with most existing continuous processes.

Very simple equipment can be utilized for handling magnetized resins, ranging from countercurrent-fluidized beds (52) to co-current entrained pipe reactors (51), as depicted in Fig. 2 and 3. Both these systems have been tested on the pilot scale.

Compared with a conventional continuous contactor of the Asahi type, the fluidized bed system requires only about 25% of the resin (52). Complexities in large scale operation can be avoided by the use of an entrained pipe reactor, the capital cost for a 5 ML/day system being about half that of the fluidized bed (51). These plants are adequate for reactions such as dealkalization, which require only one or two stages.

More recently a multistage version of the fluidized bed column has been constructed which contains perforated dividing plates and a stirrer in each compartment. It was specifically designed for the use of hot water as the regenerant (50). It has been tested in a desalination pilot plant, and a 1 ML/day demonstration unit is to be constructed in 1982.

One of the main benefits of magnetic resin systems is that, in contrast to the normal packed bed techniques, they can treat raw water that contai suspended matter, without the system clogging up. This is especially important for the treatment of wastewater, where costs can be cut by avoiding preclarification. For example, a weakly acidic magnetic resin has been used to dealkalize unfiltered primary sewage effluent which has been treated with lime (51).

For economical wastewater treatment, the regeneration effluent should contain the highest possible concentration of contaminants to ensure their economic recycling. Theoretically, the use of columns operated in a continuous countercurrent manner should produce highly concentrated effluent. However, when a highly concentrated solution is used for regeneration of the resin, the ratio of the regeneration solution to the resin should be low, as the concentration of ions to be removed from the resin never exceeds 2 eq/L of the resin. In a semicontinuous countercurr unit, the movement of the resin in the column is achieved by a hydraulic pulse (for example in the Higgins unit). The liquid moves beyond the resin, which means that some of the regeneration solution will slip into the rinse section. At the same time, the exhausted resin brings some raw water into the regeneration section (see Fig. 4). The regeneration effluent from the rinse section is pushed back into the regeneration section by the rinse water during the 'working mode'. However, at the very beginning of each working mode, the first part of the regeneration effluent is raw water brought in during the resin movement. This leads to dilution of the regeneration effluent. In theory it should be possible to partition-off the regeneration effluent, separating the first part of the liquid from the other more concentrated position. Due to the short pulse times involved, this partitioning-off may not be easy to achieve. A truly continuous countercurrent column should reduce this dilution effect, since a steady state system would ensure a proper flow balance. In fluidized bed systems backmixing has to be carefully controlled.

A range of functional types of magnetic resins can be prepared (49). The simplest forms are those in which the magnetic component, usually about 20-50% by weight, is present within a bead or granule of an otherwise conventional resin structure. The particle size is smaller at about 100 μm across, versus 300-1200 μm for conventional resins. Polystyrene derived resins having strongly acidic and strongly or weakly basic characteristics can be readily synthesized.

Even more rapidly reacting resins can be obtained by grafting the active ion-exchange polymer onto an inert magnetic polymer core, such as that obtained by dispersing the magnetic material throughout a crosslinked polyvinyl alcohol. Weakly acidic resins based on polyacrylic acid have been successfully prepared by this means (49). The exchange sites in such 'whisker' resins are more accessible, so that the rate advantage is five times that of a conventional weakly acidic resin of the same size. The weakly acidic groups can be converted to other functionalities by various synthetic steps, but the linkage of the active group to the polymer chain usually invoves a hydrolytically unstable ester or amide group (53). It is our belief that the future development of these resins will further reduce the costs of ion exchange as applied to the recycling of wastewater constituents.

Review of flowsheets for wastewater treatment with magnetic resins

The most important features of magnetic resins - fast reaction and ease of transfer during operation in a continuous system - permit simplification of plant design and size. This should make magnetic systems more competitive with existing technology.

One of their biggest potential areas of application is the removal of heavy metals from industrial wastewater and sewage. An application of magnetic resins to the recycling of wastewater components from the rinse water of metal finishing industries is illustrated in Fig. 5 for the recycling of nickel salts. The use of magnetic resins for the recycling of other types of rinse waters would be very similar. Another possible example is the removal of metals from mixed plating effluents (see Fig. 6).

Magnetic resins can also be applied to sewage treatment. They have special advantages for the treatment of liquids which contain suspended matter initially (51) or for liquids containing precipitates that form during the ion exchange operation, as for example in the Liberti process (34-36). (See Fig. 7). Even the adsorbent clinoptilolite, which is used to remove NH_4^+, can be converted to a magnetic mode, by the synthesis of a composite bead containing the finely divided adsorbent and iron oxide within an inert polymeric binder such as crosslinked polyvinyl alcohol. This is a variation on the magnetic activated carbon particles devised by the CSIRO magnetics program (54). Although the capacity of the magnetic particle is lowered because 40% of the bead's volume consists of inert material, the adsorption rates are enhanced by an order of magnitude. Also, the mechanical strength of the particles is improved considerably.

One of the most economical and practical approaches to wastewater treatment is its integration with the basic manufacturing process. One example is the integrated system employed in the manufacture of metal oxides, which could be operated with magnetic resins (see Fig. 8).

Another area where use of the magnetic resin would be beneficial is in the separation of an acid from its salts (see Fig. 9). When conventional resins are employed the process is limited by slow kinetics (20). The use of a magnetic micro-resin with its far better kinetic properties should improve the opportunities for the acid retardation process in the wastewater area. Other processes which would benefit from the magnetic approach, and where preliminary laboratory studies have been carried out, include colour removal from paper-mill bleach plant effluent (57) and the recovery or organic compounds in the food industry (58).

These few examples illustrate ways in which the unique properties of magnetic resins might be exploited for treating various wastewaters. One magnetic resin, the magnetic whisker type, in addition to tolerating suspended matter in the contacting equipment has another advantage. Because the active polymer chains which are grafted to the magnetic core are essentially uncrosslinked, resin fouling is unlikely to be a problem. As already mentioned, a weakly acidic whisker resin has been successfully used in pilot plant studies for the removal of calcium bicarbonate from lime-treated primary sewage effluent (51). Further uses are suggested in Figs. 6 and 8 (49). So far, not much has been done on the application of magnetic resins to wastewater treatment.

An extension of the magnetic concept to the removal of insoluble contaminants has been successfully developed to large scale operation (55). Whilst the removal of suspended matter could be achieved with magnetic whisker resins based on grafted cationic polymers, the process was too expensive to compete with conventional clarification using alum, iron salts or lime. However, it has been found that iron oxide particles themselves can be employed to adsorb dissolved materials, and heterocoagulate colloidal and suspended matter. Magnetite of diameter 1-10 μm has a positive surface charge at slightly acidic pH levels of 5 to 6, and effectively removes colour anions and clays, silica and microbes which are all negatively charged at that pH level. The addition of a soluble cationic polyelectrolyte increases the load that the magnetite can carry. Rapid separation of the treated water and magnetite is achieved after magnetization, and the oxide can be regenerated by raising the pH level to 11.5, whereupon the surface charge is reversed so that the like-charged impurities are released. The process has been proven for the removal of turbidity and colour from underground and surface waters (55), and more recently, attention has been focussed on wastewaters, including effluents from a steel cold rolling mill, textile dyehouse, secondary sludge treatment plant, and papermaking (56).

We believe that the unusual characteristics of the new reagent particles will guarantee their future usage.

REFERENCES

1. CALMON, C. Ion Exchange for Pollution Control, edited by C. Calmon, and H. Gold. CRC Press, 1979.
   a) Mercury removal by ion exchange, pp 201-206.
   b) Precious metal recovery, pp 213-215.
   c) Specific ion exchangers, pp 151-154.

2. NAGATA MASANORI, UCHINO YOSHITOSHI, KUBOTA NOBORU, HONDA YUKIO and TAKADA TOMOKO. Wastewater treatment in electronic parts plating plant and metal recovery - Gold recovery from plating rinse wastewater by ion exchange, Kenkyu Hokoku - Kymamoto-ken Kogyo Shikenjo, 1979, 8-13 (Chem. abstracts 94: 6282).

3. HENMI EIICH, Recovery of gold in Washings of acid gold-plating bath, Jap. Patent 7768023 (1977).

4. NAGATA MASANORI, UCHINO YOSHITOSHI and KUBOTO NOBORU, Wastewater treatment and recovery of useful materials from electronic part plating factories. I. Recovery of silver from wastewater by ion exchange, Kumamoto-ken Kogyo Shikenjo, Kenkyu Hokoku, 1976, 51, 12-19. (Chem. abstracts 90: 435549).

5. NAGATA MASONORI, HONDA YUKIO, KUBOTO NOBORU, UCHINO YOSHITOSHI and TAKATA TOMOKO, Investigations on the treatment of effluents from electronic component parts plating plants and the recovery of valuables. V. Experiments on the recovery of silver from silver plating waste waters. Kumamoto-ken Kogyo Shikenjo Kenkyu Hokoku, 1977, 52, 24-35. (Chem. abstracts: 90: 172075).

6. YABE EICHI, Treatment of plating waste solutions, Jap. Patent 7975159 (19

7. GEMERT J.T., Using and recovering silver, Industrial and Commercial Photography, 1980 April/May, 42-43.

8. MINA R., Silver recovery from photographic effluents by ion exchange methods, J. of Appl. Photogr. Eng., 1980, 6, 120-125.

9. COOLEY AUSTIN C., Ion exchange silver recovery for process EP-2 with nonregenerated bleach-fix. J. Appl. Photogr. Eng., 1981, 7, 106-110.

10. USEHIMA HIROMOTO, SHINMURA TAKEO, MURAKAMI HIROYOSHI and OKAMOTO MITSUTAMI, Recovery of selenium from wastewater from the electrolytic colouring process of aluminium and its reuse. Japn Patent 7980194 (1977)

11. KAWAMOTO MICHIO and MURAKAMI HIROYOSHI, Recovery of selenium oxyacids in wastewater. Jap. Patent 78108898 (1978).

12. UESHIMA HIROMOTO, YAMASHITA SHINTARO, MURANISHI SEIICHIRO, MURAKAMI HIROYOSHI and OKAMOTO MITSUTAMI, Recovery and recycling of selenium values from a spent aluminium electrolytic colouring bath, Jap. Patent 79155945 (1979).

13. HEIJWEGEN, C.P. and VAN ZIJLL LANGHOUT P.A. Treatment of waste liquid from the electrolytic tin-coating lines at Hoogovens Jmuiden B.V., Belg.-Ned. Tijdschr. Oppervlaktetech Mater., 1980, 24, 162-168.

14. UMARAKHUNOV M.Kh., KAPLAN A.L., SHAMSIEV S.M. and REZQEV N.U., Optimization of the separation of molybdenum from nitrate and c-rbonate ions on the anion exchanger AV-17 in chloride form. D:posited Doc., 1974, VINITI 2808-74.

15. VLASOVA V.I., PETRYAEV E.P. and AFANASEVA T.S., Treatment of arsenic-containing wastewater by ion-exchange sorption. Jzv. Vyssh. Uchelon. Zaved. Khim. Khim. Teknol. 1971, 24, 212-215.

16. GRANT P.J. and SCHEPERS J.R., Power plant water purification, chemical recovery and waste-treatment: a novel technique with multiple in plant application. Proc. Amer. Power Conf., 1978, 975-985.

17. ITO YUTAKA, Apparatus and Method for recovering boric acid from radioactive waste solutions, Jap. Patent 8000430 (1978).

18. ZAGULSKI, L., PAWLOWSKI, L. and CICHOKI, A. Methods for recovery of water and vanadium compounds from wastewater, Physicochemical Methods for Water and Wastewater Treatment, ed. L. Pawlowski, Pergamon Press, 1980, pp 229-236.

19. DE YONG, G.J. and REKERS, C.J.N. AKZO Process: for removal of mercury, J. Chromatography, 1974, 443-450.

20. VIS, J.H. Mercury recovery from waste streams by means of the Imac TMR ion exchange process. Chem. Age. India, 1980, 31, 481-487.

21. GNOT, W. and KOCIUBAS, Z. Recovery of mercury from wastewaters in the electrolytic production of chlorine. Zesz. Nauk. Politech. Slask. Chem., 1979, 57-69.

22. GRINSTEAD, R.R., NASUTAVICUS, W.A. and WHEATON, R.M. New selective ion exchange resins for copper and nickel recovery. Extractive Metallurgy of Copper, American Institute of Mining, Metallurgical and Petroleum Engineers, 1976.

23. YONES, KENNETH, C. and GRINSTEAD, ROBERT R., Properties and hydrometallurgical applications of two new chelating ion exchange resins. Chem. and Ind., 1977, 637-641.

24. GRINSTEAD ROBERT R., Copper-selective ion exchange resin with improved iron rejection. Journal of Metals, 1979, 31, 13-16.

25. JONES KENNETH C., and PYJOER, RANDALL A. Copper recovery from acidic bleaching liquors by continuous ion exchange and electrowinning. Journal of Metals, 1979, 31, 19-25.

26. JOFFE, A.Z., CHAMAEV, V.N., MARKOVA, L.M. and KULEBYAKINA, L.A. Reduction in environmental pollution by wastewater from cuprammonium fibre production. Khim. Volokna, 1979, 4, 50-51.

27. INOUE TAKASHI, MATSUOKA MAKOTO, NIIMURA TAKEO and UEJIMA HIROMOTO, Vanadium and nickel recovery from desulfurization solution. Jap. Patent 7890 653 (1978).

28. MAKITA ZENJI, HAYASHI MASAMI and UTSUGI MINORU. Recovery of high purity zinc. Jap. Patent 7761125 (1977).

29. ANDERSON, R.E. Some examples of the concentration of trace heavy metals with ion exchange resins, Report No. EPA - 902/9-74-001, US EPA, Cincinnati, 1976.

30. CHAMBERLAIN, D.G. Selective removal of zinc from cooling tower blowdown by ion exchange, Proc. Int. Water Conference, Pittsburgh, 1970.

31. CALMON, C. Specific and chelate exchangers: new functional polymers for water and wastewater treatment, Journal AWWA, 1981, 652-656.

32. WAITZ WILLIAM H. Ion exchange in heavy metals removal and recovery. Amber Highlights, No. 162, Philadelphia, 1979.

33. PAWLOWSKI, L., CICHOCKI A. and KUHNKE, T. Note on a continuous counter-current ion exchange with two step regeneration for zinc recovery from synthetic fibre plant wastewater. Water Res., 1976, 10, 781-782.

34. LIBERTI, L., LIMONI, N., PASSINO, R. and PETRUZZELLI, D. Ammonium phosphate recovery from urban sewage by selective ion exchange. Physicochemical methods for water and wastewater treatment, edit. L. Pawlowski, Pergamon Press 1980, pp 73-85.

35. LIBERTI, L. and PASSINO, R. An ion exchange process to recover nutrients from sewage. Resource Recovery and Conservation, 1981, 6, 263-273.

36. LIBERTI, L., LIMONI, N. and PETRUZZELLI, D. Nutrient removal from sewages by selective ion exchange. Quard.-wst. Ric. Acque, 1979, 38, 13/1-13/25.

37. GOLD, H. Metal finishing wastes. Ion Exchange for Pollution Control, edit. by C. Calvin and H. Gold. CRC Press 1969, p. 173.

38. SHULDT, A.A. and HUBER, R. Chromate and water recovery from tin plating operation using ion exchange technologies. Proc. Int. Water Conf. Eng. Soc. West Pa., 1978, 38, 185-199.

39. STROMQUIEST, D.M. and REENTS, A.C. Removal of cations from chromic acid solutions, Proc. 6th Ind. Waste Conf., Series 76, Purdue University, Lafayette, Ind., 1952, p. 289.

40. COSTA, R.L. Regeneration of chromic acid solutions by ion exchange. Ind. Eng. Chem., 1950, 42, 308-311.

41. PAWLOWSKI, L., KLEPACK, B. and ZALEWSKI, R. A new method of regeneration of anion exchanger used for recovering chromates from wastewater. Water Res., 1981, 15, 1153-1156.

42. KUNIN, R. Ion exchange for the metal products finisher. Prod. Finish (Cincinnati), (April-June) 1969.

43. HIGGINS, I.R., CHOPRA, R.C. and ROLAND, L.D. Continuous ion exchange achieves unique solution in pollution control. Inst. Chem. Eng. Symp. Ser., 1978, 54, 105-110.

44. ROLAND, L.D. The recovery of ammonium nitrate from fertiliser factory wastes. Physicochemical Methods for Water and Wastewater Treatment, ed. L. Pawlowski, Pergamon Press, 1980, pp 299-311.

45. ARION NICOLAE M. Treating and recovering wastewater from fertilizer manufacture, USA Pat. 4 002 455 (1977).

46. POPOVICI I.V. Technical solutions and technical advances made in Romania to control environmental pollution effects arising from fertilizer plants. UNIDO, TD/WG 175/18, Expert Group Meeting, Helsinki 26-31, 1974.

47. PAWLOWSKI, L. Pilot plant study of ammonia and water recycling in the nitrogen industry. Resource Recovery and Conservation, 1979, 4, 43-57.

48. N.V. BLESING, B.A. BOLTO, D.L. FORD, R. McNEILL, A.S. MACPHERSON, J.D. MELBOURNE, F. MORT, R. SIUDAK, E.A. SWINTON, D.E. WEISS, and D. WILLIS. Some ion exchange processes for partial demineralization. Ion Exchange Processes in Pollution Control, Society of Chemical Industry, London, 1970, pp 371-381.

49. B.A. BOLTO. Magnetic micro resins. In. Ion Exchange Processes in Pollution Control, Vol. 2, C. Calmon and H. Gold, Eds. CRC Press, Boca Raton, Fla., 1979, pp 213-225.

50. B.A. BOLTO, E.A. SWINTON, P.R. NADEBAUM and R.W. MURTAGH. Desalination by continuous ion exchange based on thermally regenerable magnetic microresins. Water Science and Technology, accepted for publication.

51. B.A. BOLTO, D.R. DIXON, A.J. PRIESTLEY and E.A. SWINTON. Ion exchange in a moving bed of magnetized resin, Prog. Water Tech., 1977, 9 833-844.

52. B.A. BOLTO, D.R. DIXON, R.J. ELDRIDGE, L.O. KOLARIK, A.J. PRIESTLEY, W.G.C. RAPER, J.E. ROWNEY, E.A. SWINTON and D.W. WEISS. Continuous ion exchange using magnetic miroresins. The Theory and Practice of Ion Exchange, Society of Chemical Industry, London, 1976, paper 27.

53. B.A. BOLTO, D.R. DIXON, and R.J. ELDRIDGE. Graft polymerization on magnetic polymer substrates. J. Appl. Polym. Sci., 1978, 22, 1977-1982.

54. D.R. DIXON and J. LYDIATE, Selective magnetic adsorbents. J. Macromol. Sci-Chem., 1980, A14, 153-159.

55. N.J. ANDERSON, N.V. BLESING, B.A. BOLTO, D.R. DIXON, L.O. KOLARIK, P.R. NADEBAUM, A.J. PRIESTLEY, W.G.C. RAPER and D.E. WEISS. Water clarification with the use of regenerable magnetic particles. Effl. Water Treat. J., 1981, 21, 359-366.

56. N.J. ANDERSON, B.A. BOLTO, D.R. DIXON, L.O. KOLARIK, A.J. PRIESTLEY and W.G.C. RAPER. Water and wastewater treatment with reusable magnetic particles. Water Science and Tech., accepted for publication.

57. B.A. BOLTO, A.J. PRIESTLEY and R.V. SIUDAK. A continuous process for decolorizing hardwood caustic bleach plant effluent using a magnetic basic ion exchange resin, Appita, 1979, 32, 373-377.

58. R.A. GRANT. Protein recovery as an effluent treatment process, Effl. Water Treat. J., 1975, 15, 616-620.

59. B.A. BOLTO and L. PAWLOWSKI. Reclamation of wastewater constituents by ion exchange. Effl. Water Treat. J., accepted for publication.

60. HATCH J. and DILLON J.A., Acid retardation, Ind. Eng. Chem. Process Design and Development, 1963, 2, 253-263.

TABLE 1

Commercially available chelate resins (31)

Manufacturer	Resin Designation	Chelate Group
Akzo Chemie	IMAC GT-73	weakly acidic
Diamond Shamrock	Duolite ES-346	amidoxime
	Duolite ES-466	iminodiacetic acid
	Duolite ES-467	amino-phosphonic
Dow	Dowex A-1	iminodiacetic acid
	Dowex XF-4195	weakly basic
	Dowex XF-4196	weakly basic
Nippon Soda	Nisson ALM-525	carbamate
Mitsubishi	Diaion CF-10	iminodiacetic acid
	Diaion CR-20	polyamine
	Diaion CR-40	etheleneimine
Rohm and Haas	Amberlite IRC-718	pyridine type
Initica	Unicellex UR-10	iminodiacetic acid
VEB Farbenfabric Wolfen	Wofatit UR-10	iminodiacetic acid
Czechoslovakia	Biosorbent M	mixture of products of a culture (mycellium)
Bayer	TP-207	iminodiacetic acid
Ayalon	Srafion NMRR	isothiouronium
Sumitomo	Sumichelate Q-10	isothiouronium

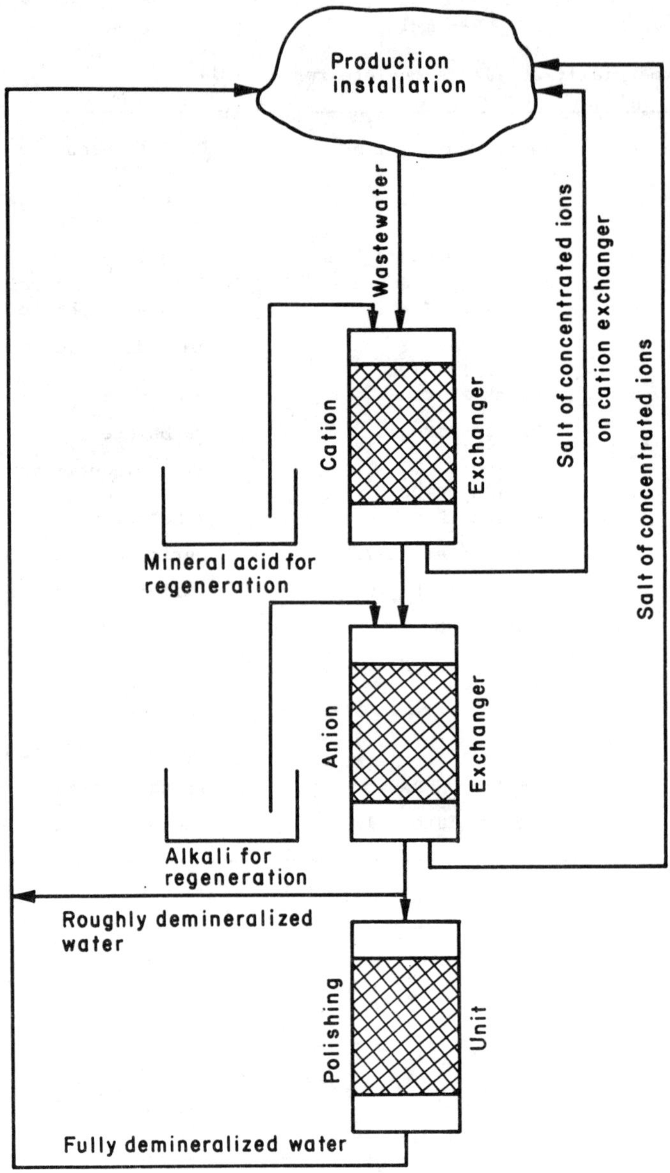

FIG. 1. FLOWSHEET FOR THE RECYCLING OF ALL WASTEWATER CONSTITUENTS BY ION EXCHANGE.

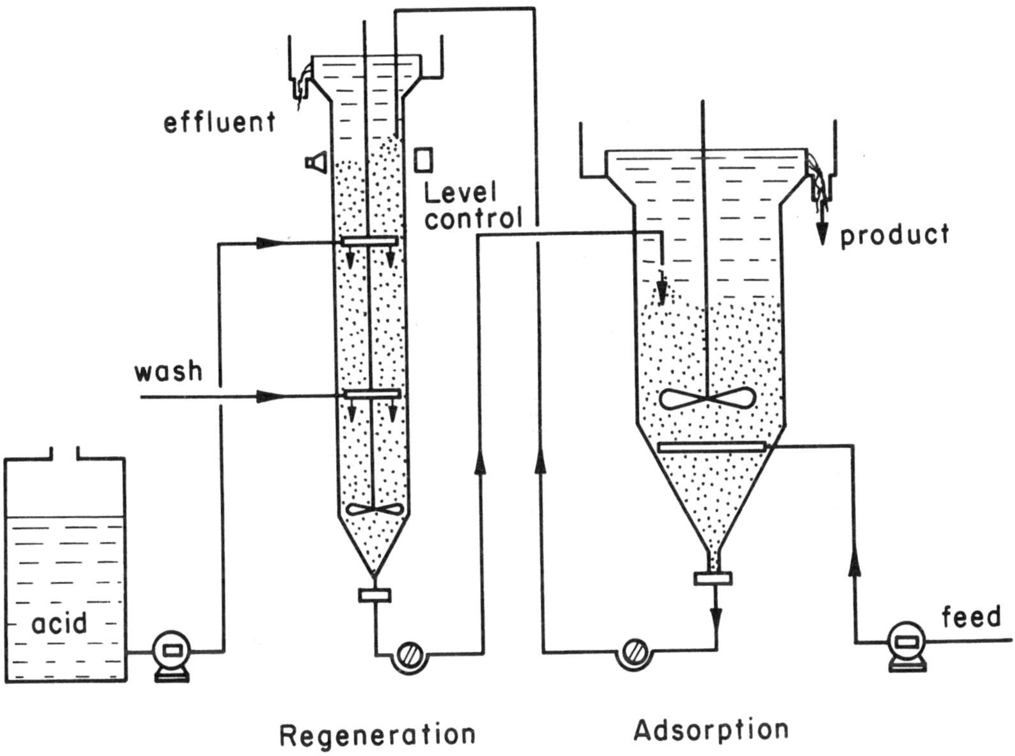

FIG. 2. FLOWSHEET FOR A MOVING BED SYSTEM WITH A FLUIDIZED BED ADSORBER UNIT.

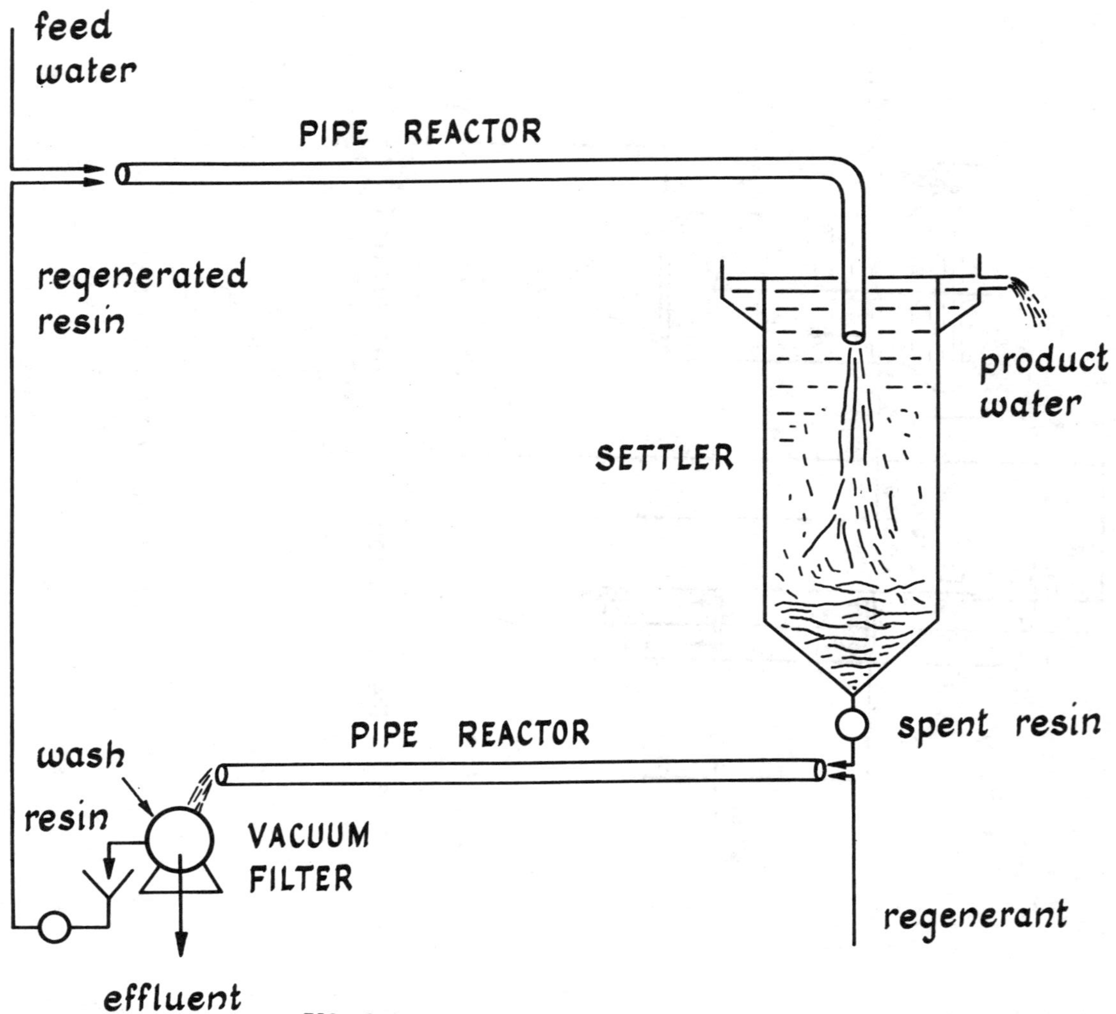

FIG. 3 GENERALIZED FLOWSHEET BASED ON ENTRAINED PIPE REACTORS

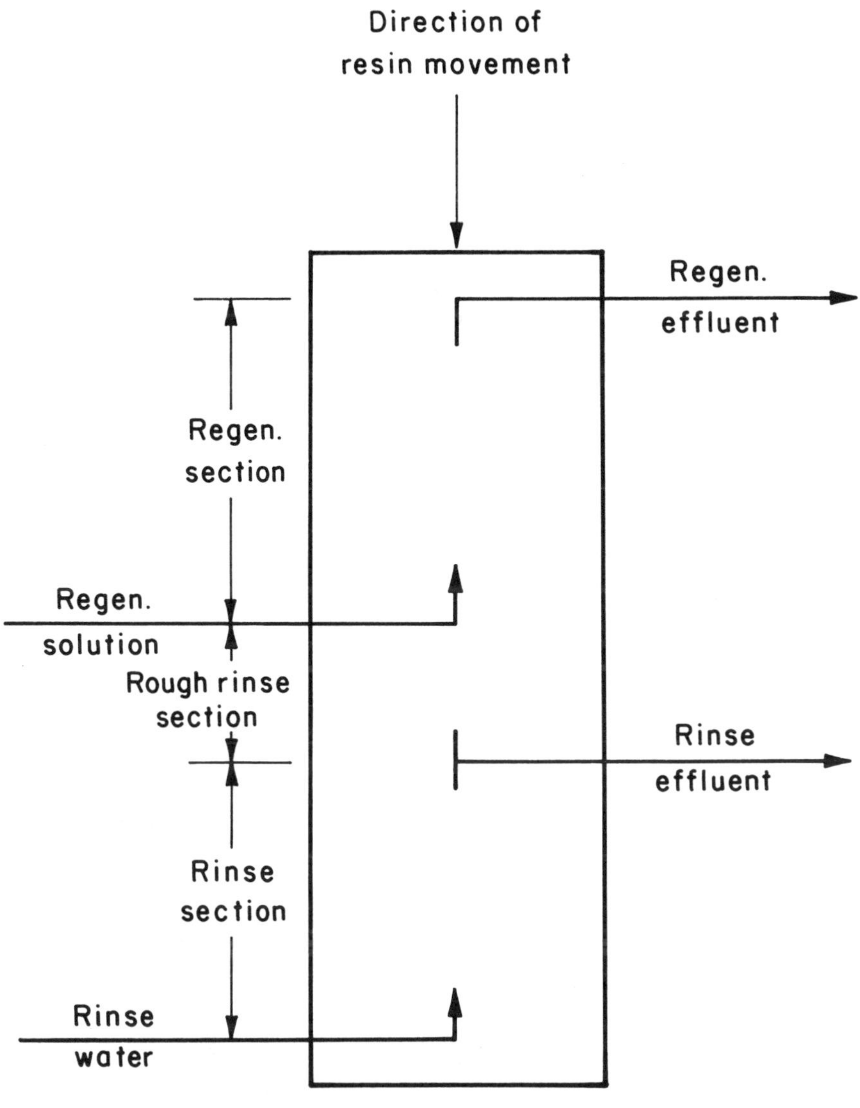

FIG. 4. SECTION OF A MOVING BED SYSTEM.

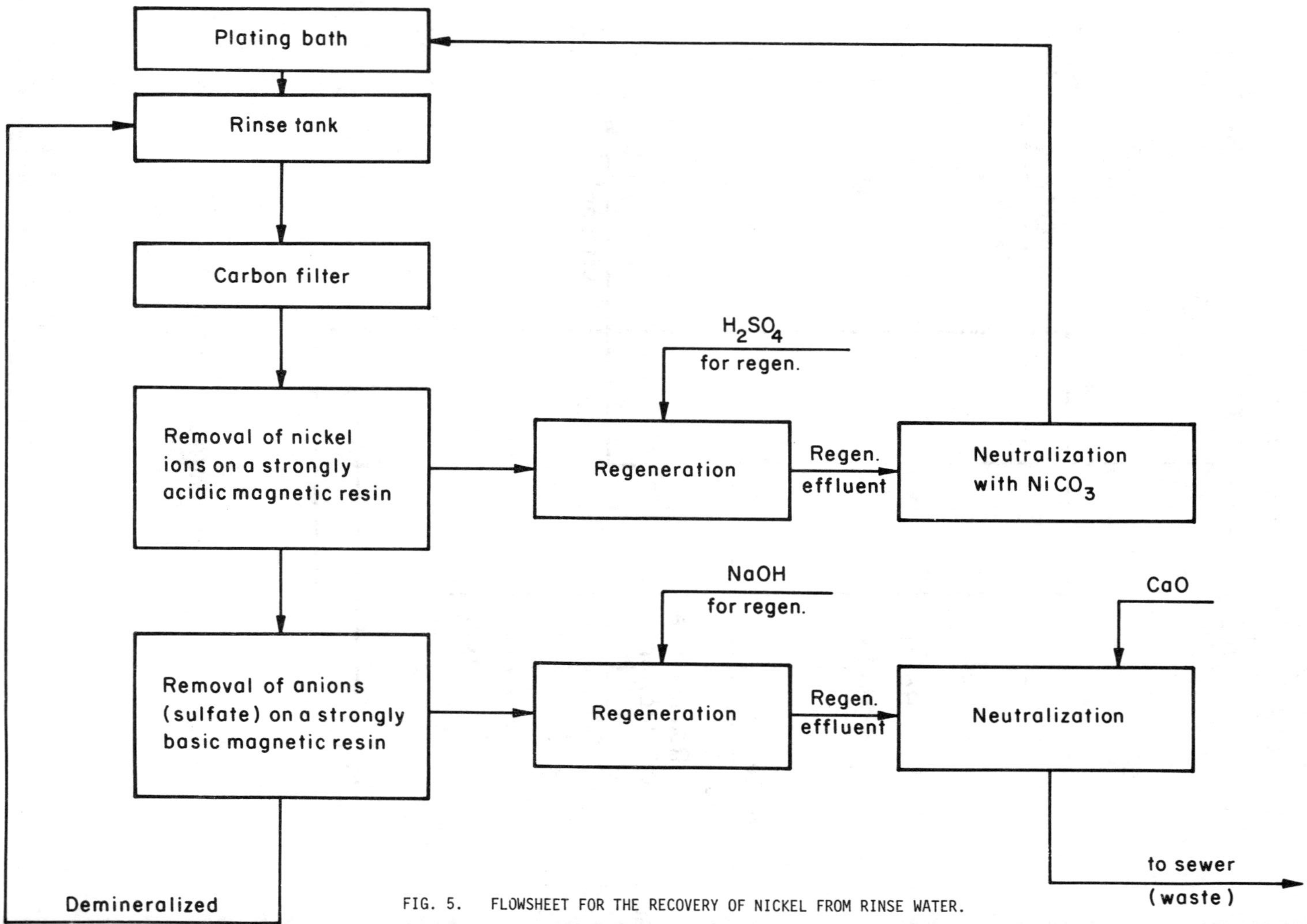

FIG. 5. FLOWSHEET FOR THE RECOVERY OF NICKEL FROM RINSE WATER.

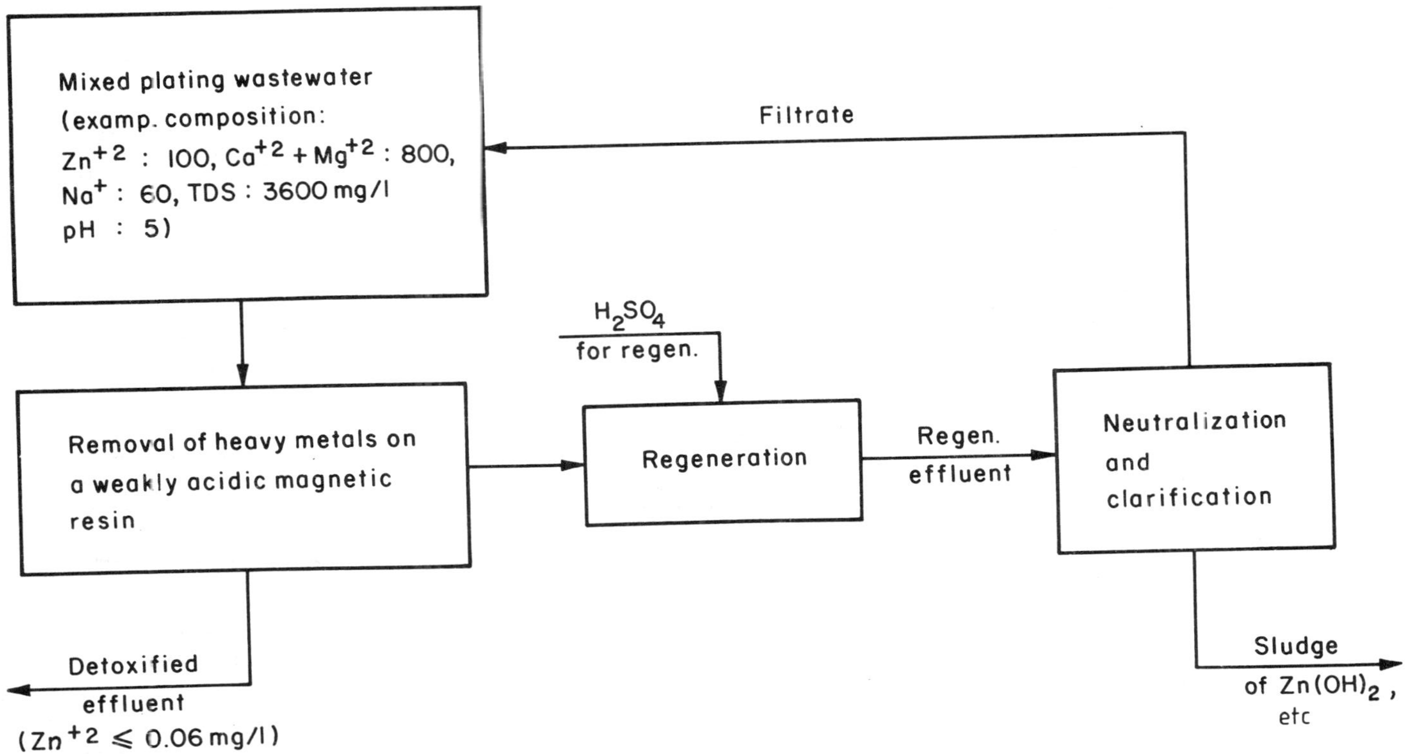

FIG. 6. FLOWSHEET FOR THE PURIFICATION OF MIXED PLATING WASTEWATER.

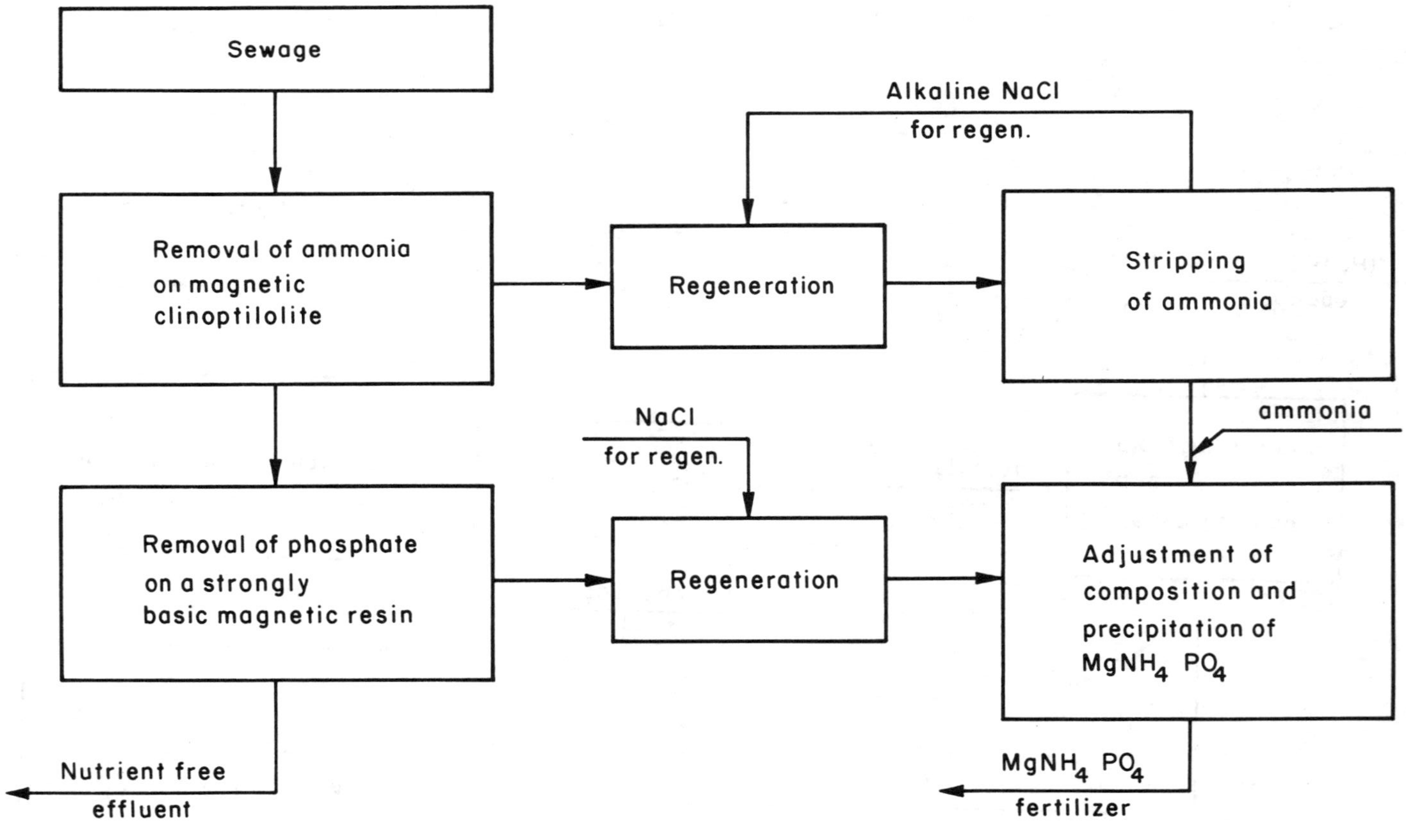

FIG. 7 FLOWSHEET FOR THE RECOVERY OF NUTRIENTS FROM SEWAGE

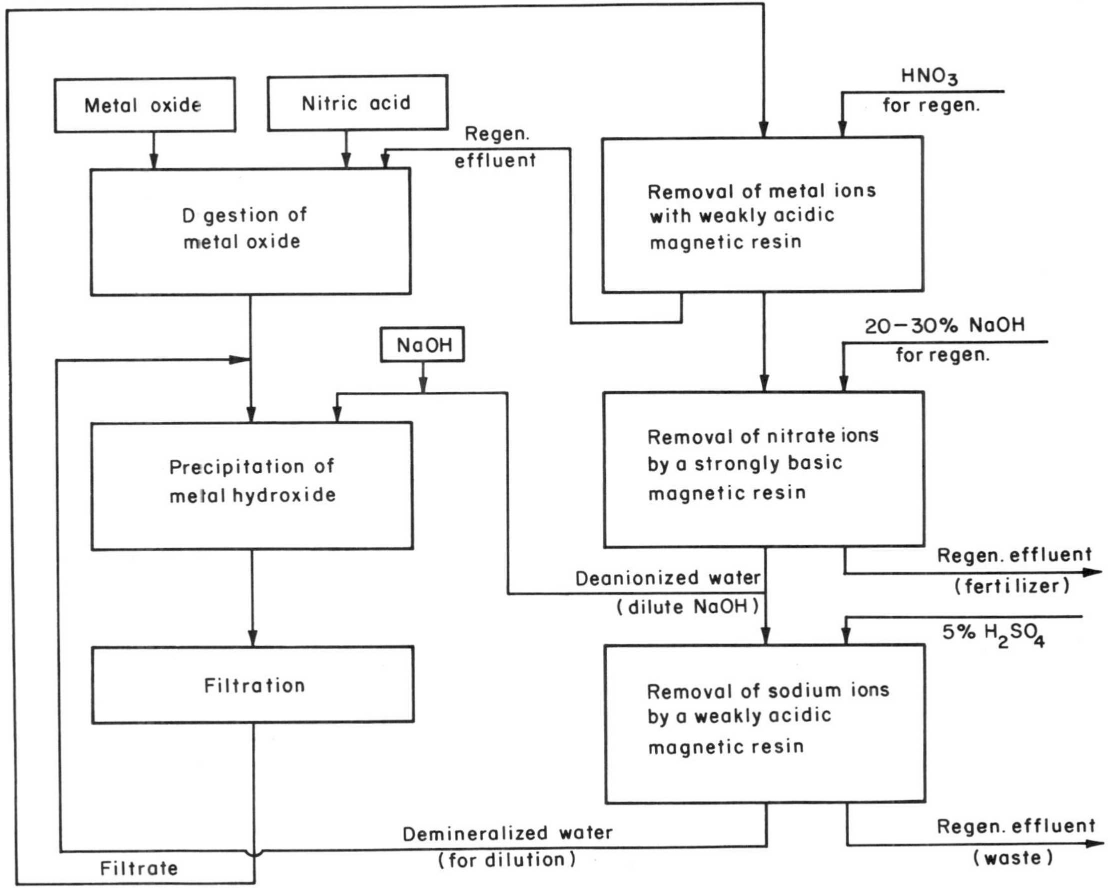

FIG. 8. FLOWSHEET FOR A CLOSED-LOOP SYSTEM OF PROCESSING OF HEAVY METAL OXIDES CATALYSTS.

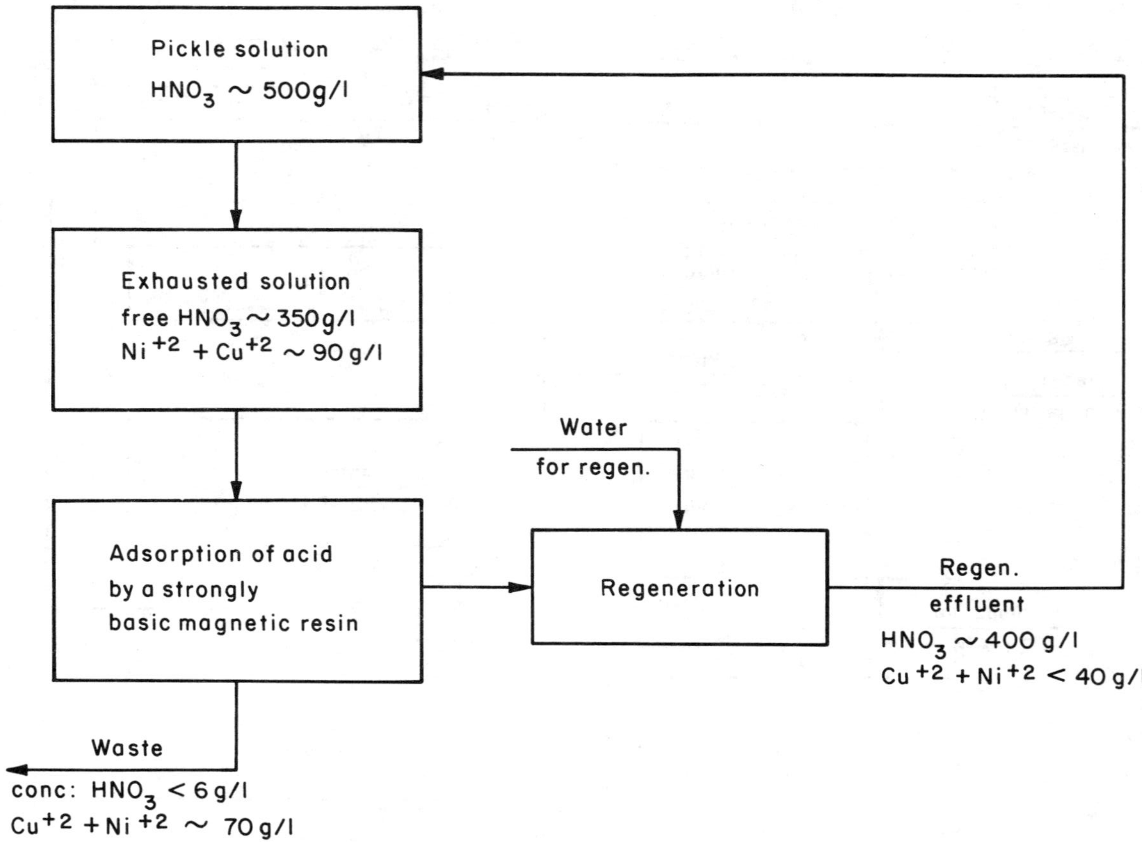

FIG. 9. FLOWSHEET FOR THE SEPARATION OF ACIDS AND SALTS BY ION EXCHANGE.

CATION EXCHANGE SOFTENING COUPLED WITH ELECTRODIALYSIS FOR HIGH RECOVERY DESALINATION

John W. Kaakinen

Bureau of Reclamation
U.S. Department of the Interior
PO Box 25007
Denver CO 80225 USA

INTRODUCTION

Recovery of a maximum volume of desalted product water from a given saline feed water source is a goal in desalting where there is a shortage of feed water or the disposal of the desalting reject brine volume is a problem. In many instances of inland brackish water desalting, brine disposal by such methods as deep well injection or evaporation in ponds can be the dominant cost. Thus, by minimizing the brine volume to be disposed of and maximizing the product water volume, high recovery desalting may be economically feasible in water-scarce localities of the world. The Bureau of Reclamation, U.S. Department of the Interior, has been investigating high recovery desalting for possible application in the arid southwestern United States. The Bureau is not only investigating methods of providing the benefits of fresh water from brackish water, but it has been studying methods to limit the amount of dissolved solids entering the Colorado River because of the detrimental effects of salinity on irrigation and other uses of Colorado River water. One possible solution for keeping saline waters from entering the Colorado River is to utilize them for industrial or utility cooling in place of fresh water.

This paper describes field experiments of two processes, IX (ion exchange) as a pretreatment coupled with ED (electrodialysis) desalting. Many brackish waters require pretreatment to keep calcium carbonate, calcium sulfate, and sometimes strontium and barium sulfates from precipitating on membranes or on cooling tower surfaces as the dissolved solids are concentrated. Cation exchange provides an attractive, cost effective process for this pretreatment when the reject brine from a desalting unit or blowdown from a cooling tower can provide the sodium for regenerating the cation exchanger. ED is attractive for high recovery brackish water desalting because it can produce higher brine concentrations and, thus, for a given feed water salinity can achieve higher product water recoveries than RO (reverse osmosis), because RO equipment is limited by allowable applied pressure required to overcome the osmotic pressure of the brine. ED is generally cheaper and less energy intensive for brackish water desalting than commercially available distillation processes.

Purpose and Outline

The purpose of the IX and ED experiments was to demonstrate the applicability and to obtain feasibility data for IX and ED at two test sites. Different IX operating conditions and procedures were tested in an effort to obtain optimum IX operation at each site. The IX data were analyzed to determine maximum values for calcium resin capacity divided by IX cycle time, which is inversely proportional to the amount of IX resin required for an IX plant design. But the use of the ED reject brine for IX regeneration can lead to severe gypsum scaling in the IX resin and IX effluent piping during regeneration, which can cause severe operational problems as described later in this paper. Thus, a dominant dependent variable was the intensity of this gypsum scale in the IX column, which could be minimized with certain sets of IX operating conditions and procedures that did not correspond with those yielding the minimum resin requirements for design purposes. While the experimental ED unit was operated primarily to make concentrated brine for IX regeneration in the IX experiments, useful and interesting ED operational data for high recovery desalting were collected. Also, work has just begun at the Bureau of Reclamation to develop IX models which can greatly decrease the amount of IX field testing required at other desalting and cooling water sites that are under study.

This paper contains the following sections. This introduction includes a description of the two test sites, water compositions, and a general process description. The experimental methodology section includes a description of pilot plant equipment, procedures, and experimental strategy. The results section describes much data and findings of the IX and ED experiments at both test sites and presents results of a simple IX equilibrium model. These are followed by the Conclusions and Recommendations. The acknowledgments, references, tables, and figures are at the end of this report.

Field Test Sites at Yuma and LaVerkin Springs

The first test site was the YDTF (Yuma Desalting Test Facility) near Yuma, Arizona. Most testing at the YDTF was for the purpose of obtaining design data for the YDP (Yuma Desalting Plant), which is now under construction as one of the largest membrane desalination plants in the world with a maximum design capacity of 4.2 m^3/s [1]. The design recovery of the YDP is 70 percent. An 82-km-long canal will convey the waste reject brine to the Gulf of California. In a competitive bidding procedure that gave approximately equal weight to cost and technical design, two manufacturers of spiral wound RO equipment were selected over other manufacturers of RO and ED equipment to provide the desalting units for the YDP. Part of the congressional authorization [2] for the YDP provided for studying various means of replacing the 30 percent reject brine stream, which is lost for beneficial use. This paper describes the results of experiments at Yuma aimed at obtaining feasibility design [3, 4] data for increasing the product water recovery of the YDP design to at least 90 percent.

Raw feed water for the YDTF and YDP is irrigation return flow pumped from wells in the Wellton-Mohawk Valley where 30 000 hectares are irrigated with Colorado River water containing approximately 860 mg/L of TDS (total dissolved solids). The return flow contains about 3000 mg/L of TDS primarily because of evapotranspiration. Conveyance of the irrigation return flow is in an open canal, which results in windblown dust, aquatic weeds, and algae in the raw saline water. Raw and lime-treated water compositions are listed in table 1. The lime treatment also clarifies the raw water. Partial lime-treated water is adequate for RO and ED at 70 percent desalting recovery.

To avoid membrane scaling, much higher recoveries would require additional removal of calcium for RO and ED and greater silica removal for RO but not ED because ED does not concentrate un-ionized silica. Calculations using methods developed by Marshall and Slusher [5] indicate that calcium concentrations should be reduced to less than about 35 mg/L at 90 percent recovery and 17 mg/L at 95 percent recovery to avoid gypsum scale. (An alternative pretreatment possibility to avoid gypsum precipitation would be to remove sulfate with anion exchange, but this approach was considered less favorable than cation exchange removal of calcium with the Yuma water compositions.) Some more recent ED experiments have shown that membrane scaling by supersaturated calcium sulfate can be avoided with polarity reversal [6]. High pH lime-treated water in table 1 includes silica removal for high recovery using RO; this water was used in the IX-ED experiments because the IX experiments were to apply to pretreated water for RO as well as ED, because adding additional RO equipment for high recovery in the YDP would be more economical than tail-end ED when the original YDP will contain only RO equipment. The final IX report [7] and a previous paper [8] describe the IX experiments in more detail than in the present paper.

Typical water compositions at LaVerkin Springs are given in table 1. LaVerkin Springs is located in southwestern Utah along the Virgin River, which flows into the Colorado River. The springs have a water temperature of about 40 °C. Tests at LaVerkin Springs were for the purpose of obtaining desalting data for feasibility estimates. High desalting recovery would be very important because onsite reject brine disposal would probably be part of any desalting project here. In addition to IX and ED, partial lime softening in a reactor clarifier system and in a spiractor, lime-soda softening, and RO were other processes tested at LaVerkin Springs [9]. Only the IX and ED experiments are described here.

IX Process Description

Using reject desalting brine to regenerate a fixed-bed cation exchange softener requires different procedures [10, 11, 12, 13] than when using common NaCl regenerant solution (table 2) [14]. These differences are due to the presence of supersaturated calcium sulfate in the regeneration effluent and the limited volume of regenerant (reject brine) available per IX cycle. This supersaturated calcium sulfate is a consequence of the high concentrations of calcium eluting from the cation exchange resin during regeneration plus the high concentrations of sulfate in the desalting reject regenerant. In order to avoid gypsum (calcium sulfate dihydrate) scaling of the resin, the regeneration is carried out with upward flow, countercurrent to the flow direction of the exhaustion or service mode, generally with a flow rate high enough to fluidize the resin bed in the nature of a backwash. (Similar methods have been developed for sulfuric acid regeneration of cation exchange resins.) [15] Because the precipitation of large gypsum particles is not immediate due to the kinetics of gypsum formation [16], the used regenerant can be removed from the resin bed before a gypsum scale can form in the bed. Studies have shown that downflow, concurrent regeneration, using desalting reject brine, is not so successful [10, 17].

Reusing the reject brine regenerant is an innovation [12] which offsets the limited volume of fresh desalting reject made per IX cycle, V_F, which is determined by the desalting recovery, R, and volume of exhaustion product, V_E, per cycle according to:

$$V_F = (1-R)\ V_E \tag{1}$$

Somewhat less than half of the sodium in the regenerant is generally absorbed in a single pass through a strong acid cation exchange resin, and much less than half if a relatively fast regeneration flow rate is used. The supersaturated used regenerant is desupersaturated through contact with gypsum crystallites in an agitated tank [12] or a spiractor [17]. After stopping the agitator in a tank, the gypsum precipitate is separated from the regenerant solution by gravity settling. This separation removes much calcium and sulfate ions and, therefore, lowers the TDS of the used regenerant. The supernatant becomes the recycled regenerant and is transferred to a storage tank. Because recycled regenerant has a lower concentration of sodium, a lower TDS concentration, and a higher calcium concentration compared to the fresh desalting reject brine, it is used immediately prior to the higher quality fresh regenerant.

The individual modes of a general IX cycle using brine regeneration (table 2) have the following purposes:

- Exhaustion (also called service) - feed water is softened downflow by the cation exchange resin as a pretreatment for the desalting units.

- Drain 1 - feed water is removed from the resin bed to avoid subsequent excessive dilution of recycled regenerant.

- Regeneration 1 - recycled regenerant from a storage tank as described in the previous paragraph is used for an upflow backwash and partial regeneration. This backwash removes suspended particles filtered from the feed water during exhaustion and reclassifies the resin beads by particle size. Regeneration 1 effluent is the major process waste stream containing the desalting reject after its use as IX regenerant. Regenerant 1 volume per cycle should match regenerant 3 volume for system steady state if there are no other regenerant losses.

- Regeneration 2 - stored recycled regenerant is also passed through the resin bed upflow, but the effluent is returned to the recycling system for further use as recycled regenerant.

- Regeneration 3 - fresh desalting reject provides the highest quality solution used to regenerate the IX resin, and the effluent is recycled.

- Drain 2 - the excess regenerant is removed from the resin bed to lower the subsequent rinse volume.

- Rinse - the remaining regenerant is flushed downflow from the resin to provide a high quality effluent at the beginning of the exhaustion mode of the following cycle.

EXPERIMENTAL EQUIPMENT AND METHODOLOGY

Equipment

A simplified flow diagram of the pilot plant equipment used for the high recovery ion exchange-electrodialysis experiments is shown in figure 1. High lime-softened, clarified water was provided by a solids-contact internal solids recirculation reactor-clarifier at Yuma [8] and by an inline reactor-flocculator-clarifier system with external solids recycle at LaVerkin Springs [9]. The clarifier effluent at each site was filtered through beds of granular anthracite coal over silica sand. Sulfuric acid was added to the filter influent to maintain a pH of 7 in the filter effluent. Gaseous chlorine at Yuma and sodium hypochlorite solution at LaVerkin Springs were used to maintain a chlorine residual of about 0.5 g/m^3 through the lime treatment and filtration. Dechlorination with sodium sulfite solution was applied just prior to the cation exchanger to protect the resin and subsequently the ED membranes from chlorine attack. At Yuma, there was no further chlorination of the IX product-ED feed, but this scheme changed at LaVerkin Springs as discussed later.

The IX pilot plant contained two transparent acrylic columns, 2.5 m high and 0.34 m inside diameter, each charged with about 0.10 m^3 of cation exchange resin. The resin used at Yuma was Amberlite 200 macroreticular type. Resin bead size was 0.4 to 0.6 mm in diameter. Bed depth was about 1 m. Perforated plates covered by a fine plastic screen at the top and bottom of the column acted as flow distributors and kept resin levels from leaving the column. To control the IX process, a microprocessor specially designed by the Bureau selected the sequence of some 35 electric motoractivated valves and 4 pumps. The length of each step (hereafter referred to as a mode) in an IX cycle was controlled by either the time duration set in the microprocessor clock or through operators' manual override, using as criteria the monitored volume of water throughput or the calcium breakthrough concentration in the exhaustion effluent.

The ED pilot plant was an Ionics, Inc., Aquamite V model with a single Mark II stack (fig. 2). The Aquamite V skid contained the pumps, flow control devices, variable voltage transformer, rectifier, and associated monitoring and control devices. The stack contained the following components at each site:

	Yuma	LaVerkin Springs
Electrical stages	2	2
Total hydraulic stages	4	6
Cell pairs/stage	75-50-50-75	45-45-40-40-45-45
Total cell pairs	250	260
Cation membrane model No.	CR61 AZL183	CR61 CZL183
Anion membrane model No.	AR103 PZL183	AR204 SXZL183

Each cell pair in the ED consisted of one anion membrane, one cation membrane, and a tortuous path spacer between each membrane. Each stack component measured 508 by 457 mm, overall. The effective area of each membrane was 0.144 m^2. The product TDS concentration at LaVerkin Springs was relatively high at approximately 1 g/L, but lower product TDS could have been achieved with additional stages.

The membrane models used at Yuma were standard brackish water types, Ionics, Inc.'s most commonly used membranes. The membrane types used at LaVerkin Springs were specially selected for higher current efficiencies, which is important for demineralizing high brine concentrations. Both membrane types at Yuma and the cation membrane at LaVerkin Springs were comprised of cross-linked copolymers of vinyl monomers. The anion membrane type used at LaVerkin Springs was made from acrylic material.

As seen in figure 2, the dilute stream flowed once through the stack. The brine was recirculated internally to equalize the flow rates and pressure differential on each side of a membrane. Desalting recovery and brine concentration were easily set by adjusting the brine makeup flow. Part of the dilute (IX product) provided the electrode rinse at Yuma, but the electrode rinse is shown in figure 2 as originating from the IX feed water; this was done at LaVerkin Springs to avoid wasting IX product and to maintain the correct ratio of reject brine regenerant volume to IX product volume for the IX experiments.

Polarity reversal occurred once every 15 minutes at Yuma and every 30 minutes at LaVerkin Springs. With polarity reversal, the dilute and concentrate streams exchange identity in the stack; i.e., the dilute stream becomes the concentrate stream and the concentrate stream becomes the dilute stream. This reversal was accomplished with MOV (motor-operated valves) on the Aquamite V connected to the inlet and outlet of the stack and by reversing the electrical polarity of each electrode. Following polarity reversal, the product was automatically diverted to waste until the new dilute stream had lowered to a set-point conductivity. In addition, the ED system contained a special modification to divert the concentrate to waste following polarity reversal until a set-point brine conductivity was reached, which then automatically caused the flow of brine to be sent to the IX fresh regenerant storage tank.

Procedures

Operators on duty 24 hours per day measured and set flow rates, measured tank volumes; manually titrated samples of the IX effluent to determine the calcium breakthrough point of the IX; made numerous other measurements, readings, and observations which they recorded on data sheets and in their log; and collected samples for analysis by the YDTF chemical laboratory. Chemical analyses of samples collected during IX data cycles were made using atomic absorption spectroscopy and other standard analytical techniques.

Desalting recoveries were calculated from TDS concentrations by:

$$R = \frac{C'_F - C'_E}{C'_F - C'_P} \tag{2}$$

where R is the desalting recovery as a fraction, C'_F is the TDS concentration of the desalting reject (fresh IX regenerant), C'_E is the TDS concentration of the desalting feed (IX exhaustion product), and C'_P is the desalted product TDS concentration.

Equation 2 was derived from mass balances of water, TDS, and the normal definition of recovery (product flow divided by feed flow). Recoveries were calculated from salinity measurements alone using equation 2, not from flow rate measurements, which are generally subject to greater experimental error than TDS concentrations. At Yuma, these brine concentrations were determined operationally by evaporation at 103 °C in the chemistry laboratory, usually daily because more than a week was required to obtain summation-of-ions data for TDS. TDS measured by both methods were in agreement. Using this procedure, reject brine TDS concentrations of 20, 35, and 50 g/L correspond to calculated recoveries of 85.5, 91.8, and 94.3 percent, respectively, when the feed TDS concentration is 3.3 g/L and the product TDS concentration is 0.427 g/L, the approximate values at Yuma. At LaVerkin Springs, the TDS concentrations of the ED reject were 40 g/L and 92 g/L corresponding to 80 and 92 percent recoveries for an ED feed TDS concentration of 8.7 g/L and a product concentration of 1.0 g/L of TDS.

For a closed-loop process (no imported regenerant chemicals), the new regenerant volume available per cycle would normally be limited and equal to the volume of ED reject brine generated per cycle on the average according to equation 1. To operate the IX closed loop at YDTF, the fresh regenerant volume selected for use was determined by the volume theoretically made, V_F calculated using equation 1. An iterative, trial and error approach toward a balance of volumes was necessary because the exhaustion volume, V_E, is affected somewhat by the amount of regenerant being used. The procedure used is as follows: In each experimental run, an initial fresh regenerant volume was selected by experience and insight. After at least three similar IX cycles to approach an equilibrium for V_E using this selected fresh regenerant volume, a theoretical V_F available was calculated using the measured V_E and R in equation 1. This calculated V_R was compared to the actual volume used, and if these two values did not agree within 10 percent, a new fresh regenerant volume was selected and the process repeated until agreement between the volume used and V_F was achieved.

The limitation in fresh regenerant volume generated by the ED per IX cycle was overcome by recycling regenerant as discussed in concept previously. The recycling system, consisting of T-6 and T-5 shown in figure 1, functioned as follows. Regenerant effluent supersaturated with calcium sulfate was placed in T-6. During the regeneration, a residual of gypsum crystals from previous cycles was maintained in suspension in T-6 using an agitator. This agitation promoted rapid desupersaturation of the calcium sulfate. At least 1 hour prior to the next use of recycled regenerant, the agitator was turned off to allow the crystals to settle to the bottom of T-6. The clear supernatant from T-6 was transferred by gravity flow to T-5 for temporary storage immediately prior to its use as recycled regenerant. The effluent from T-5 was also filtered immediately prior to use as regenerant during some of the later experiments. The gypsum solids in the bottom of T-6 needed to be drained out infrequently, approximately monthly, because their accumulation rate was small compared to the volume of T-6.

The ED was operated on IX softened water supplying the need for reject brine regenerant for the IX experiments. The recovery and brine concentration were determined by the IX experimental design. ED operating conditions and samples for chemical analysis were collected after steady state had been reached following a change in polarity.

Experimental Strategy

The IX experiments at Yuma were carried out in three phases: phase 1, exploratory experiments; phase 2, response-surface experimental design; and phase 3, additional experiments. The purpose of phase 1 was to screen process variables and to establish the ranges for the independent control variables for use in phase 2. In phase 2, a systematic experimental design was followed to establish the relationships between dependent and independent variables. Phase 3 was necessary to obtain additional data and answer some new questions resulting from phase 2.

A Box-Behnken experimental design [18] was followed in phase 2. The basic design for 4 independent variables consisted of 3 levels, a high, a low, and a midpoint, for each independent variable and 27 experimental IX data runs or observations. Data were analyzed by multiple regression [19] using Hewlett-Packard desk top computer models 9825 and 9835 with multiple regression software.

At LaVerkin Springs, time constraints did not allow a statistical experimental design. Instead, features of the optimum process conditions found at Yuma and process modifications were tested at LaVerkin Springs, which are discussed later with the LaVerkin Springs IX results.

Dependent Variables

Important dependent variables selected for these experiments were the resin capacity q_{Ca}, the cycle time, t_c, and the intensity of gypsum scale in the IX column during and following regeneration. The specific resin capacity, normally used in characterizing IX resins and processes, is the number of equivalents of ions, (in this case, Ca^{++}), which are absorbed in an IX cycle per volume of resin used (in liters) [14]. The specific resin capacity divided by the cycle time, q_{Ca}/t_c is inversely proportional to the amount of working resin required for a design. However, the intensity of gypsum scaling was a dominant consideration in selecting optimum sets of IX operating conditions and procedures. The average calcium leakage concentration from the resin bed during exhaustion was not directly important as a dependent variable in these experiments because it was always comfortably less than the concentrations where gypsum scaling would occur in the desalting concentrate.

Independent Variables

The major independent variables selected at Yuma were the TDS concentration of the fresh regenerant (reject brine TDS concentration related to desalting recovery by equation 2), the fresh regenerant flow rate, the recycled regenerant flow rate, and the recycled regenerant volume. An uncontrolled independent variable, which did not come to light until the middle of phase 2, was the fresh regenerant temperature. These variables, plus several other control variables which emerged after initial experimental results, are discussed further in the results for Yuma and LaVerkin Springs. The IX feed water composition was relatively constant, although there were minor fluctuations in the seasonal canal composition and in the daily operation of Train IV reactor, particularly in the reaction zone pH which affected the calcium concentration in the IX feed water.

Other variables were not independent in the high recovery experiments. The volume of fresh regenerant used was set by the desalting recovery (fresh regenerant concentration) and the resulting exhaustion throughput volume according to equation 1. The maximum calcium breakthrough allowed before termination of exhaustion was also tied to desalting recovery, which affects the allowable calcium concentration in the reject brine before gypsum precipitation ensues [5]. At Yuma, calcium breakthrough concentrations used were 4.5, 3.0, and 1.5 meq/L with fresh regenerant concentrations of 20, 35, and 50 g/L of TDS, respectively. At LaVerkin Springs, the calcium breakthrough concentrations used were 12 meq/L for a 40-g/L TDS reject brine concentration and 6 meq/L for a 92-g/L TDS reject brine concentration. These maximum calcium breakthrough concentrations were approximately equal to the average concentrations that would have caused gypsum precipitation in the reject. This was a conservative approach since the average concentration of calcium in the exhaustion effluent was always much less than the maximum breakthrough calcium concentration.

RESULTS AND DISCUSSION

Phase 1 Exploratory IX Experiments at Yuma

In phase 1, the control variable levels were established for the phase 2 response surface experiments. The 50-g/L high fresh regenerant concentration was the highest practical, with the ED membranes and stack configuration used at Yuma. A 24-L/min regenerant flow rate corresponded to about 50 percent bed expansion with water temperatures of 15 °C. Other variable limits were based on process practicalities, especially duration of regeneration. Other control variables established for phase 2 included two values for the maximum allowable calcium concentration before terminating exhaustion (run only with 20 g/L fresh regenerant concentration) and the type of backwash water, IX feed water versus recycled regenerant.

Several methods of automatic termination of the exhaustion mode were evaluated in phase 1, but none were found satisfactory compared to manual Ca^{++} titrations and exhaustion termination by the operators.

During phase 1, the regenerant recycling techniques previously described in the procedures were established. In a typical cycle which used 50 g/L of TDS reject brine and a large volume of recycled regenerant, the data showed that 68 percent of the calcium ion in the used regenerant was removed during the recycling process. This compares to a 72 percent theoretical removal of gypsum under these conditions at chemical equilibrium. The TDS also decreased by 11 g/L through this gypsum removal. Less calcium ion and TDS removal than this occurred during regenerant recycling with lower regenerant concentrations because the degree of calcium sulfate supersaturation was less.

Phase 2 Response-surface Experiments at Yuma

A summary of the data from the phase 2 experiments is given in table 3. The order of each data cycle was random with the following exception. Regenerant concentrations were grouped in blocks to save some of the time required to make new tank volumes of fresh and recycled regenerant in equilibrium with the IX process, which took over 1 week for fresh regenerant concentration of 50 g/L of TDS, for example.

One of the most important findings in phase 2 was the influence of regenerant temperature upon gypsum scaling of the resin. The regenerant temperature was not controlled in these experiments but varied with the ambient temperature. During phase 1, which was run in late 1978 when regenerant temperatures were 15 to 20 °C, there was no visually observable gypsum scaling in the resin bed under any combination of the control conditions run. This lack of gypsum scaling influenced the selection of a relatively low range of fresh regenerant flow rates for use in phase 2. As ambient temperatures rose in late spring and summer of 1979, the regenerant temperatures also increased to as high as 36 °C. Severe gypsum scaling of the resin bed and sometimes of effluent piping occurred at high temperatures, especially during combinations of high regenerant concentration, low fresh regenerant flow rate, and low volumes of recycled regenerant. This effect of temperature was further verified using laboratory-sized IX columns in which the regenerant temperature was controlled; resin scaling was observed with 30 °C regenerant temperatures but not with 15 °C regenerant temperatures with other control variables being comparable. It is notable that, in all cases, the gypsum scale could be removed and the resin rejuvenated to its original appearance and resin capacity after regeneration with a 3-percent NaCl solution. Thus, gypsum scale of resin was completely reversible, which is not generally true for gypsum scaling of RO and ED desalting membranes.

It is somewhat surprising that the importance of temperature on gypsum precipitation in the resin bed during reject brine regeneration has not been reported previously in any of the numerous studies of this IX process. This is apparently because these previous studies were carried out inside laboratories and with water which had relatively constant temperatures. However, basic kinetics studies have shown that the temperature definitely affects rates of gypsum precipitation [16]. In a related application, temperature is taken into account to avoid gypsum scale when sulfuric acid is used to regenerate cation exchange resins [15].

Other phase 2 findings in table 2 indicate that there was some improvement in resin capacity between using calcium breakthrough points of 1.5 meq/L (run numbers with C suffix) and 4.5 meq/L, which should be expected. There was no significant difference in resin capacity between using an IX feed water backwash (run numbers with B suffix) and using the same relatively low volume of 240 L of recycled regenerant. Note that run 3.11B, which used a feed water backwash and a fresh regenerant concentration of 50 g/L of TDS, was the only phase 2 run that was not self-sustaining. That is, the IX product-ED feed volume was not sufficient to provide enough desalting reject volume at 94.3 percent recovery to regenerate the IX in a closed-loop process. Apparently, the use of recycled regenerant tended to suppress the peak calcium elution concentration and, thus, the level of calcium sulfate supersaturation as compared to the use of fresh regenerant alone.

Phase 3 Experiments to Mitigate Gypsum Scaling of Resin at Yuma

The gypsum scaling of the resin bed in phase 2 during reject brine regeneration led to special phase 3 experiments (table 4) to find methods for mitigating the scale. Measures tested included higher fresh regeneration flow rates, addition of SHMP (sodium hexametaphosphate) scale inhibitor to the regenerant solutions, and air mixing (declassifying) the resin bed between regeneration and exhaustion.

An effective method found for mitigating the resin scaling at Yuma was the use of a high fresh regenerant flow rate sufficient to expand the resin bed by about 50 percent in regeneration 3 plus use of large volumes of recycled regenerant in regeneration 2 (table 1). These findings agree with those of others using a different composition feed water [12].

Alternatively, the addition of 100 mg/L of SHMP to the regenerant (runs 3.18.21E, 4.04.07E, and 4.06.07F) also eliminated resin scaling when lower fresh regenerant flow rates were used. However, q_{Ca} was still not as high as with the high fresh regenerant flow rates alone (tables 3 and 4). Regenerant recycling was also severely hampered by the presence of SHMP in the used regenerant as the recycled regenerant contained higher concentrations of calcium compared to other runs.

The air mixing (run 4.05.09B) to destratify the resin bed caused the lowest calcium resin capacity of any IX run in the test program, mainly due to a high calcium leakage which caused an early termination of exhaustion using the established breakthrough concentration criteria. Therefore, this procedure is not recommended for the present IX application.

Gypsum Scaling Intensity at Yuma

When using reject brine as regenerant, the regeneration effluent was probably always supersaturated in calcium sulfate because high concentrations of calcium ion were eluted into used regenerant containing high concentrations of sulfate. A gypsum scale accumulation on the resin can be largely avoided using the methods described previously. However, these data do not clearly indicate what effect the presence of gypsum precipitation in the bed has on the IX performance as measured by q_{Ca}/t_c. This lack of information on the effect of gypsum scaling led to the following additional analysis of the existing IX data.

A new dependent variable called the scaling intensity was developed as a semiquantitative grade given to each IX run based on operators' observations of resin scale as regularly recorded in their log.

- A value of 10 was assigned when symptoms of severe calcium sulfate scaling of the resin were noted. These included cementation of resin beads, flow channeling, and sometimes the formation of cavities in the top 300 to 600 mm of the resin bed; the formation of volcano-like cones on the top surface of the bed during regeneration with fresh desalting brine; and cloudiness of the regeneration effluent in the column above the resin and settling of white calcium sulfate crystals on the top of the resin bed.

- A value of 5 was assigned when these symptoms were less severe and confined to the top 300 mm or less of the resin bed.

- A value of zero was assigned when no symptom of gypsum scaling was observed. A tabulation of these scaling factor values is contained in the last columns of tables 3 and 4.

Correlation coefficients were calculated for the variables for 28 runs in tables 3 and 4. What was most interesting about these statistical analyses was that there was only a -0.21 correlation coefficient between the scaling intensity and q_{Ca}/t_c. In other words, there was only a weak relationship observed between overall performance of the IX as measured by the q_{Ca}/t_c and the observed presence of gypsum scale. Thus, while extremely severe gypsum scale should be avoided in the IX column because flow distribution may be harmed, the presence of some gypsum scale in the resin bed apparently does not affect overall IX performance.

Microbiological Growth Causing High Plugging Factors at Yuma

High plugging factors (filter plugging index), including some values of over 90 percent, were measured in the IX product water. Conversely, partially softened IX feed water had consistently very low plugging factors. No steps were attempted to lower the high plugging factors in the IX product because they had no immediate measurable effect on ED performance (but see ED results). Scanning electron micrographs plus chemical analyses for amino acids and polysaccharides indicated the presence of microbiological growth [20] on the surface of the cation exchange resin and in the ED feed water but not in the IX feed water. This growth was attributed to dechlorination of the IX feed water, necessary to prevent gradual oxidation and deterioration of the resin. To protect the ED membranes from chlorine attack, the feed to the ED should also not contain a chlorine residual. Disinfection of IX product will maintain low plugging factors in the desalting feed as demonstrated elsewhere [17].

Statistical Analysis of Yuma IX Data

Multiple regression analysis [19] was performed on the data from 28 observations in phases 2 and 3. With calcium resin capacity as the dependent variable, only rejected regenerant volume was found to be a significant independent variable using the F test and a 95 percent confidence limit. A scattergram and a linear fit are given in figure 3.

Different independent variables became important when the resin capacity divided by the cycle time q_{Ca}/t_c, inversely proportional to the design resin requirement, was used as the dependent variable in a multiple regression analysis.

The regenerant brine TDS concentration, the fresh regenerant flow rate, and the recycled regenerant volume were found to significantly affect q_{Ca}/t_c and t_c by itself. This is illustrated by an equation including each of the components of cycle time:

$$t_c = \frac{V_E}{G_E} + \frac{V_E\ (C'_E - C'_P)}{G_F\ (C'_F - C'_P)} + \frac{V_R}{G_R} + t_o \tag{3}$$

where t_c is the cycle time in minutes,

G_E is the exhaustion flow rate in L/min,

V_E is the exhaustion throughput volume of feed water per cycle in L,

C'_E, C'_P, and C'_F are the TDS concentrations of ED feed, product, and reject brine (IX fresh regenerant) in g/L, respectively,

G_F is the fresh regeneration flow rate in L/min,

V_R is the volume in L of recycled regenerant in addition to backwash used per cycle,

G_R is the rejected regeneration flow rate, and

t_O is other time used per cycle, almost always 25 minutes in the present experiments, for the backwash, rinse, and drains.

The term V_E $(C'_E - C'_P)/C'_F - C'_P)$ in equation 3 equals the fresh regenerant volume according to equations 1 and 2.

Equation 3 illustrates how the time required for a step in the IX cycle will decrease as the flow rate used in that step is increased. However, mass transfer considerations will generally result in a decrease in resin capacity and V_E for the same volume of throughput as a flow rate is increased. Besides mass transfer limitations, pressure drop in the bed in downflow and allowable bed expansion in upflow are hydraulic considerations affecting the design flow rates and resin bed dimensions.

Multiple regression analysis of the Yuma IX data done to determine the effects of fresh reject brine TDS concentration (C'_F related to R by equation 2), G_F, V_R, and G_R on cycle time. Results in figure 4 show a fit of t_c as a function of $1/G_F\ C'_F$ and V_R/G_R, which correspond to terms related to fresh regeneration and recycled regeneration times in equation 3. (Because C'_F is much greater than C'_P in equation 3, the term $1/G_F\ C'_F$ is proportional to fresh regeneration time.) The relatively good fit of the two terms to t_c illustrates the applicability of equation 3 to the Yuma data. The plot in figure 4 could not be made using equation 3 without experimental data because V_E is a function of G_F, C'_F, V_R, and G_R.

Consideration of the dependencies of resin capacity and cycle time with the dependent variables and a high value for resin capacity divided by cycle time led to a set of recommended cycle operating conditions for Yuma. Another important consideration was whether resin scaling by gypsum occurred, for some conditions which gave the highest values for q_{Ca}/t_c were eliminated because of the gypsum scaling. A set of recommended operating conditions is given in table 5, which corresponds closely to data for cycle 4.04.42. Table 6 gives chemical compositions of water sample collected during cycle 4.04.42. Plots of the exhaustion and regeneration concentrations for calcium, magnesium, and sodium are given in figures 5 and 6. This recommended cycle did not have the highest q_{Ca}/t_C of all cycles run, but because gypsum scaling in the resin was well controlled under the conditions of table 5, it was selected over some other cycles with higher q_{Ca}/t_C. The selection of the recommended IX cycle conditions of table 5 did not consider the benefits of a higher exhaustion flow rate. All IX data were collected with a G_E of about 0.3 bed volumes per minute. A higher G_E would decrease t_C according to equation 3. A higher G_E would also decrease q_{Ca} but to a lesser degree than the decrease in t_C, which is suggested by the low Ca leakage during exhaustion (fig. 5). Thus, q_{Ca}/t_C would increase and the design resin requirement would decrease with a higher G_E.

IX Process Changes Tested at LaVerkin Springs

At LaVerkin Springs, four changes were made to the IX pilot plant equipment and procedures based on results at Yuma.

1. The Amberlite 200 macroreticular cation exchange resin used at Yuma was replaced at LaVerkin Springs with a common gel-type cation exchange resin, Dowex HCR (8 percent cross linked). The gel-type resin was selected for LaVerkin Springs because:

   a. The higher physical strength of the macroreticular type is unnecessary in the present process application.

   b. The availability of the gel type is greater and its cost is lower.

   c. The capacity of the gel type is roughly 10 percent higher.

   d. Sulfate is largely excluded from the interior of the bead of gel resin so that gypsum formed around the smooth bead surface is easily removed. Conversely, the macroreticular type has pores allowing bulk flow of sulfate containing solution to enter the bead wherein gypsum may form during regeneration and be more difficult to remove. (While this potential for gypsum formation inside a macroreticular resin bead exists, no clear evidence of this was noted in the IX data from the Yuma experiments.)

2. A heat exchanger was added to allow heating the IX regenerant to the maximum expected summertime water temperature of about 25 °C. The shell-side heating fluid for the heat exchange was excess raw well water, which had a temperature averaging about 40 °C. The regenerant was heated because the higher summertime temperatures at Yuma were found to accelerate the rate of gypsum precipitation in the IX column.

3. A system was installed on the IX pilot plant for maintaining a packed resin bed during high upflow regeneration flow rates [21, 22]. The system utilized a regenerant effluent collector consisting of a cross of perforated pipe wrapped with fine mesh plastic screen installed about 50 mm below the top of the settled resin bed. Upon completion of a 10-minute backwash with recycled regenerant, the upward flow through the column was stopped, and the resin allowed to settle while the solution in the column was drained to the level of this midcolumn collector. Just prior to further regeneration, a small flow of compressed air was applied to the top of the column and the air allowed to exit the column through the midcolumn collector. This downward air flow, maintained at 50 to 100 kPa in the column during regeneration, was sufficient to cause the upward flowing regenerant to leave through the midcolumn collector and to maintain a packed bed with maximum upward flows (limited in these experiments by the capacity of the regenerant pump used) that were up to 50 percent greater than the flow rates that were normally used for a 50-percent-expanded fluidized bed during regeneration.

4. An additional secondary chlorination system was installed to maintain a chlorine residual of 0.5 g/m^3 to control microbiological growth in the IX product water piping and storage tank. A dechlorination system was also added to remove this free chlorine residual just prior to the ED to prevent chlorine attack of the ED membrane.

IX Results at LaVerkin Springs

The IX experiments at LaVerkin Springs were designed to obtain the following data:

1. Verification that the most successful IX operating conditions found at Yuma (a fluidized bed during regeneration and the use of a substantial volume of recycled regenerant preceding regeneration with fresh ED reject brine) would also be successful with the water compositions at LaVerkin Springs.

2. The evaluation of additional possible IX process improvement suggested by results from Yuma including the use of gel-type cation exchange resin, rather than a compacted bed during upflow regeneration, addition of SHMP to the regenerant to inhibit gypsum scale formation, and special rinse modes designed to minimize resin scaling.

Time limitations did not allow a statistical experimental design for data analysis by multiple regression as was done at Yuma. However, the number of IX data cycles completed at LaVerkin Springs was sufficient to achieve the above objectives and to obtain feasibility design data on IX performance for LaVerkin

Springs. The numerical results of the LaVerkin Springs IX experiments are summarized in table 7.

The dependent variables included calcium resin capacity, q_{Ca}, the calcium resin capacity divided by cycle time, q_{Ca}/t_c (both in table 7), and the qualitative severity of gypsum scaling during and following regeneration. The effect of each independent variable was as follows.

Regeneration TDS Concentration. Both q_{Ca} and q_{Ca}/t_c were higher with a fresh regenerant concentration, C'_F = 92 g/L of TDS (92 percent recovery ED reject brine) than they were with C'_F = 40 g/L of TDS (80 percent recovery reject brine). With C'_F = 40 g/L, no gypsum scale accumulated on the resin when recycled regenerant was used prior to fresh regenerant and a fluidized bed was used during each regeneration step. But with C'_F = 92 g/L, substantial amounts of gypsum scale remained in the IX column at the end of regeneration at all flow and temperature conditions. (However, when SHMP was used, the gypsum scale did not form as discussed below.)

Use of Recycled Regenerant. The use of recycled regenerant increased q_{Ca} and made gypsum scaling less severe. With C'_F = 92 g/L, resin scaling was very severe plus q_{Ca} and q_{Ca}/t_c were lower when recycled regenerant (V_R = 0) and SHMP were not used. In comparing recycled regenerant volume V_R = 1600 L with V_R = 800 L, the values of q_{Ca} were about the same but q_{Ca}/t_c was the highest at V_R = 800 L, indicating that the 800 L should be close to the optimum V_L, with the other operating conditions at LaVerkin Springs. Thus, increase in V_R from 800 to 1600 L caused a proportionally greater increase in t_c than in q_{Ca}.

Packed Bed and Regeneration Flow Rate. The maximum regeneration flow rates (G_F for fresh regenerant and G_R for recycled regenerant) possible with a fluidized bed and 50 percent bed expansion were about 24 L/min at the lowest water temperatures because the fluidized bed reached the top of the IX column with 50 percent expansion. A packed bed allowed upflow rates of 36 L/min at the maximum flow capacity of the regenerant pump used. Assuming a 0.3-voids fraction in the 115-L bed, the residence volume for water in the packed bed was estimated to be 34.5 L, and the residence volume in the 50 percent expanded bed was 92 L. Thus, the minimum regenerant residence time was 92 L/24 L/min = 3.8 min for the fluidized bed and only 34.5 L/36 L/min = 0.96 min for the packed bed. Qualitatively, it was observed that gypsum scaling of the resin with C'_F = 92 g/L was less with the packed bed than with the fluidized bed, which apparently was due to the lower residence time allowing gypsum crystal growth in the packed bed. Accumulation of gypsum scale in the midcolumn collector piping caused IX operation to cease several times while scale was removed and severely scaled piping was replaced. The data in table 7 show no consistent difference in q_{Ca} whether a packed bed or a fluidized bed was used, but the q_{Ca}/t_c was less for the packed bed because the faster regeneration resulted in lower t_c.

Regenerant Temperature. No consistent differences in q_{Ca} or q_{Ca}/t_c were observed whether or not the heat exchanger was used to warm the regenerant to a maximum estimated summer temperature of 26 °C. The gypsum scaling in the IX column was worse at higher temperatures, which agrees with results found at Yuma.

SHMP Added to Fresh Regenerant. The addition of 100 mg/L of sodium hexametaphosphate to the fresh regenerant while a packed bed was used resulted in complete control of gypsum scaling in the IX column and piping. Recycled regenerant was not used in cycles with SHMP addition. While using SHMP the values for q_{Ca} were somewhat lower but q_{Ca}/t_c was about the same in comparison to cycles not utilizing SHMP but with recycled regenerant. SHMP dosages less than 100 mg/L probably would also be effective, but lower dosages were not tested.

Alternative Rinse Modes. Two special rinse modes were tested (the last two cycles listed in table 7) to determine their effect on gypsum scale retention in the column. All rinse water was IX feed, and the rinses were done immediately after regeneration. The first special rinse procedure tested was an upflow rinse followed by a normal downflow rinse. Gypsum scale was not visually alleviated using this procedure. The second special rinse method included three steps: a 1-minute downflow rinse, 10-minute simultaneous downflow and upflow rinses with the column effluent leaving through the midcolumn collector, and a final 1-minute downflow rinse. This second procedure was somewhat effective in reducing visible gypsum scale in the resin bed, but there still were problems with a scale buildup in the effluent collector piping.

Most Successful IX Conditions at LaVerkin Springs. Cycles L.22.15, L.23.19, and L.24.11 were the most successful at LaVerkin Springs in terms of high q_{Ca}/t_c and low resin scaling by gypsum. Operating conditions for these runs are given in table 7.

If some resin scaling can be tolerated, then the conditions of cycle L.22.15 including the use of recycled regenerant and a packed resin bed, would be preferred and the cost of SHMP would be avoided. A buildup of gypsum scale in the regeneration effluent piping can be avoided if they are flushed with feed water during each rinse and exhaustion, which can best be accomplished by using common piping for regeneration effluent and exhaustion feed, where possible.

In cycles L.23.19 and L.24.11, the addition of SHMP to the fresh regenerant eliminated any visible resin scaling in the column or piping. Because no recycled regenerant was used in these cycles, no regenerant recycling system would be necessary in using these cycle conditions. A more economical dosage of SHMP required to prevent gypsum scaling would have to be established through experiment for each regenerant effluent water composition.

Ion Exchange Equilibrium Calculations

A resin capacity at theoretical equilibrium was calculated for the various feed and regenerant compositions encountered at Yuma and LaVerkin Springs. The equilibrium composition of the cation exchange resin for Ca-Mg-Na was calculated from the solution concentrations by solving the following three equations simultaneously [23, 24]. Any K present was treated as Na by adding the K normality to the Na normality.

$$\frac{Y_{Ca}}{X_{Ca}}\left(\frac{X_{Na}}{Y_{Na}}\right)^2 = \frac{K_{Ca/Na}\,Q}{C_o} \tag{4}$$

$$\frac{Y_{Mg}}{X_{Mg}}\left(\frac{X_{Na}}{Y_{Na}}\right)^2 = \frac{K_{Mg/Na}\,Q}{C_o} \tag{5}$$

$$Y_{Ca} + Y_{Mg} + Y_{Na} = 1 \tag{6}$$

Where Y_{Ca} is the equivalent fraction of resin in the calcium form

Y_{Mg} is the equivalent fraction of resin in the magnesium form

Y_{Na} is the equivalent fraction of resin in the sodium form

X_{Ca} is the equivalent fraction of calcium in the bulk solution

X_{Mg} is the equivalent fraction of magnesium in the bulk solution

X_{Na} is the equivalent fraction of sodium in the bulk solution

$K_{Ca/Na}$ is the selectivity coefficient of calcium over sodium

$K_{Mg/Na}$ is the selectivity coefficient of magnesium over sodium

Q is the total resin capacity in equivalents per liter of resin

C_o is the total normality of the solution.

A combination of equations 4, 5, and 6 yields

$$Y_{Na} = \frac{(1 + 2A)^{1/2} - 1}{A} \tag{7}$$

where

$$A = \frac{2}{X_{Na}^2} \frac{Q}{C_o} (K_{Ca/Na} X_{Ca} + K_{Ca/Mg} X_{Mg})$$

Given X_{Na}, X_{Mg}, and X_{Ca}; Y_{Na} was first calculated from equation 7; then Y_{Ca} was calculated from equation 4; and, finally, Y_{Mg} was calculated from equation 6. For each component i, $q_i = Y_i Q$.

The following data were used from manufacturers' literature for Amberlite 200 (1.75 eq/L) used at Yuma and Dowex HCR (2.0 eq/L) used at LaVerkin Springs. Values for $K_{Ca/Mg} = 2.0$ and $K_{Mg/Na} = 1.2$ were from Klein, et al. [24], which were obtained in laboratory experiments with nearly the same solution normality range for Dowex 50W-X8, which is an older model designation for essentially the same resin now designated HCR according to Dow Chemical Company. The values of Klein agree with other published values [25]. The same K values were assumed for Amberlite 200 because no other applicable experimental data were found in the literature.

Equilibrium calculation results for some average water compositions at both test sites are given in tables 8 and 9. By comparing results from table 8 with experimental calcium resin capacities at Yuma from tables 3 and 4, it is seen that, at most, just over half of the equilibrium capacity was achieved in the Yuma field experiments. Also, the equilibrium capacity (table 8) increased with recovery but not the experimental resin capacity (tables 3 and 4). An explanation may be that the volume of fresh regenerant per cycle decreased with increased recovery, which would cause the resin to be further from equilibrium with the regenerant solution because of mass transfer limitations and a shorter contact time with the same regeneration flow rate (see equation 3). It may also be that the regenerant volumes used would not be sufficient to fully regenerate the resin because of the unfavorable equilibrium of the regeneration and column dynamics as indicated by a $K_{Ca/Na}Q/C_o$ near 10. A favorable or self-sharpening concentration profile should occur in a column during regeneration when $K_{Ca/Na}Q/C_o$ is less than 1.0 [24].

In contrast to the calculation results for Yuma, the equilibrium q_{Ca} values for LaVerkin Springs (table 9) are close to the field experimental values (table 7). This agreement seems almost fortuitous considering the deviation between calculated and experimental values for q_{Mg} and the much poorer agreement between calculated and experimental q_{Ca} and q_{Mg} for the Yuma data. It is true that during regeneration the formation of gypsum precipitate and the formation of ion pairs containing Ca or Mg would tend to cause greater conversion of the resin to the sodium form than predicted by assuming (equations 4 through 6) that Ca and Mg are fully ionized in solution [26]. The gypsum scaling at LaVerkin Springs was much greater at 92 g/L of TDS regenerant brine concentration than it was at Yuma. On the other hand, the qualitatively observed amount of gypsum scale at LaVerkin Springs was actually less at 40 g/L of TDS regenerant concentration than it was at Yuma. More study is required to better explain the results of these equilibrium calculations and to develop more realistic models for cation exchange softening with regeneration using waste reject brine.

ED Experimental Results

Two ED equipment problems had to be solved at Yuma before completely successful operation was achieved. Both problems can be attributed to the unusually high brine concentrations and the standard ED design. The first problem was rapid erosion of the bronze impeller of the brine recirculation pump during highest recovery operation, during which time the brine pH also dropped significantly. Substitution of a pump with plastic impeller and case and a Hastelloy shaft completely solved the pump corrosion problem. The second problem was shorting through insulation on the stack electrodes. Ionics, Inc., replaced the standard electrodes with ones specially encapsulated with plastic for better insulation in critical nonflow areas. There were minimal ED operational and maintenance problems during the remainder of testing at Yuma and at LaVerkin Springs following these two equipment modifications.

Typical ED operating conditions for the brine concentrations tested are given in table 10, which are mean values for positive and negative polarities. Chemical analyses of corresponding feed, product, and brine samples are listed in table 11.

Cell pair resistance (R_1) for the first stage is shown plotted in figures 5 and 6. Corrections to 25 °C from temperature t were made using $(0.6053)\ (1.020285)^t$, which was supplied by Ionics, Inc. Figure 7 shows R_1 decreasing with brine concentration, which is expected since solution electrical conductance decreases with increasing ionic concentration. The scatter at 35 g/L of TDS reflects the increase in R_1 with time at Yuma. R_1 at LaVerkin Springs was less than R_1 at Yuma because of the higher feed and product water concentrations at LaVerkin Springs.

Figure 8 demonstrates an increase in R_1 with operating time at Yuma but not at LaVerkin Springs. Much of the data scatter is because three brine concentrations are included, but a similar plot using only 35 g/L of TDS gave essentially the same result. As mentioned previously, slime-producing microbiological growth occurred at Yuma in the IX resin and IX product water tank because the IX feed was dechlorinated and additional disinfection was not used. It is generally known that membrane fouling will cause cell pair resistance to increase. However, no buildup of slime was ever observed on the membrane surfaces of the ED unit, suggesting that organic material may have combined internally with the anion membrane. No high pH cleaning of the stack was attempted. If microbiologically produced organic material at Yuma was the cause for the increased R_1, the chlorination of the IX product and dechlorination just prior to the ED at LaVerkin Springs effectively mitigated the problem as shown in figure 8. The raw well water at LaVerkin Springs also contained less organic material than the raw canal water at Yuma. It is also believed that the aliphatic membranes used at LaVerkin Springs are not as sensitive to organic materials as the anion membranes operated at Yuma.

Current efficiency is shown decreasing with brine concentration in figure 9. This is expected because current efficiency is largely a function of ionic concentration difference across a membrane. The higher current efficiency of the membranes used at LaVerkin Springs is clearly demonstrated in figure 9.

Direct current energy consumption per volume of product water versus brine concentration is shown in figure 10. These energy consumption values were calculated from instantaneous readings such as those in table 10 and include corrections to 25 °C, a factor of 0.94 for the rectifier efficiency, and a factor of 0.90 to account for an estimated 10 percent of operating time with off-specification water following polarity reversal. These data show as the brine concentration increases how the decreased current efficiency has a greater effect than the decreased cell pair resistance upon power consumption, which is expected since the power equation contains current squared times resistance to the first power [6]. The direct current energy consumption at LaVerkin Springs was reasonable considering the large numbers of ions that were moved during demineralization.

CONCLUSIONS AND RECOMMENDATIONS

The successful IX and ED pilot plant experiments at Yuma and LaVerkin Springs have demonstrated that the coupled IX-ED combination has great potential for achieving high product water recoveries in desalting brackish (2000 to 15,000 mg/L of TDS) water, which is particularly applicable to inland sites. The IX easily produced product of sufficiently low calcium concentration to avoid gypsum scaling in the ED.

The IX could also be coupled equally well with other processes producing a concentrated reject such as a reverse osmosis desalting unit or wet cooling tower.

In these experiments, the ED successfully produced higher desalting recoveries (up to 95 percent) with the feed water tested and higher brine concentrations (up to 100 g/L) than is possible with present RO technology, a prime competitor of ED, and the energy-consumption of the ED desalting was not unreasonable considering the amount of salt that had to be removed.

There was a different set of optimum IX operating conditions (see table 1) for fresh regenerant TDS concentration, C_F; fresh regeneration flow rate, G_F; volume of recycled regenerant, V_R; recycled regenerant flow rate, G_R; and other described independent variables for each selected dependent variable including specific resin capacity for calcium removal, q_{Ca}; IX cycle duration, t_c; q_{Ca}/t_c which is inversely proportional to the volume of resin needed in an IX plant design; and intensity of gypsum scaling in the IX column and effluent piping. Increased regenerant temperatures were found to substantially increase the kinetic rate of gypsum formation during regeneration.

If the predominant dependent variable is the prevention of gypsum scale accumulation in the IX system, then a substantial volume of recycled regenerant (roughly 10 bed volumes) and high regeneration upflow rates (sufficient for 50 percent fluidized bed expansion) are recommended. Alternatively, a resin holddown procedure with collection of regenerant effluent right in the top of the packed bed allowed even higher upflow rates than were possible with a fluidized bed and resulted in a several-fold decrease in the residence time of the supersaturated calcium sulfate in the IX column and a decrease in the amount of gypsum scale formed in the column and effluent piping per cycle. Because accumulating gypsum scale gradually caused severe flow blockages only in regeneration effluent piping that was not flushed with IX feed water, it is recommended for design that regeneration effluent piping be flushed with IX feed water following regeneration during each cycle, which may be most easily accomplished by designing common piping for the feed and the regenerant effluent. These design measures were successful in controlling gypsum scale accumulation in the IX at Yuma and with 80-percent desalting recovery at LaVerkin Springs.

Yet, even with the preceding design measures, including use of recycled regenerant, the rate of gypsum scale formation at LaVerkin

Springs during regeneration with 92-percent recovery reject brine was so rapid that a gypsum-cemented resin bed and flow blockages in the regeneration effluent collectors occurred irrespective of the high calcium resin capacities obtained. Only the addition of sodium hexametaphosphate (a 100-mg/L dosage was used but not optimized) to the regenerant completely eliminated operational problems. The cost of purchasing sodium hexametaphosphate can be somewhat offset by savings in not providing and operating a system for recycling regenerant, which the experimental results indicate can be omitted if sodium hexametaphosphate is used.

The experimental results indicated that only the most severe gypsum scaling, which occurred with the highest desalting recoveries (highest C_F and lowest V_F) and when no recycled regenerant was used ($V_R = 0$), adversely affected IX performance. In fact, some cycles with fairly severe resin scaling yielded relatively high values for the performance parameter, q_{Ca}/t_c. However, in every case of resin bed scaling with both the gel and macroreticular types of resin, the original resin capacity returned through fluidized bed regeneration and dissolution of the gypsum with a sodium chloride solution. Gel-type resin is recommended over macroreticular type because the gel type seemed to repel gypsum scale better, has a higher capacity, and is cheaper.

The equilibrium calculations yielded results which were not consistently in agreement with the experimental column results. More laboratory equilibrium data for the selectivity coefficients for Ca-Mg-Na are needed. Much more work on modeling the column dynamics, regeneration with gypsum formation, and recycled regenerant use to optimize q_{Ca}/t_c could lead to a better design procedure for optimum IX softening with reject brine regeneration.

Design and operation of an ED unit for high recovery operation are demonstrably feasible technically provided that membranes with high current efficiencies are used, the stack and electrodes have adequate insulation to prevent shorting due to extremely high brine concentrations, and the materials of components such as pumps are selected to resist the high TDS concentration and lowered pH of the recirculating brine stream. The high performance of the aliphatic anion exchange membrane in the LaVerkin Springs experiments supports the value of its further development and possible application where its advantages are desired. Disinfection of the IX product-ED feed, for example by chlorination and dechlorination, should ensure that organic biproducts of microbiological growth in the IX product tank and piping do not form which may foul the membranes.

Regarding an interesting research area not studied in the present experiments, colloidal fouling of desalting membranes may be decreased to some extent through cation exchange softening [27]. Nearly complete softening causes an increase in the double-layer thickness and the effective electrical charge of a colloid. This should cause colloids to be more stable rather than coagulate during concentration in a desalting unit. Possible benefits in lower colloidal fouling of RO and ED membranes through the use of IX pretreatment need to be verified by experiment.

DISCLAIMER

The information contained in this report, stated or implied, regarding commercial products of firms may not be used for advertising or promotional purposes and is not to be construed as an endorsement of any product or firm by the Bureau of Reclamation.

ACKNOWLEDGMENTS

The author wishes to acknowledge the following individuals:

Mr. Roy Eisenhauer of the Bureau of Reclamation was project engineer of the LaVerkin Springs test program;

Dr. Charles Moody of the Bureau provided technical suggestions to the IX process and data analysis;

Mr. Paul Laverty, PRC (Planning Research Corporation), Yuma Desalting Test Facility, was the contractor's engineer in charge of operating the IX-ED system at Yuma;

Mr. Ron Hudson, formerly of PRC, was the chief IX-ED operator;

Mr. Rick Amalfi, formerly of PRC, was primarily responsible for the chemical analyses at Yuma;

Ms. Cindy Hoeft and Ms. Shirley McAdams of PRC did the original data reduction.

Mr. David Elyanow of Ionics, Inc., designed the two high recovery ED stack configurations, did troubleshooting, and suggested improvements in the ED data analysis.

Mr. Gerhardt Klein and Dr. Theodore Vermeulen of the University of California provided much information on IX modeling during lively discussions with the author.

REFERENCES

[1] Taylor, Ivyl G., and Lorentz A. Haugseth, "Yuma Desalting Plant Design," Desalination, vol. 19, pp. 505-523, 1976.

[2] Public Law 93-320, Colorado River Basin Salinity Control Act, June 24, 1974.

[3] Colorado River Basin Salinity Control Project, Reject Stream Replacement Study, Yuma Desalting Plant High Recovery, draft report submitted by the Chief Design Engineer to the Chief, Division of Planning Technical Services, April 9, 1980.

[4] Kaakinen, John W., Roy J. Eisenhauer, and Cornelis Van Hoek, "High Recovery in the Yuma Desalting Plant", Desalination, vol. 23, pp. 357-366, 1977.

[5] Marshall, William L., and Ruth Slusher, "Aqueous Systems at High Temperature," Journal of Chemical and Engineering Data, vol. 13, pp. 83-93, 1968.

[6] Elyanow, David, Earnest Sieveka, James Mahoney, "The Determination of Supersaturation Limits in an EDR Unit With Aliphatic Anion Membranes," Presented at the Ninth Annual Conference, National Water Supply Improvement Association, Washington, D.C., May 31-June 4, 1981.

[7] Kaakinen, John W., and Paul E. Laverty, Cation Exchange Pretreatment Studies for High Recovery - Yuma Desalting Plant, REC-ERC-82-11, Bureau of Reclamation, Denver, Colorado, in preparation.

[8] Kaakinen, John W., and Paul E. Laverty, "Cation Exchange Following Lime Softening as Pretreatment for High Recovery in the Yuma Desalting Plant," presented at Water Reuse Symposium II, August 23-28, 1981, Washington, D.C.

[9] Eisenhauer, Roy, LaVerkin Springs On Site Pretreatment for Desalting Test Evaluation, REC-ERC Series, Bureau of Reclamation, Denver, Colorado, in preparation.

[10] Haugseth, Lorentz A., and C. D. Beitelshees, Evaluation of Ion-exchange Pretreatment for Membrane Desalting Processes, Report No. REC-ERC-74-26, Bureau of Reclamation, Denver, Colorado, December 1974.

[11] Van Hoek, Cornelis, John W. Kaakinen, and Lorentz A. Haugseth, "Ion Exhange Pretreatment Using Desalting Plant Concentrate

for Regeneration," paper in proceedings of First Desalination Congress of the American Continent, Mexico City, Mexico, October 24-29, 1976, Elsevier Scientific Publishing Company, Amsterdam, The Netherlands.

[12] Sephton, Hugh H., and Gerhard Klein, "A Method of Using Irrigation Drainage Water for Power Plant Cooling," paper presented at the First Desalination Congress of the American Continent, Mexico City, Mexico, October 24-29, 1976, Elsevier Scientific Publishing Company, Amsterdam, The Netherlands.

[13] Klein, G., S. Cherney, E. L. Ruddick, and T. Vermeulen, "Calcium Removal From Sea Water by Fixed-Bed Ion Exchange," Desalination, vol. 4, pp. 158-166, 1968.

[14] Kunin, Robert, Ion Exchange Resins, second edition, John Wiley and Sons, Inc., New York, 1958.

[15] Kunin, Robert, "Helpful Hints in Ion Exchange Technology (Cont.)," Amber-hi-lites, No. 131, Rohm and Haas Company, Philadelphia, Pennsylvania, November 1972.

[16] Liu, Sung-Tsuen, and George H. Nancollas, "Kinetics of Crystal Growth of Calcium Sulfate Dihydrate," Journal of Crystal Growth, vol. 6, pp. 281-289, 1970.

[17] Mindler, Albert B., and Steve T. Bateman, Increased Product Water Recovery by Reverse Osmosis Using Interstage Ion Exchange Softening and a Spiractor, Office of Water Research and Technology, Contract No. 14-34-0001-7554, Permutit Company, Princeton, New Jersey, June 20, 1978.

[18] Strategy of Experimentation, revised edition, E. I. du Pont de Nemours and Company, Wilmington, Delaware, 1974.

[19] Draper, N. R., and H. Smith, Applied Regression Analysis, John Wiley and Sons, Inc., New York, New York, 1966.

[20] Winters, Harvey, "In-plant Microfouling in Desalination," Proceedings of International Congress on Desalination and Water Reuse, Nice, France, October 1979, Elsevier Scientific Publishing Company, Amsterdam, The Netherlands.

[21] Dorfner, Conrad, Ion Exchangers: Properties and Applications, pp. 136-138, Ann Arbor Science Publishers, Inc., Ann Arbor, Michigan, 1972.

[22] Barraque, C., "Counter Current Regeneration Processes," Presented at Liberty Bell Corrosion Conference, IX1980, Pittsburgh, Pennsylvania.

[23] Vermeulen, Theodore, Gerhard Klein, and Nevin K. Heister, "Adsorption and Ion Exchange," Chemical Engineers' Handbook, fifth edition, sec. 16, McGraw-Hill Book Company, New York, New York, 1973.

[24] Klein, Gerhard, Milagres Villena-Blanco, and Theodore Vermeulen, "Ion-Exchange Equilibrium Data in the Design of a Cyclic Sea Water Softening Process," I&EC Process Design and Development, vol. 3, pp. 280-287, July 1964.

[25] McIlhenny, W. F., "Sea Water Softening by Ion Exchange as a Saline Water Conversion Pretreatment," Saline Water Research and Development Report No. 62, Office of Saline Water, Washington, D.C., May 1962.

[26] Helfferich, Friedrich, Ion Exchange, McGraw-Hill Book Company, New York, 1962.

[27] "Pretreatment Considerations for Reverse Osmosis," Permasep Technical Bulletin, No. 401, E. I. du Pont de Nemours and Company, Wilmington, Delaware, March 1, 1976.

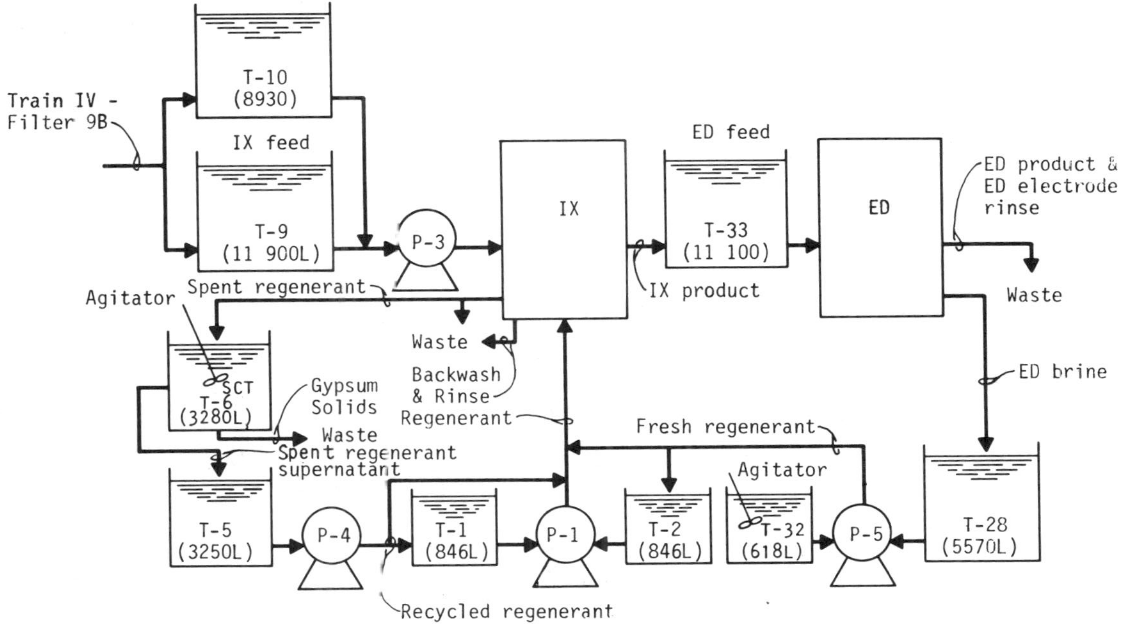

Pumps

P-1	Regenerant
P-3	IX feed
P-4	Recycled regenerant transfer
P-5	Fresh regenerant transfer

Processes

IX	Ion exchanger
ED	Electrodialyzer
SCT	Solids contact tank (T-6) (Calcium sulfate precipitator for recycling regenerant)

Tanks

T-1	Recycled regenerant (metering)
T-2	Fresh regenerant (metering)
T-5	Recycled regenerant
T-9	IX feed
T-10	IX feed
T-28	ED brine
T-32	Sodium chloride solutioning
T-33	IX product/ED feed

Figure 1. - Pilot plant flow diagram.

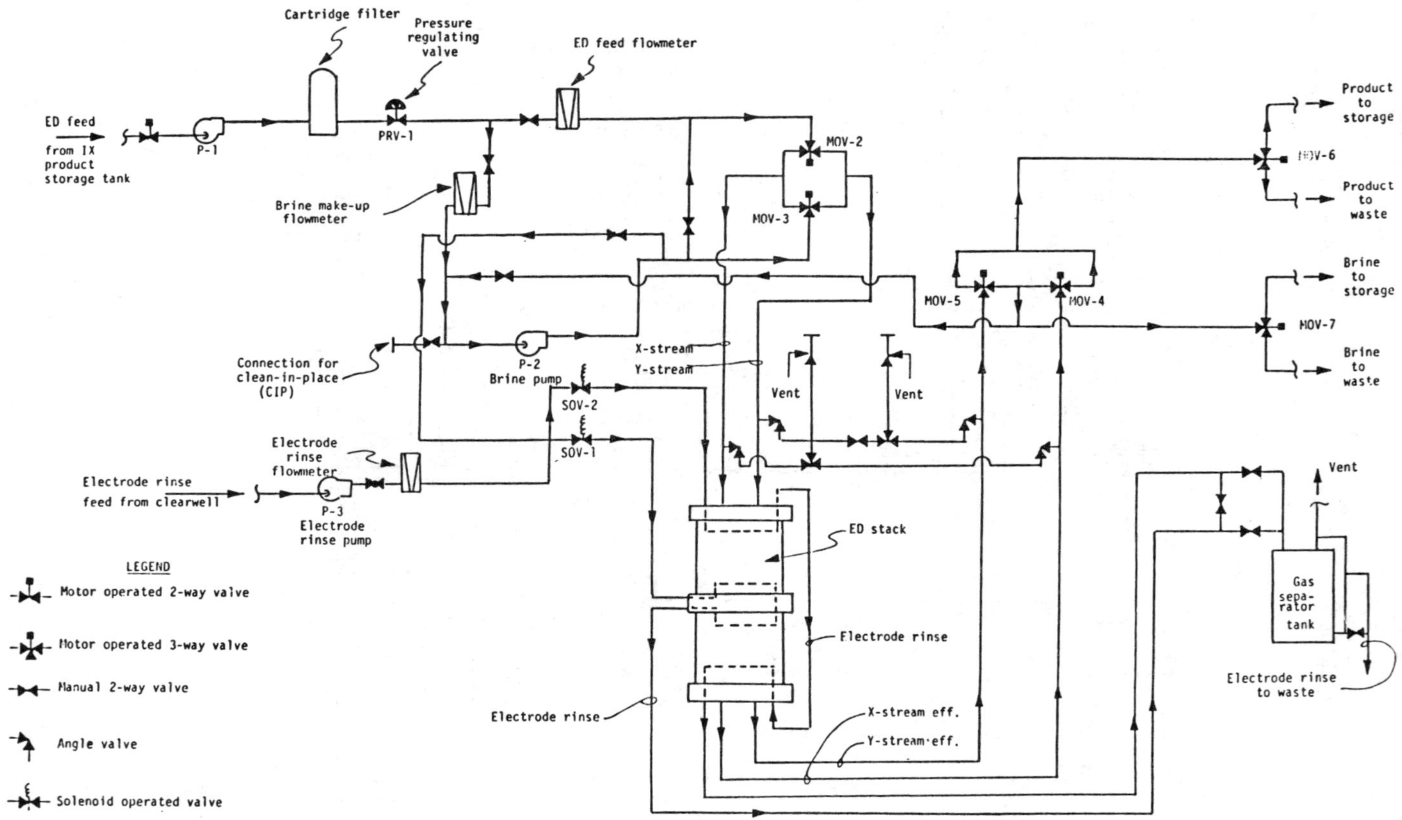

Figure 2. - Ionics Aquamite V electrodialysis demineralizer, simplified flow diagram.

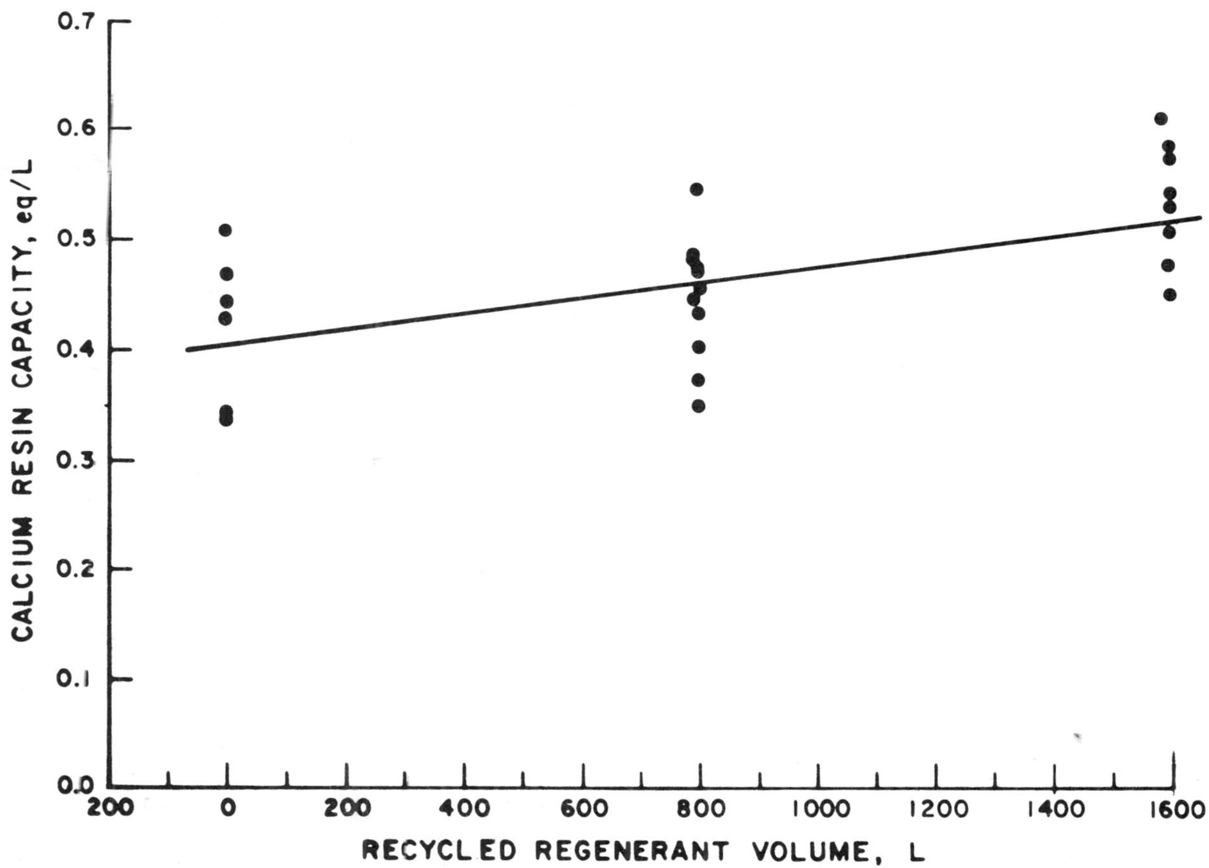

Figure 3. - Calcium resin capacity versus recycled regenerant volume at Yuma.

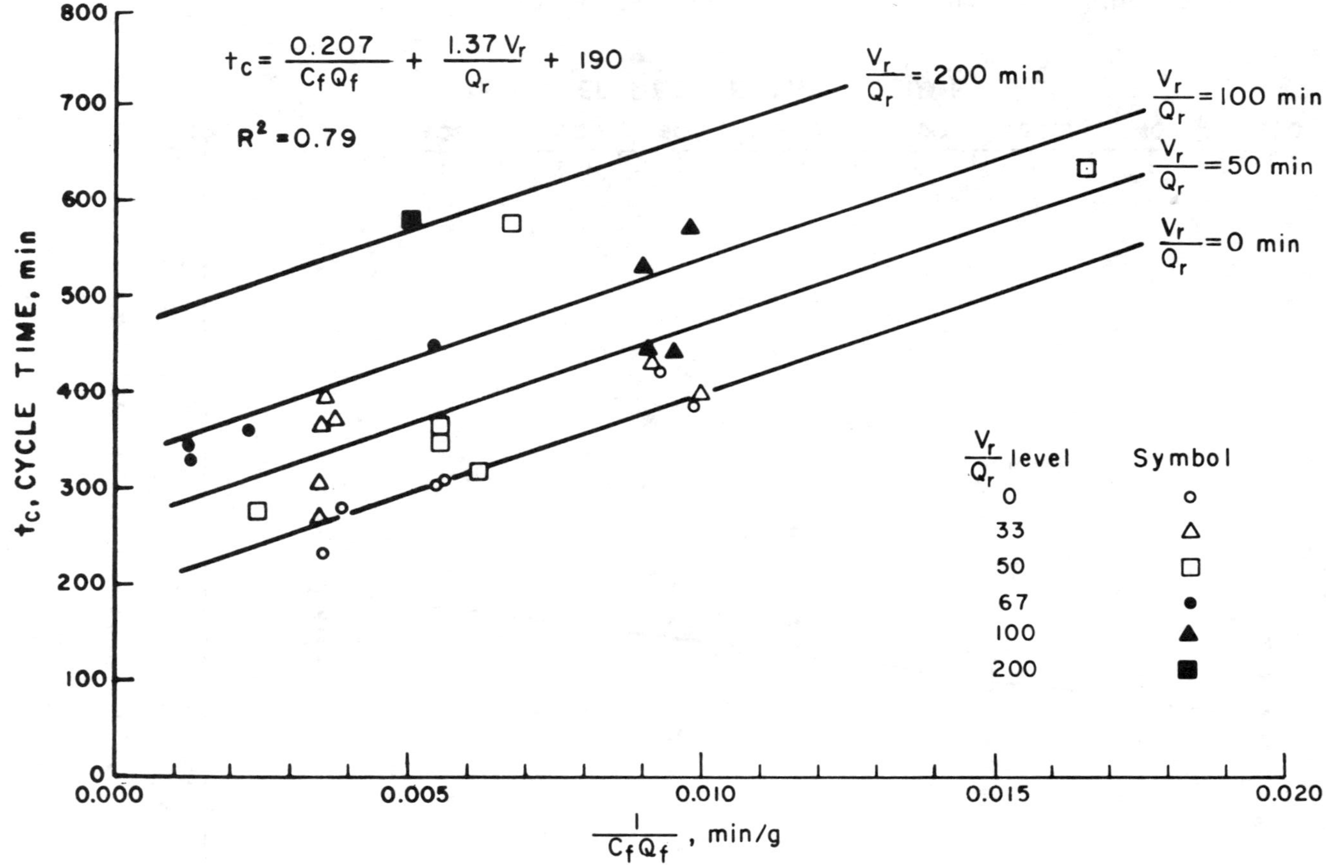

Figure 4. - Regression fit of IX cycle time to regeneration operating variables.

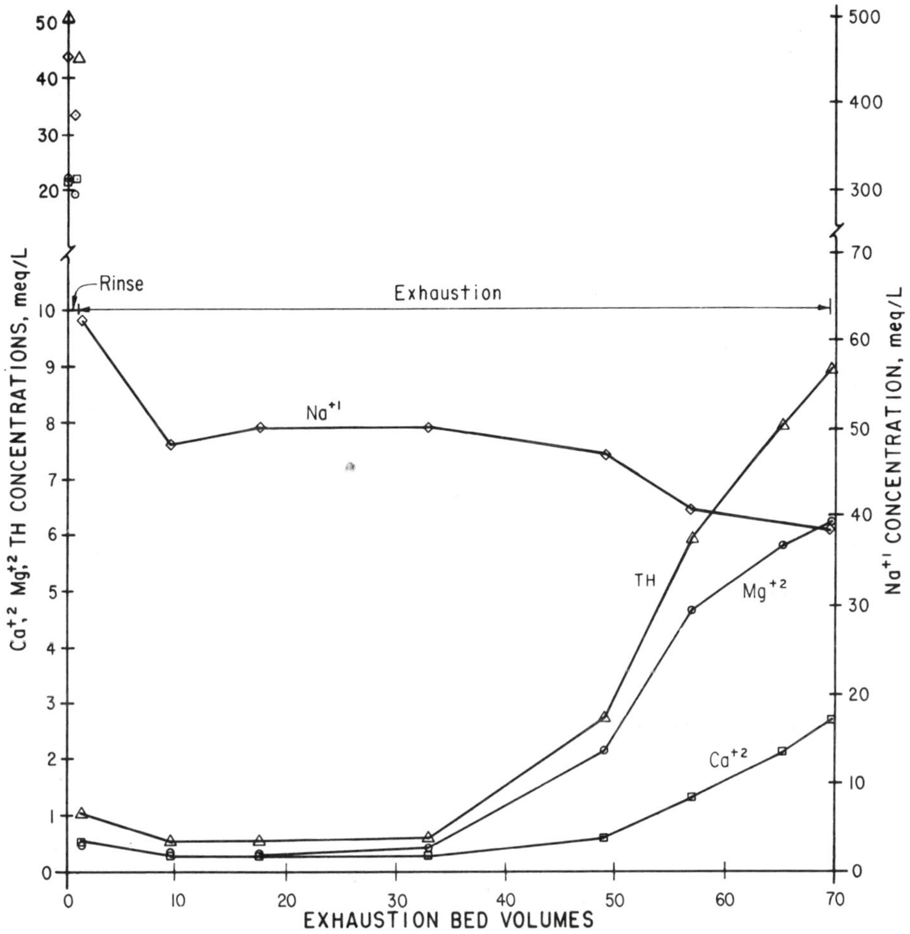

Figure 5. - Major cations in rinse and exhaustion effluent of cycle 4.04.42.

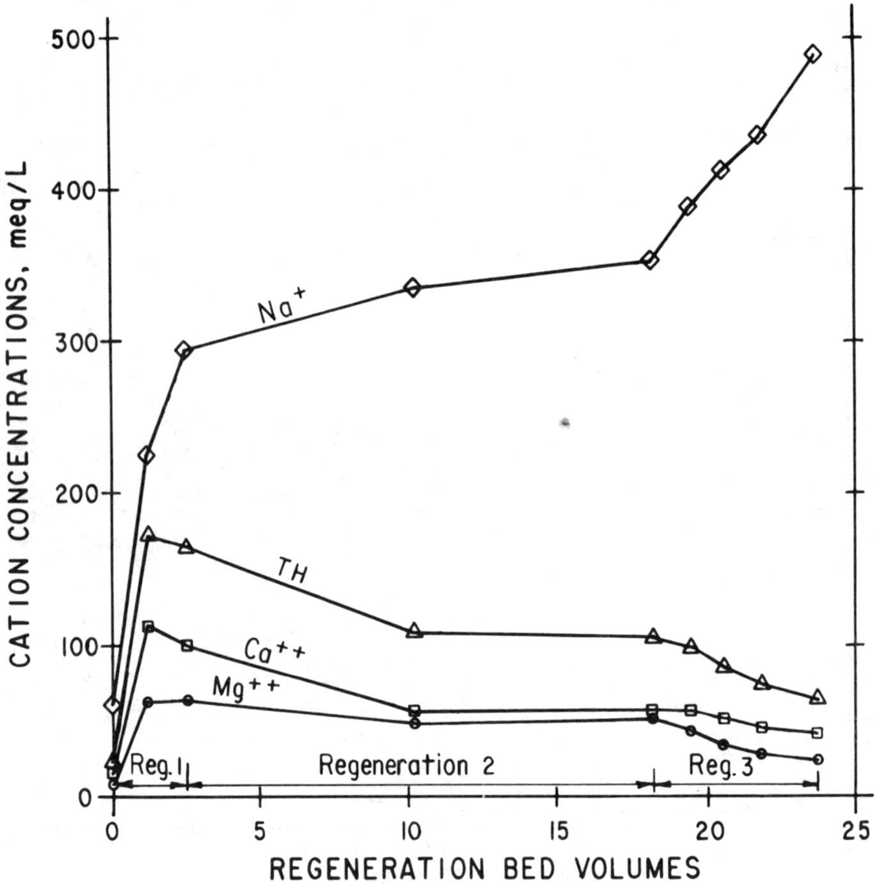

Figure 6. - Major cations in regeneration effluent of cycle 4.04.42.

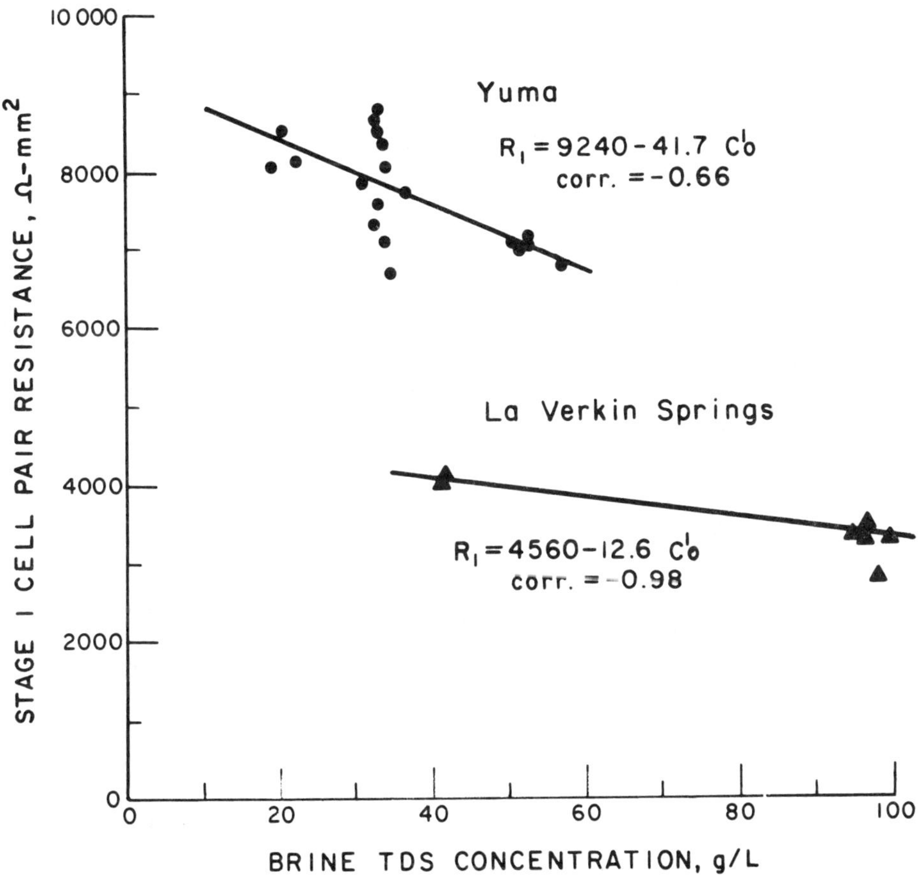

Figure 7. - First stage cell pair resistance (25 °C) versus brine TDS concentration.

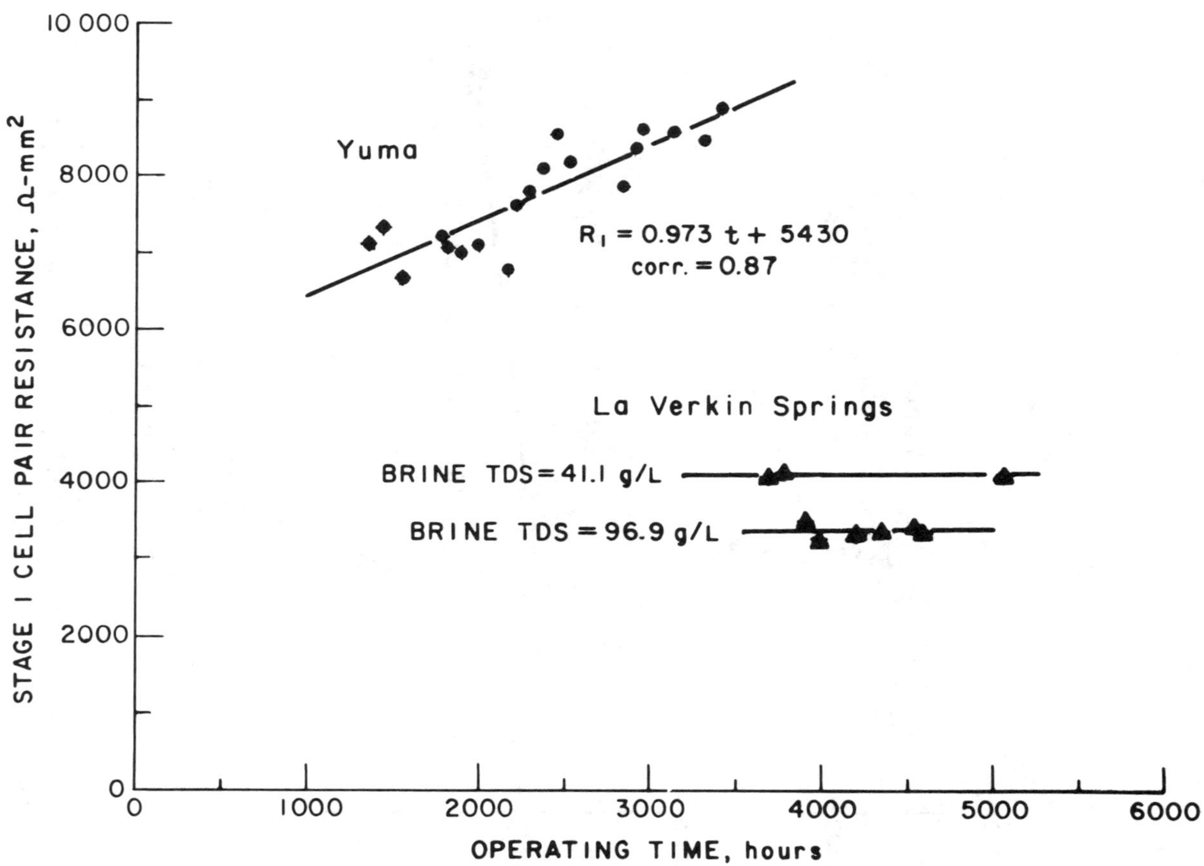

Figure 8. - First stage cell pair resistance (25 ^{o}C) versus operating time.

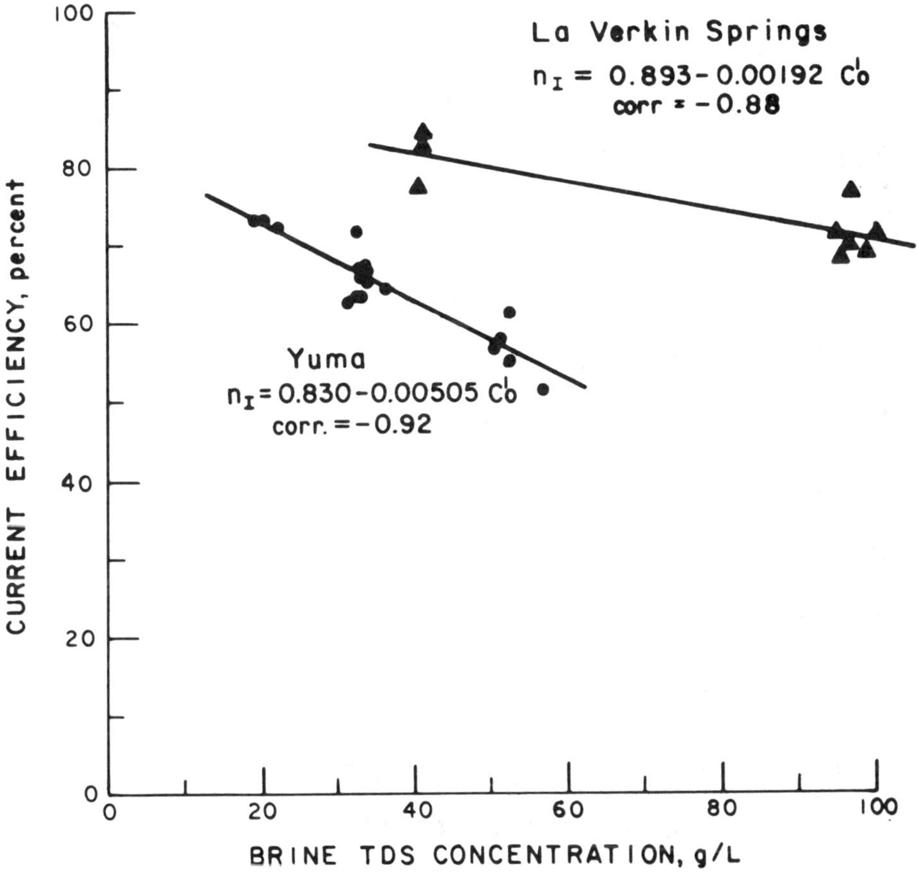

Figure 9. - Electrodialysis current efficiency versus brine TDS concentration.

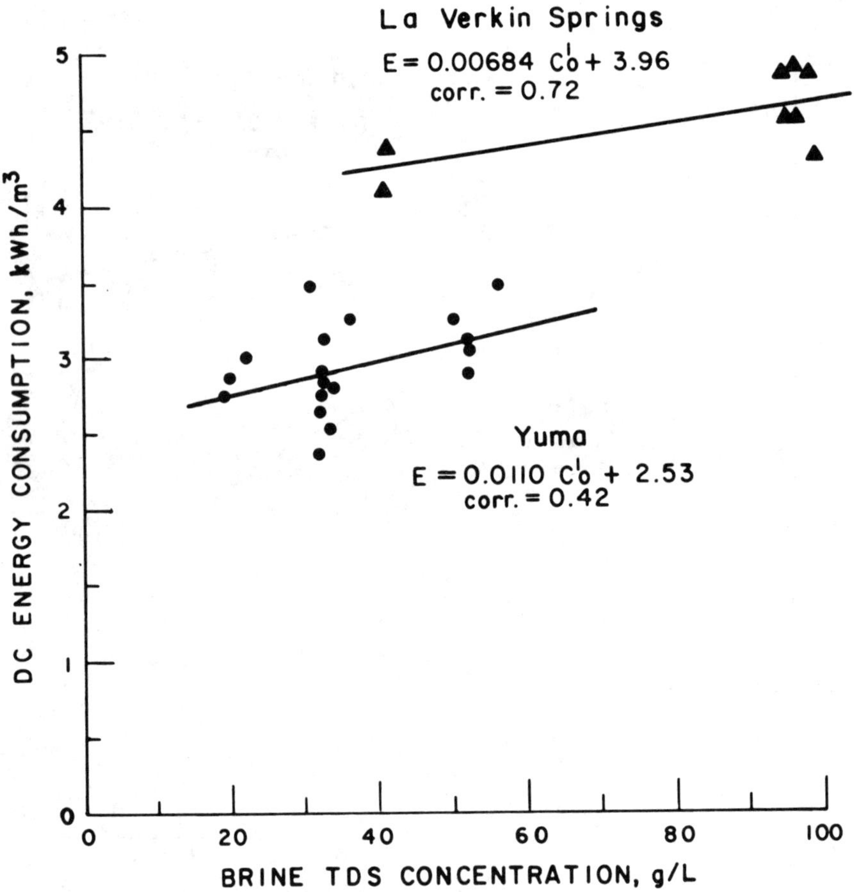

Figure 10. - Electrodialysis direct current energy consumption (25 ^{o}C) versus brine TDS concentration.

Table 1. - Typical raw and partially pretreated water compositions at Yuma and LaVerkin Springs. The partially treated water underwent lime treatment to the pH specified, filtration, and stabilization with sulfuric acid.

Component	Wellton-Mohawk Canal Water, Yuma			LaVerkin Springs	
	Raw Canal	Partial lime treatment (pH = 9.5)	High lime treatment (pH = 10.4)	Raw well	Lime treatment (pH = 9.8)
pH units	7.8	8.2	7.2	6.0	7.5
silica g/m^3	29	19	5	30	12
calcium g/m^3	228	134	169	790	430
magnesium g/m^3	94	84	58	160	128
sodium g/m^3	916	895	897	2300	2329
potassium g/m^3	9	9	9	210	211
iron g/m^3	0.3	<0.1	<0.1	0.09	0.06
strontium g/m^3	3.7	2.4	2.9	12.0	6.0
bicarbonate g/m^3	417	32	19	1270	47
sulfate g/m^3	904	908	899	1980	1980
chloride g/m^3	1160	1160	1160	3480	3470
free carbon dioxide g/m^3	0	0	0	800	0
TDS g/m^3	3750	3240	3210	9430	8590

Table 2. - Comparison between IX softening cycles with NaCl regeneration and desalting reject brine regeneration. The direction of flow in each mode is downflow except backwash, regeneration 1, regeneration 2, and regeneration 3, which are upflow

Mode	Input	Output
A. NaCl regeneration used in standard cation exchange softening:		
Exhaustion	Feed	Product
Backwash	Feed	Waste
Regeneration	NaCl solution	Waste
Drain	Vent or air	Waste
Slow rinse	Feed	Waste
Fast rinse	Feed	Waste
B. Regeneration with desalting brine for high recovery pretreatment: 1/		
Exhaustion	Feed	Product
Drain 1	Vent or air	Waste 1/
Regeneration 1	Recycled regenerant	Waste
Regeneration 2	Recycled regenerant	Used regenerant
Regeneration 3	Fresh desalting reject	Used regenerant
Drain 2	Vent or air	Waste 1/
Rinse (slow)	Feed	Waste 1/

1/ Except for regeneration 1, waste streams could be recycled by combining them with other process streams to minimize the net wastage of water.

Table 3. - Results of phase 2 response-surface experiments at Yuma.

Cycle No.	Date	Fresh Regen. Conc. (g/L-TDS)	Fresh Regen. Flowrate (L/min)	Recycled Regen. Flowrate (L/min)	Recycled Regen. Volume (L)	Fresh Regen. Temp. (°C)	Resin Capacity (eq-Ca/L)	Resin Capacity (eq-TH/L)	Exhaustion Duration (min.)	Cycle Duration (min.)	Exhaustion fraction of Cycle Time	q_{Ca}/t_c* meq/L-min	Visual Scaling Intensity
3.01.32C	1/27/79	22.3	8.00	14.6	800	12.5	0.328	0.511	176	350	0.50	0.94	0
3.02.12C	2/1/79	20.4	3.15	14.5	800	15.2	0.421	0.639	203	298	0.36	0.74	0
3.01.15	2/8/79	20.6	7.18	14.5	800	15.7	0.543	0.646	322	575	0.56	0.94	0
3.02.08	2/14/79	20.0	3.01	14.8	800	19.5	0.471	0.579	226	630	0.37	0.73	0
3.03.28	3/3/79	33.9	5.85	7.96	1600	16.7	0.570	0.794	238	579	0.41	0.98	0
3.04.23	3/12/79	33.5	5.44	8.0	0	20.1	0.506	0.657	182	303	0.60	1.67	0
3.05.15	3/17/79	32.8	5.51	15.6	800	19.0	0.466	0.667	204	367	0.56	1.27	0
3.06.13	3/21/79	33.3	5.52	23.7	1600	18.8	0.582	0.731	246	446	0.55	1.30	0
3.07.04	3/22/79	32.9	5.42	24.0	0	18.9	0.466	0.612	189	307	0.62	1.52	0
3.07.04B	3/23/79	34.3	5.49	N/A**	N/A**	21.5	0.513	0.690	180	296	0.60	1.72	0
3.08.77	5/2/79	51.3	3.16	16.2	800	22.0	0.346	0.520	158	318	0.50	1.09	10
3.09.10	5/4/79	52.0	8.02	16.5	800	27.0	0.469	0.616	171	277	0.62	1.69	10
3.10.15	5/9/79	50.3	5.59	16.6	1600	22.3	0.526	0.766	208	396	0.52	1.32	10
3.11.12	5/12/79	51.8	5.49	16.0	0	24.2	0.342	0.537	152	231	0.67	1.50	10
3.11.--B	N/A	50.0	5.5	N/A**	N/A**		run not self-sustaining so never completed successfully						10
3.26.12D	5/26/79	52.1	5.48	7.82	800	28.0	0.456	0.527	185	367	0.50	1.24	10
3.27.25D	6/2/79	53.4	5.42	24.2	800	28.0	0.369	0.497	162	270	0.60	1.37	10
3.12.09	6/6/79	35.7	3.09	8.06	800	28.7	0.399	0.536	175	444	0.39	0.90	5
3.13.08	6/8/79	33.9	2.96	24.6	800	31.5	0.398	0.544	183	394	0.46	1.01	5
3.14.10	6/11/79	33.0	5.47	16.3	792	31.9	0.442	0.565	190	346	0.55	1.28	10
3.15.08	6/13/79	33.7	7.93	8.06	792	35.8	0.477	0.581	193	374	0.52	1.28	10
3.16.06	6/15/79	36.1	8.01	24.3	791	33.5	0.483	0.594	193	306	0.63	1.58	10
3.17.08	6/18/79	20.2	5.49	17.3	1595	28.4	0.473	0.597	231	528	0.44	0.89	5
3.18.09	6/21/79	20.0	5.38	16.0	0	27.8	0.441	0.541	213	418	0.51	1.05	5
3.18.13B	6/25/79	20.6	5.47	N/A**	N/A**	30.2	0.392	0.471	176	349	0.52	1.15	0
3.19.09	6/28/79	19.4	5.42	8.77	300	35.0	0.429	0.518	180	440	0.41	0.98	0
3.20.10	7/2/79	19.6	5.58	23.1	305	32.0	0.452	0.557	200	430	0.46	1.04	0
3.18.21E	7/9/79	21.8	5.40	N/A**	N/A**	30.3	0.380	0.488	189	351	0.54	1.08	0
3.21.56	8/2/79	33.9	3.00	16.1	1588	36.5	0.607	0.737	245	570	0.43	1.06	5
3.22.12	8/6/79	33.8	7.68	16.0	0	33.2	0.426	0.537	186	278	0.67	1.53	5
3.23.33	8/22/79	34.4	5.51	16.2	794	29.5	0.655	0.815	336	546	0.62	1.20	5
3.24.11	8/25/79	34.6	7.92	15.5	1573	30.8	0.690	0.867	346	561	0.62	1.23	0
3.25.10	8/29/79	33.3	3.04	16.0	0	29.5	0.336	0.468	211	383	0.55	0.88	0

*Calcium resin capacity/cycle time. **N/A = not applicable.

Table 4. - Results of phase 3 experiments at Yuma.

Cycle No.	Date	Fresh Regen. Conc. (g/L-TDS)	Fresh Regen. Flowrate (L/min)	Recycled Regen. Flowrate (L/min)	Recycled Regen. Volume (L)	Fresh Regen. Temp. (^oC)	Resin Capacity (eq-Ca/L)	Resin Capacity (eq-TH/L)	Exhaustion Duration (min.)	Cycle Duration (min.)	Exhaustion Fraction of Cycle Time	q_{Ca}/t_c* meq/L-min	Visual Scaling Intensity
4.01.39	9/9/79	33.4	10.17	N/A**	N/A**	29.5	.292	.383	131	194	0.68	1.51	5
4.01.51	9/11/79	33.7	8.25	N/A**	N/A**	24.0	.317	.438	134	208	0.64	1.52	5
4.02.27	9/16/79	32.3	24.4	23.6	1600	30.9	.446	.606	217	331	0.66	1.35	0
4.02.34	9/18/79	34.4	12.8	23.2	1600	28.0	.539	.679	225	361	0.62	1.49	0
4.02.42	9/20/79	33.4	24.0	23.8	1600	28.7	.503	.625	231	346	0.67	1.45	0
4.03.13B	9/22/79	32.7	5.40	N/A**	N/A**	30.4	.262	.353	110	197	0.56	1.33	5
4.04.07E	9/24/79	32.8	5.56	N/A**	N/A**	29.2	.322	.394	130	215	0.60	1.50	0
4.05.09B	9/25/79	32.1	5.66	N/A**	N/A**	33.3	.187	.244	82	152	0.54	1.23	0
4.06.07F	9/26/79	33.1	5.60	15.8	800	30.0	.514	.654	210	378	0.55	1.36	0

*Calcium resin capacity/cycle time.
**N/A = not applicable.

Table 5. - Recommended IX cycle with 35 g/L of TDS reject brine regenerant, based on a resin BV (bed volume) of 100 L and bed depth of 1 m, Amberlite 200 resin

Mode name	Input	Output	Throughout volume (BV)	Duration (min)
Exhaustion	IX feed	RO feed	68.0	228
Drain 1	Air vent	RO feed	0.4	2
Regeneration 1	Recycled reg.	Waste	2.5*	10
Regeneration 2	Recycled reg.	Reg. recycle	15.8	68
Regeneration 3	RO reject	Reg. recycle	5.4	23
Drain 2	Air vent	Reg. recycle	0.6	3
Rinse	IX feed	Reg./feed recycle	1.3	10

* Must use a volume and flow rate,which vary with water temperature, to maintain a 50 percent bed expansion for 10 min.

Total cycle time = 344 min
Fraction of cycle time in exhaustion = 0.66
Resin capacity = 0.47 meq/L Ca^{+2}
Time-weighted resin capacity = 0.31 meq/L Ca^{+2}
Calcium removal efficiency = 90 percent

Table 6. - Composite water compositons during cycle 4.02.42

Constituent	Regeneration 1 and 2 influent	Regeneration 1 effluent	Regeneration 2 effluent	Regeneration 3		Rinse and exhaustion influent	Rinse effluent	Exhaustion effluent
				Influent	Effluent			
TDS mg/L 1/	27 984	19 742	28 827	33 631	31 508	3 079	21 507	3 254
Silica mg/L	2.4	2.4	2.6	3.4	2.8	3.5	3.0	3.5
Calcium mg/L	1 030	1 590	1 490	167	1 000	168	332	21.8
Magnesium mg/L	602	600	738	254	484	46.4	194	31.0
Sodium g/L	8 230	4 660	7 730	11 500	9 530	841	7 070	1 100
Potassium mg/L	57	36	53	83	60	7.4	44	11.8
Strontium mg/L	25	25	31	3.6	25	2.1	5.0	0.4
Carbonate mg/L	97.6	68.3	122.0	119.6	146.4	26.8	89.3	29.8
Sulfate mg/L	5 500	4 120	6 080	9 420	8 220	930	5 800	950
Chloride mg/L	12 440	8 640	12 580	12 080	12 040	1 054	7 970	1 106

1/ Calculated by summing the concentrations of the components.

Table 7. - Results of LaVerkin Springs ion exchange experiments.

Cycle designation	Notes	1980 date	Fresh regen. conc. (TDS g/L)	Fresh regen. flow rate (L/min)	Recycled regen. flow rate (L/min)	Recycled regen. volume (L)	Fresh regen. temp. (°C)	Resin capacity (eq-Ca/L)	Exhaustion duration (min)	Cycle duration (min)	Exhaustion fraction of cycle time	Resin capacity divided by cycle time (meq/L-min)
80 percent recovery reject brine regeneration. Exhaustion termination concentration 6 meq-Ca/L												
L.01.97		1-29	41.3	12.1	NA	0	15.5	0.585	120	216	0.559	2.71
L.02.02	H	2-1	39.8	12.1	24.0	1591	11.8	0.761	152	317	0.480	2.40
L.03.35	PH	2-14	39.4	11.9	NA	0	25.9	0.681	153	253	0.599	2.69
L.04.11	PH	2-17	39.8	12.0	24.2	1573	25.5	0.860	189	370	0.510	2.32
92 percent recovery reject brine regeneration. Exhaustion termination concentration 12 meq-Ca/L												
L.05.54	H	2-28	92.0	26.0	NA	0	26.8	0.341	73	118	0.619	2.89
L.10.14	H	3-8	93.4	23.6	23.1	1594	26.4	1.019	187	308	0.609	3.31
L.12.72	PH	3-14	92.8	23.7	23.8	1610	25.0	0.996	201	318	0.631	3.13
L.17.17	PH	3-18	93.3	33.6	32.2	1602	25.7	1.082	196	296	0.662	3.66
L.18.13	P	3-21	91.3	23.6	24.1	1592	17.6	1.089	196	321	0.610	3.39
L.19.27	P	3-27	90.8	36.3	34.7	1597	13.6	1.012	188	283	0.660	3.58
L.20.09		3-29	91.9	23.7	23.6	1603	13.6	1.065	198	316	0.630	3.37
L.22.15	P	4-4	90.9	35.8	33.0	793	17.5	0.991	178	250	0.710	4.35
L.23.19	PS	4-16	90.0	24.0	NA	0	22.5	0.964	177	228	0.780	3.86
L 24.11	PHS	4-18	90.1	23.2	NA	0	29.5	0.866	191	243	0.786	3.56
L.25.22	PR1	4-9	92.8	30.6	33.1	791	19.4	0.998	193	282	0.680	3.54
L.26.14	PR2	4-12	92.8	33.0	33.1	798	19.4	1.103	200	278	0.750	3.97

Notes:

H - heat exchanger used to heat regenerant to summertime temperatures.
P - resin bed packed during upflow regeneration. In other cycles the resin bed was fluidized during regeneration.
S - SHMP (sodium hexametaphosphate) added to regenerant to prevent gypsum scaling.
R1 - special upflow rinse followed by a downflow rinse.
R2 - simultanteous upflow and downflow rinses passing out of the column through the collector in top of the resin bed.
NA - not applicable.

Table 8. - Calculated equilibrium resin compositions at Yuma
$Q = 1.75$ eq/L, $K_{Ca/Na} = 3.0$, $K_{Mg/Na} = 1.2$

Cation i	IX feed C_o = 49.0 meq/L			85.5 percent recovery recycled regenerant C_o = 297 meq/L			85.5 percent recovery fresh regenerant C_o = 318 meq/L		
	Ca	Mg	Na	Ca	Mg	Na	Ca	Mg	Na
C_i meq/L	8.40	3.82	36.8	42.2	27.5	228	6.5	14.2	299
x_i	0.172	0.078	0.751	0.142	0.093	0.765	0.020	0.045	0.935
y_i	0.721	0.131	0.149	0.518	0.135	0.347	0.175	0.153	0.672
q_i eq/L	1.261	0.229	0.260	0.907	0.236	0.607	0.306	0.268	1.176
q_i (feed) - q_i (regenerant) eq/L				0.354	-0.007		0.955	-0.039	
				91.8 percent recovery recycled regenerant C_o = 460 meq/L			91.8 percent recovery fresh regenerant C_o = 531 meq/L		
C_i meq/L	8.40	3.82	36.8	51.5	49.5	361	8.4	20.9	504
x_i	0.172	0.078	0.751	0.112	0.108	0.781	0.016	0.039	0.945
y_i	0.721	0.131	0.149	0.405	0.156	0.440	0.107	0.107	0.785
q_i eq/L	1.261	0.229	0.260	0.708	0.272	0.769	0.188	0.188	1.374
q_i (feed) - q_i (regenerant) eq/L				0.553	-0.043		1.073	0.041	
				94.3 percent recovery recycled regenerant C_o = 636 meq/L			94.3 percent recovery fresh regenerant C_o = 824 meq/L		
C_i meq/L	8.40	3.82	36.8	68.5	124	445	6.0	26.6	595
x_i	0.172	0.078	0.751	0.108	0.195	0.697	0.007	0.032	0.961
y_i	0.721	0.131	0.149	0.333	0.241	0.426	0.039	0.070	0.890
q_i eq/L	1.261	0.229	0.260	0.582	0.421	0.746	0.069	0.123	1.558
q_i (feed) - q_i (regenerant) eq/L				0.679	-0.192		1.192	0.106	

Table 9. - Calculated equilibrium resin compositions at LaVerkin Springs
$Q = 2.0$ eq/L, $K_{Ca/Na} = 3.0$, $K_{Mg/Na} = 1.2$

Cation i	IX feed C_o = 139 meq/L			80 percent recovery recycled regenerant C_o = 584 meq/L			80 percent recovery fresh regenerant C_o = 637 meq/L		
	Ca	Mg	Na	Ca	Mg	Na	Ca	Mg	Na
C_i meq/L	21.7	10.6	107	55.2	47.2	503	18.4	23.2	624
x_i	0.156	0.076	0.768	0.094	0.081	0.825	0.029	0.036	0.935
y_i	0.639	0.125	0.236	0.366	0.126	0.508	0.171	0.086	0.742
q_i eq/L	1.277	0.250	0.473	0.733	0.252	1.016	0.343	0.173	1.485
q_i (feed) - q_i (regenerant) eq/L				0.544	-0.002		0.934	0.077	
				92 percent recovery recycled regenerant C_o = 1234 meq/L			92 percent recovery fresh regenerant C_o = 1475 meq/L		
C_i meq/L	21.7	10.6	107	47.9	100.3	1131	21.4	41.7	1471
x_i	0.156	0.076	0.768	0.112	0.081	0.880	0.015	0.028	0.957
y_i	0.638	0.125	0.236	0.405	0.114	0.749	0.053	0.041	0.906
q_i eq/L	1.277	0.250	0.473	0.708	0.229	1.498	0.106	0.082	1.812
q_i (feed) - q_i (regenerant) eq/L				1.004	0.021		1.171	0.168	

Table 10. - Typical high recovery electrodialysis performance at Yuma and LaVerkin Springs.

	Yuma			LaVerkin Springs	
	7-7-79	6-4-79	6-18-79	2-11-80	3-10-80
Operating time (hour)	7,902.8	8,219.3	8,376.8	9,786.5	10,210.7
Brine concentration (g/m^3)	51 680	33 361	19 406	41 600	96 500
Feed temperature (°C)	25.0	29.5	27.0	23.8	25.1
Dilute flow rate (L/min)	34	34	34	22.0	21.9
Brine makeup flow rate (L/min)	0	1.35	4.4	3.7	0.2
Product flow rate (L/min)	32.7	33.0	32.0	20.0	20.7
Brine flow rate (L/min)	1.65	3.2	5.88	4.3	1.6
Brine pH (units)	4.54	5.24	6.45	6.3	5.5
Electrical stage 1					
Current (amperes)	23.6	24.0	21.6	26.2	31.9
Voltage (volts)	153	154	155	109	104
Specific cell pair resistance ($ohm-mm^2$)	6980	7570	8050	4090	3270
Electrical stage 2					
Current (amperes)	9.6	8.6	7.4	15.6	18.1
Voltage (volts)	122	123	124	104	98
Specific cell pair resistance ($ohm-mm^2$)	13,440	16,570	18,470	6520	5400
Feed cationic concentration (eq/m^3)	51.2	55.3	55.1	147.2	148.9
Product cationic concentration (eq/m^3)	7.10	6.06	6.25	18.2	20.5
Current efficiency (percent)	58.1	66.1	73.7	84.0	69.5
Energy consumption (kWh/m^3)	2.88	3.11	2.73	4.31	4.86

Table 11. - Chemical analysis of typical electrodialysis stream samples at Yuma and LaVerkin Springs.

Site	Yuma			Yuma			Yuma			LaVerkin Springs			LaVerkin Springs		
Date (percent recovery)	5-7-9 (94.6)			6-4-79 (90.9)			6-18-79 (83.9)			2-11-80 (80.2)			3-10-80 (91.7)		
Sample stream	Feed	Product	Brine	Feed	Product	Brine	Feed	Product	Brine	Feed	Product	Brine	Feed	Product	Brine
pH (units)	6.8	5.6	4.4	6.9	5.0	4.8	7.2	4.8	6.4	7.9	7.5	7.4	7.3	7.4	6.8
TDS (calculated) (g/m^3)	3251	494	51 680	3443	435	33 361	3487	431	19 406	9140	1150	41 600	9180	1280	96 500
Conductivity at 25 °C (4 S/cm)	5413	814	62 830	5797	736	43 276	5718	696	27 008	1360	205	5090	1370	225	9670
Silica (g/m^3)	3.5	3.6	3.5	4.0	4.2	6.5	4.6	1.8	6.5	12	12	17	16	19	26
Calcium (g/m^3)	7.1	1.2	117	9.6	0.7	109	18.5	1.7	128	60	2.4	310	35	5.0	310
Magnesium (g/m^3)	19.7	1.6	35.4	15.6	1.1	190	30.5	2.7	210	54.9	3.9	275	51.9	3.1	506
Sodium (g/m^3)	1126	158	18 140	1224	136	11 500	1180	136	6570	3110	402	13 900	3170	449	33 000
Potassium (g/m^3)	11.2	1.6	188	10.0	0.7	101	12.4	1.2	68	176	12.5	860	196	19.6	1960
Total iron (g/m^3)	<0.10	<0.10	<0.10	<0.10	<0.10	<0.10	<0.10	<0.10	<0.10	0.06	0	0.42	0.04	0	0.72
Total manganese (g/m^3)	<0.30	<0.30	<0.30	<0.30	<0.30	<0.30	<0.30	<0.30	<0.30	<0.05	<0.05	0.13	<0.05	<0.05	0.30
Strontium (g/m^3)	0.7	0.6	2.3	0.7	0.6	1.8	0.3	0.1	1.5	1.5	0	0	0.3	0	2.3
Bicarbonate (g/m^3)	14.6	2.4	17.1	19.5	5.9	12.2	20.5	3.9	41.5	46.5	22.0	82.4	40.3	26.2	68.9
Carbonate (g/m^3)	0	0	0	0	0	0	0	0	0	0	0	0	0	0	0
Sulfate (g/m^3)	970	215	15 100	960	205	9340	1020	213	5050	2040	196	9620	2040	190	22 900
Chloride (g/m^3)	1114	114	18 080	1200	81.2	12 100	1200	71	7330	3660	513	16 600	3650	580	37 800
Total anions (eq/m^3)	51.53	7.21	824.85	54.17	6.66	536.1	55.43	6.50	312.6	146.4	18.88	668.6	145.9	20.72	1545.0
Total cations (eq/m^3)	51.26	7.73	802.70	55.28	6.07	523.9	55.09	6.26	311.2	147.0	18.24	666.00	149.0	20.50	1542.0

METHODS OF REDUCING CONSUMPTION OF SULPHURIC ACID IN REGENERATION OF STRONG CATION EXCHANGERS IN WATER DESALINATION

B.A. Hendry

Department of Chemical Engineering
University of Cape Town
Pvt.Bag, Rondebosch 7700, South Africa

ABSTRACT

If calcium is present on a strong cation exchanger limited concentrations of sulphuric acid regenerant can be used to avoid $CaSO_4$ precipitation which would plug fixed bed contactors. This adversely affects water recovery and economy. At low concentrations calcium selectivity is enhanced and regeneration is less efficient requiring a greater excess of sulphuric acid than if HCl or HNO_3 were used offsetting the cost advantage of the cheaper H_2SO_4.

A new ion exchange flowsheet has been developed which overcomes these limitations and puts $CaSO_4$ precipitation to use to gain a number of major advantages. The loaded cation containing a mixture of cations is first treated with a $CaCl_2$ solution (the waste from the lime regenerated weak anion resin step in a desalination process) to convert it almost entirely to the Ca^{++} form. In the subsequent sulphuric acid regeneration step precipitation of $CaSO_4$ is controlled (by promoting nucleation through recirculating seed crystals) to produce a soft sludge and prevent formation of hard scale. The fine precipitate is disengaged in a settler and the H_2SO_4 solution recycled to the regeneration unit. This results in using less than 10% excess acid giving major acid and neutralization agent (limestone) cost savings. A cost breakdown is given. The only waste from the sulphuric acid regeneration is solid $CaSO_4$ with small amounts of other cations and unused H_2SO_4. Water recovery is increased in this process.

INTRODUCTION

For desalination of water of low salinity such as required for municipal supply one may choose from two main ion exchange proccesses: (a) the DESAL process utilizing weak function resins, and (b) the common strong cation-weak anion sequence.

Desalination duty

A typical analysis of an effluent (secondary treated domestic sewage) is given below in Table 1. The typical requirement in such a case is to reduce the total dissolved solids to about 100 to 500 mg/ℓ in order to control accumulation of salinity in a water reuse situation.

TABLE 1

Typical water to be desalinated

Constituent	mg/ℓ	meq/ℓ
Na	241	10,5
Ca	44	2,2
Mg	43	3,6
NH_4	0,6	-
Cl	376	10,7
SO_4	105	2,2
NO_3	26	1,9
COD	70	-

Similar analyses are encountered with borehole water except for COD which is lower and Ca and Mg which are higher. Thus the typical desalination duty required is approximately 10 to 15 meq/ℓ.

Stoichiometry of regeneration with sulphuric acid and equilibrium

If an ion exchange process is being used which involves strong cation resin the quantity of sulphuric acid required for regeneration may be as much as three times the desalination duty, whereas the DESAL process would require roughly stoichiometric amounts of sulphuric acid. The difference is due to the equilibrium relationships between the cations on the strong function resin and the regenerating solution.

Figure 1 illustrates the Na/H exchange equilibrium for a strong cation resin (Zerolit 625) on an ion fraction diagram. If a quantity of resin, R equivalents and a quantity of liquid, L equivalents are mixed in a batch contact and allowed to equilibrate the mixture will follow the operating line with slope - R/L. This is derived from a material balance for the batch process between time o and time t:

$$y_oL + x_oR = y_tL + x_tR$$

$$y_t = y_o - \frac{R}{L}(x_t - x_o)$$

Continuous countercurrent stagewise process

Similarly, for a countercurrent, continuous process (process model in Figure 2(a)) an operating line between the conditions at the top of a column and the bottom is shown in Figure 2(b). Here the slope is R/L where R and L are now defined as equivalents per hour (product of the total ionic strength of the resin and liquid phases and their respective volumetric flow rates in the process). The operating line is derived from the material balance:

Entering			Leaving	
y_bL	$+ \; x_tR$	=	y_rL	$+ \; x_bR$
(bottom)	(top)		(top)	(bottom)

Using these operating lines and the equilibrium line a graphical construction (or a stage by stage calculation using the equations of the lines) may be performed either to determine the number of stages for a given set of feed and product compositions or the compositions of any leaving stream for given column stages and other fixed operating conditions. If the operating line touches the equilibrium line an infinite number of stages would be required and the conditions cannot be met.

Concentration effect on Ca/H equilibrium

Figure 3 gives the effect of total regenerant solution concengration on the Ca/H equilibrium line. This shows that at lower concentrations it is more difficult to get the calcium off the resin and a smaller value of R/L (i.e. greater stoichiometric excess of sulphuric acid) is needed which results in a greater fraction of unutilized acid. If a high R/L is used the regenerated resin composition will not be as good and the desalinated product water composition will be poorer. Furthermore the resin working capacity will be reduced requiring a greater resin circulation rate and a large consumption of regenerant and wash water.

Advantages of increasing sulphuric acid concentration

The above has illustrated one advantage of using high sulphuric acid concentrations in terms of acid consumption. This equally applies to any acid. Water desalting for municipal use is a high volume operation making a low value product and the water recovery has an extremely important effect on the product water cost. For example if it costs 50 cents/kilo-litre to desalt a given water and the water recovery is only 50% the product cost is 100 c/kℓ. Only when the water recovery is over ±90% does the incentive to increase it diminish. The key to high water recovery is attaining high waste solution concentrations. This is the most important reason for seeking to use sulphuric acid at high concentrations.

Ca content of feedwater and $CaSO_4$ scale formation limitations

Unfortunately because of $CaSO_4$ scaling of the walls of the CCIX columns we are normally limited to using concentrations below about 0,75 N when treating the feed water of Table 1. The calcium content of the feed water determines what regeneration conditions are feasible since, due to the greater resin selectivity for this ion than Na, all the Ca is removed from the feed water even when only ±60% of the Na is removed. (Note that in a CCIX contactor a steady-state composition profile is established. Whereas in a fixed bed operation each ion breaks through according to a characteristic pattern). Hence twice as much Ca in the feed water implies twice as much $CaSO_4$ produced in the regenerator and this demands roughly double the solution flow rate to control its concentration below the point at which scaling of column surfaces occurs. To achieve twice the solution flow, twice the amount of water is needed and hence the water recovery drops. In addition to the cost of the large excesses of sulphuric acid and the low water recovery which are incurred due to presence of Ca in the feedwater there is the extra cost of neutralising and disposing of the cation regeneration waste solution. Ideally, if the waste solution concentration can be increased any potential value of the salts contained in it can be more easily realised and the potential for practising chemical recovery is greater.

Advantages of CCIX contacting

Numerous advantages for the use of continuous ion exchange processes are apparent and these are realizable only on relatively large installations (ref. Slater). An important advantage may be the reduced initial resin cost due to better resin utilisation. Against this must be weighed the cost of resin replacement due to physical breakage which is usually (not necessarily) greater than in a fixed bed operation. Certain contactor configurations are less damaging to the resin. The author believes the fluidized bed Cloete-Streat contactor (as adapted by National Institute for Metallurgy, South Africa and used in the South African uranium mines and in pilot plant desalination of wastewater at the University of Cape Town, Chemical Engineering Department), is an excellent choice in terms of resin breakage. This very versatile contactor can be used in multiple-column flowsheets and is flexible both in design and operation as well as being operationally stable.

Operating features of the CCIX contactor

The most important features which aid attaining high waste solution concentrations are the in-column dilution of the regenerant. Details of the CCIX contactor are given in Figure 4 and typical four-column flow sheet in Figure 5. Wash water is applied to the lower stages and desorbs the acid regenerant from the macroporous resin (2=% DVB).

The sulphuric acid is injected as a 5N solution and is distributed and diluted to say 0,5N. Thus all the rinse water is used directly in the column for regenerant solution make up. Depending mainly on the resin sorption capacity and relative wash water and resin flow rates the column can be operated at any desired value of spent regenerant concentration. Operation is intermittent and the average resin flow rate is determined by the column cycle time which is the period of liquid up-flow after which the resin is drawn down one stage. The resin leaving the lowermost stage takes a quantity of unused wash water with it. This is disengaged on a sieve and returned to the product water tank. It thus enters the column at a given up-flow rate and some of it leaves each cycle (the shorter the cycle time the greater the average reverse flow of liquid). The difference or net up-flow rate of wash water then determines the spent regenerant concentration. The flow rate during the up-flow period determines the bed expansion of the fluidized beds of resin in each stage and this parameter can be varied between 100% typical in a load column and less than 10% in a regeneration column. The U.C.T. cation regeneration column has been operated at wash flow rates giving only just noticeable bed expansion.

Ion exchange resins having high ion exchange capacity and low pore sorption capacities would allow lower wash water net flows and greater waste concentration to be attained. (The fluidization properties may be adjusted to some degree by choice of bead sizes.) The effects on desorption kinetics of various operating conditions such as low flow rate and high concentrations can be offset by increasing the number of countercurrent contact stages. Capital cost of course ultimately limits the attainable performance. Usually, however, the wash stages are more efficient than the regeneration stages desorption kinetics being faster than ion exchange kinetics and their design does not become critical.

Optimum operating conditions to reduce H_2SO_4 consumption and avoid $CaSO_4$ scaling

The flexibility of operating conditions is considered to be the major advantage of the fluidized bed CCIX process. In order to determine the best operating conditions for the feedwater in Table 1 which would give the lowest acid costs and maximum waste concentration a number of tests were carried out. These are reported in detail elsewhere (Hendry). Only the results are given here. It was found that altering the stage efficiency by doubling the liquid flow rate in the regeneration column had a less important effect than changing the R/L ratio and concentration. Table 2 gives parameters tested and the results.

TABLE 2

Summarized Performance of Cation Regeneration Techniques

Nom.R/L	Conc. N	Eff.Working Capacity meq/mℓ	Acid Consumption Factor meq/meq
4-column system, using sulphuric acid only			
0,8	0,125	0,46	4,4
	0,50	0,59	3,4
	1,00	0,69	2,9
0,5	0,125	0,59	5,4
	0,50	0,78	4,1
	1,0	0,83	3,9
5-column system (suophuric acid at nom.R/L = 0,8; 0,125 N with brine or seawater rinse in intermediate column)			
0,25 (NaCl brine)	0,30	1,17	1,17
0,25 (Seawater)	0,30	0,80	2,5
5-column system (sulphuric acid recycle with nom.R/L = 0,25; 1,0N and $CaCl_2$ feed to intermediate column)			
0,25	1,00	0,54	1,08

For each operating condition the acid consumption factor was measured. This is the ratio of milliequivalents of acid applied per meq of effective working capacity achieved. The best results were obtained at R/L = 0,8 and with 0,5 N sulphuric acid which gave acid consumption of 3,4 meq/meq. Using 1,0 N sulphuric acid led to scale formation, sloughing of the scale and blockage of the resin transfer orifices in the interstage separator plates. At this optimum condition the regenerated resin contained only X_H = 0,37 and the load column product water Y_H = 0,71 (71% desalination). At R/L = 0,5 and 1,0 N the acid consumption was only slightly higher at 3,9 meq/meq with regenerated resin X_H = 0,52 and load column Y_H = 0,98 (98% desalination). However, this condition gave some slight scaling problems and is not considered safe from a stability point of view. In order to attain about 90%

desalination a good choice would be R/L = 0,65 and 0,75 N. This was tested and gave excellent results with an acid consumption factor of 3,6 meq/meq.

To obtain greater degrees of desalination would require lower R/L ratios to be used with corresponding increased acid consumption. The above tests were performed on water containing 50 mg/ℓ of calcium and 200 mg/ℓ sodium. At R/L = 0,8 : 0,5 N the water recovery of the cation system was 93%. If the total feedwater TDIS were doubled by increasing only the sodium the water recovery would drop to 86% since the waste concentration would remain the same. However, if the calcium content were raised to 150 mg/ℓ and the sodium reduced (i.e. no increase in TDIS) the waste solution would have to be diluted about three times to prevent $CaSO_4$ precipitation (all the calcium would be removed from the resin in each case due to selectivity of the resin) and water recovery would drop to 79% with a corresponding rise in the unit cost of product water.

Automatic control of CCIX process with changing feedwater parameters

Variations of the feedwater TDIS may be expected when treating effluent from a wastewater plant. The specified product TDS may then be controlled by altering the resin cycling rate to adjust the R/L ratio and thus maintain a constant product TDS. In order to minimize the consumption of regenerants the lowest resin cycling rate should be used consistent with maintaining desalination performance. If operating conditions in the regenerator are also linked to the resin cycling rate then the regenerant consumption can be minimized. This can be done automatically by monitoring the performance of the cation and anion sections and using a microcomputer control system. All the necessary data collection, calculations and feed-forward and feed-back control of cycle times or other operating variables can be carried out. Conductivity and pH measurements provide simple reliable measurements which provide the needed data and plant control is accomplished using relay operated solenoid-activated or variable position valves as required. The complete time sequencing and operating interlocking logic is carried out under softwave control by the microcomputer. Thus a relatively complex multi-column CCIX plant can be operated with ease by one operator and even a part-time operator in smaller installations. A visual display screen provides messages on plant operating conditions and alarm situations for the operator's attention; while manual control is possible for maintenance operations.

On the U.C.T. pilot plant conductivity of the feedwater and anion product water and pH of cation and anion product and of regenerant solutions is monitored. High anion product conductivity indicates low conversion in either the cation or anion systems.

The pH tells which system is at fault. Low pH indicates anion system conversion is low while correct pH indicates not enough NaCl is converted to HCl in the cation load column. Results monitored over about 12 cycles are analysed and the trends determined with weighted predictions to future cycles being made. Adjustments to cycle times are made on the basis of these predictions. The method and decision variables of these predictions can be varied to provide the necessary damping and process control stability. Where changes in input parameters such as feed TDS follow some recognisable or gradually changing pattern a successful or optimum correction procedure can also be "learned" by the microprocessor system, so that prior corrective action taken in anticipation may be adapted automatically from experience. Although this has not been applied on the CCIX process it may be desirable when treating sewage plant effluents where flow-changes at times during the day and week or the seasons may alter the feed water salinity. The technique has been applied, developed by our Civil Engineering Department and a unit built by us for control of a balancing dam at the inlet to a large activated sludge plant for load control. An important point to note is that the capital cost of microprocessor hardware for these purposes is not much greater than for corresponding fixed logic electromechanical timer sequencing systems.

New ion exchange techniques to reduce acid consumption

The above methods of optimizing the operation of the CCIX plant are fairly obvious requirements and to what extent they are put into practice depends on various economic factors and good engineering judgement. For example, very small plants may not justify sophisticated microprocessor controls. Also although the discussion above relates to the CCIX process there are many cases in which fixed bed operations will be preferred and the micro processor system can be adapted to the control of such plants equally well and this technology is being applied more and more in all kinds of plants. Further optimization than that readily available today yields diminishing returns. It is only by improving the methods of regeneration of the strong cation resin that any further major reductions in the acid consumption factor can be achieved when using sulphuric acid. One aim is to operate at higher acid concentrations. The use of an antiprecipitant agent to prevent scaling and promote a soft sludge which can be washed out of the fluidized beds has been considered but not tested by us. Pretreatment of the loaded cation resin with NaCl brine and seawater results in an improvement of the acid consumption factor to 1,71 and 2,5 meq/meq. Salt is unfortunately quite expensive (sulphuric acid + R85/ton; NaCl + R60/ton - R1,0 S.A. = $US) so that seawater is the method of choice if the plant is at the coast. These techniques are particularly important if the feedwater Ca content is high.

Recycling of cation regenerant solution

By treating the loaded cation resin with $CaCl_2$ solution (obtained from regeneration of the weak anion resin with $Ca(OH)_2$) the resin can be effectively converted to the Ca form eliminating most of the Na and Mg ions. When this is treated with sulphuric acid, precipitation of $CaSO_4$ must be promoted and controlled so as to prevent scale formation. This can be done in two ways.

(a) Recycled solution is first treated in a crystallizer with $CaSO_4$ crystals to reduce the $CaSO_4$ ion product to the saturation level. Then the solution is settled and returned through the column at a high flow rate (± 100% bed expansion of the resin) to take advantage of the nucleation delay (Hendry) with metastable $CaSO_4$ solutions and removed to the crystallizer before scaling can occur. The wash water flow is reduced drastically (incipient fluidization in the wash stages) so as not to dilute the recirculating regenerant and provides only enough water input to make up for the loss of solution with the precipitated $CaSO_4$ slurry discharged from the settler. A water recovery of 98% has been attained in this way discharging, in effect, a solid waste from the regeneration column. Careful control of the flow rates and cycle times is necessary, but 1,0 N H_2SO_4 has been used successfully in the pilot plant. The acid consumption factor is reduced to only 1,08 or almost to the stoichiometric minimum.

(b) Since the referenced paper was published the above technique has been simplified and improved by injecting the $CaSO_4$ seed slurry directly into the column operated at bed expansions of ± 20% and without the high recycle flow rates. No scaling was obtained and 2N H_2SO_4 was used. The external crystallizer is not needed and the settler is a much smaller unit. A flowsheet is shown in Figure 6. The $CaSO_4$ crystals were extremely fine and disengaged easily from the fluidised resin beds. In these methods the purpose of replacing most of the cations with Ca is to avoid accumulation of soluble sulphates of Na, Mg, etc. in the recirculating regenerant which would reduce the working capacity of the resin. Conversion to the Ca form need only be sufficient to control this since the soluble sulphates are bled off with the $CaSO_4$ slurry and therefore reach only an equilibrium concentration.

Cost of treatment

The cost for sulphuric acid to remove 500 mg/ℓ of TDS from the feedwater of Table 1 using the optimum R/L = 0,5, conc. = 0,5 N is estimated at 12,1 c/kℓ of feedwater. The corresponding lime cost is 2,0 c/kℓ and the cost of limestone to neutralise the acid waste before disposal is 3,1 c/kℓ. Total cost is 17,2 c/1ℓ. When using the acid recycle system with $CaSO_4$ seeding the acid cost drops to 5,1 c/kℓ and the neutralisation cost is only 0,3 c/kℓ. Total chemical cost is then 7,4 c/kℓ. The extra capital cost of equipment for the recycle is estimated at 2 to 3 c/kℓ. Thus the acid saving far outweighs the extra capital cost. The process is of particular significance if a very hard borehole water is to be treated which would cost no more than if it contained no calcium at all.

Finally the reduced cost of disposing of the cation waste solution by e.g. evaporation, is very significant. The $CaSO_4$ may have some value as a soil conditioner.

Other approaches to reducing sulphuric acid cost

Because H_2SO_4 is more expensive than lime in some situations (H_2SO_4 R85/ton, lime R55 ton) consideration has been given to using a strong anion resin to convert the NaCl to NaOH and regenerating the resin with solid lime. The NaOH could be removed on a weak cation resin regenerated stoichiometrically with H_2SO_4.

Precipitation of $CaSO_4$ in the anion regeneration could again be put to use to promote the regeneration reaction by removing SO_4 ions from the solution. The resin would be converted to the SO_4 form using the spent H_2SO_4 solution (similar to acid recycle) to remove the other anions. Two problems would occur:

(a) $CaSO_4$ precipitation on the Ca(OH) particles would retard the dissolution rate of the lime. It would be necessary to remove the slurry to a mill to liberate the lime particles and return the slurry to the column.

(b) The low $Ca(OH)_2$ solubility is only about twice that of $CaSO_4$ so that the OH ion fraction can never be greater than about 0,66 so that removal of SO_4 from the strong anion resin will not be complete and the working capacity would be low. Although interesting this method is likely to incur high capital costs and no work on it is being considered.

CONCLUSION

A number of unconventional ideas have been presented here. The sulphuric acid recycle system works very well and offers major cost savings. The success of these methods as with use of lime in the CCIX process depends largely on the effectiveness of a stable and flexible fluidized bed contactor which can handle the suspended solids of regenerants and by-products precipitating. The field of by-product recovery and purification is wide open to the CCIX process. The formation of solid wastes from an ion exchange process becomes feasible. $CaSO_4$ although an industrial by-product of many industries can be pure enough from the CCIX plant to be useful as a soil conditioner with no deleterious contaminants. Another interesting possibility is formation of a slow release $Ca(PO_4)(SO_4)$ fertilizer. It is believed that innovation will yield other potentially useful applications.

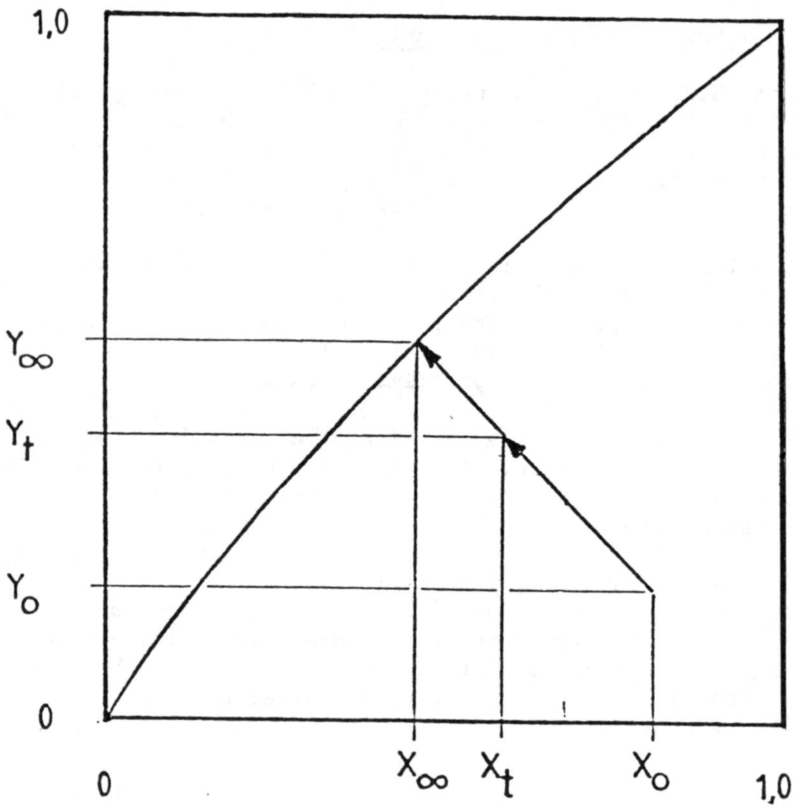

FIGURE 1 BATCH CONTACT PROCESS

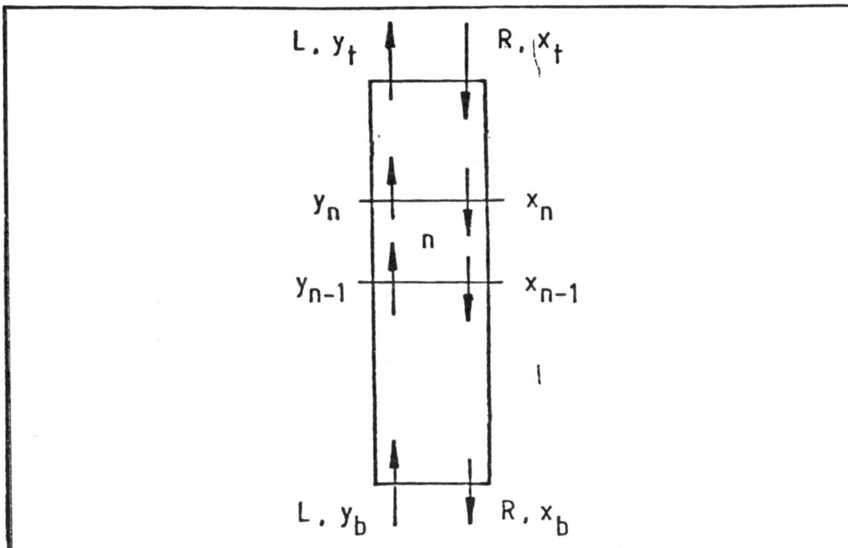

FIGURE 2a CCIX – CONTACTOR: PROCESS MODEL

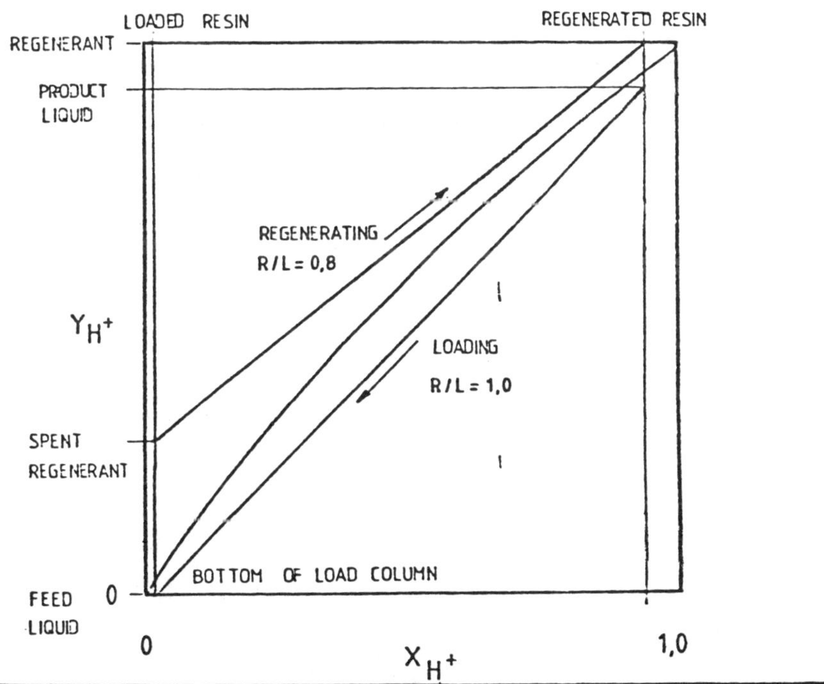

FIGURE 2b OPERATING LINES FOR Na/H EXCHANGE

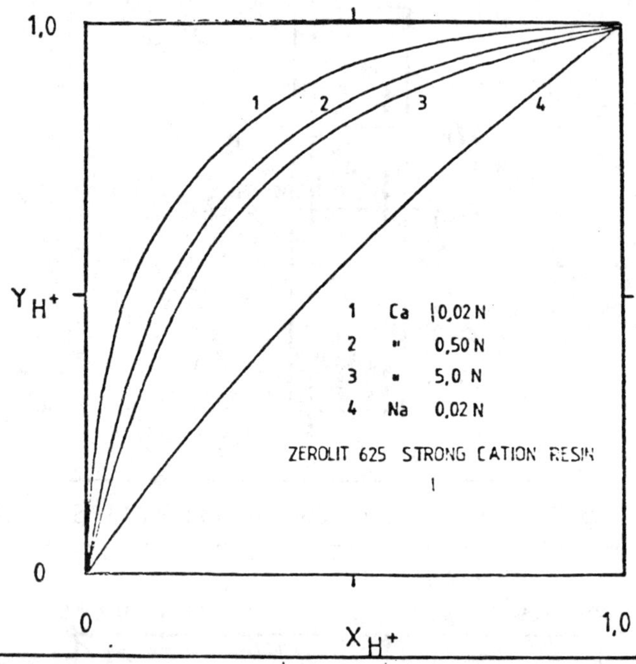

FIGURE 3 EQUILIBRIUM COMPOSITIONS – CATION

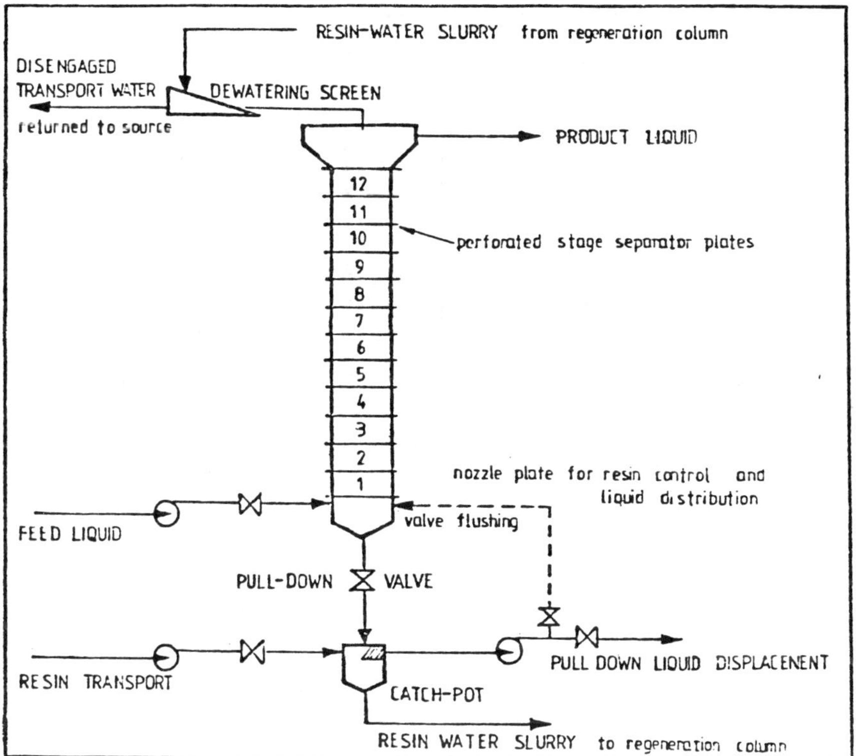

FIGURE 4 ESSENTIAL FEATURES OF CCIX COLUMN

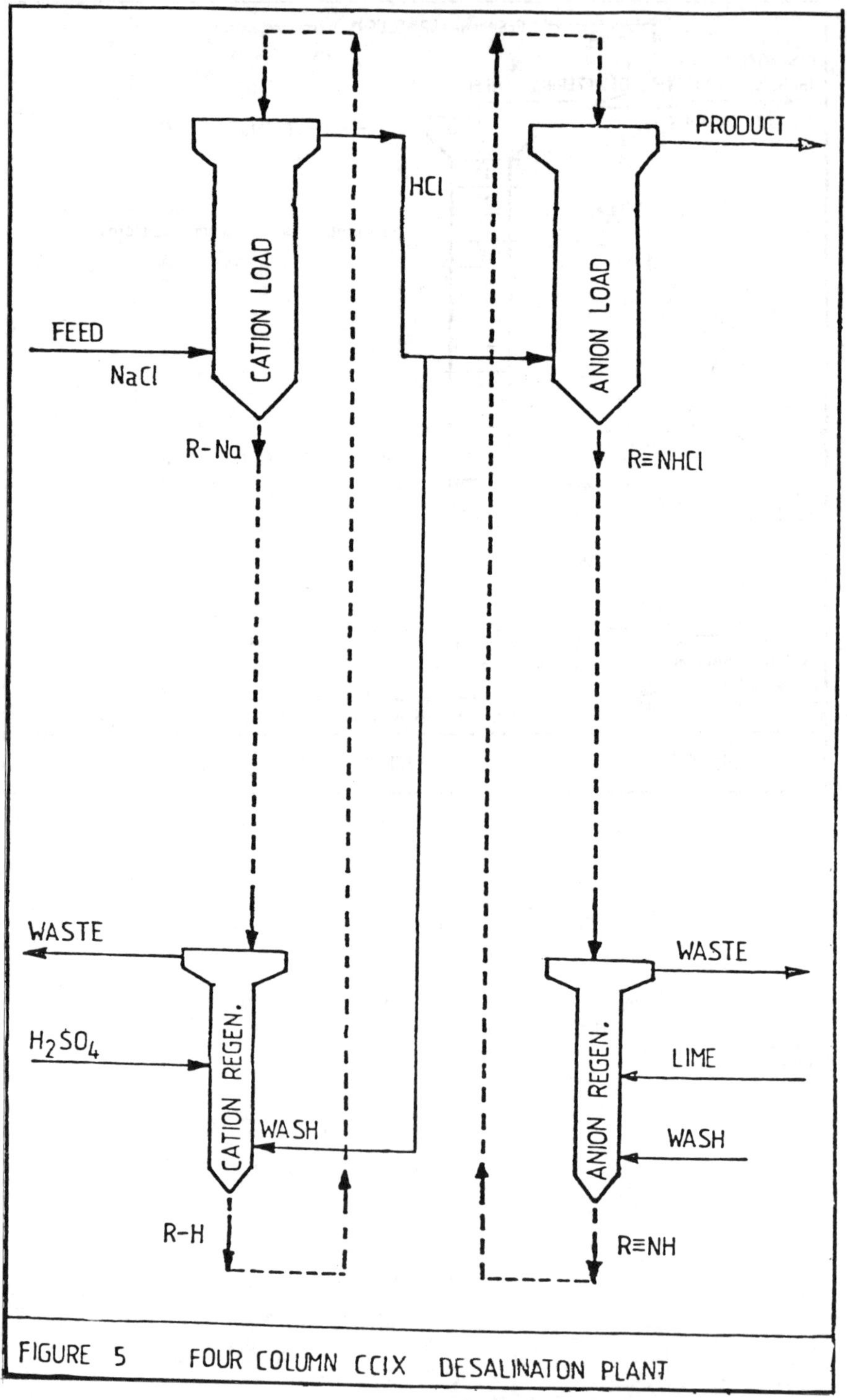

FIGURE 5 FOUR COLUMN CCIX DESALINATON PLANT

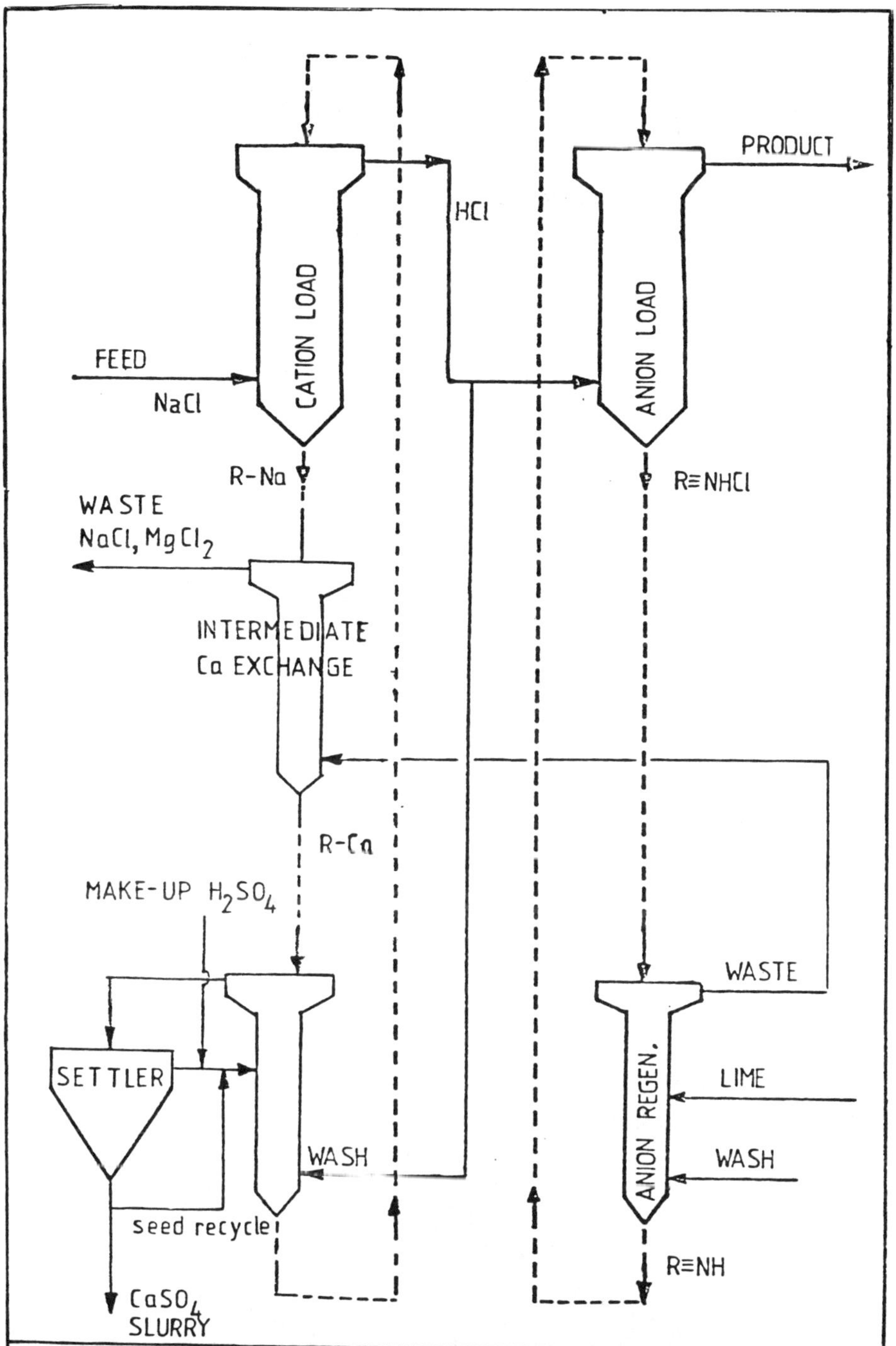

FIGURE 6 FIVE COLUMN SYSTEM WITH H_2SO_4 RECOVERY

REFERENCES

Hendry, B.A., "Continuous Counter Current Ion Exchange for Desalination and Tertiary Treatment of Effluent and other Brackish Waters." Presented at 11th Biennial Conference of the IAWPR, Cape Town, South Africa. 29 March to 2 April 1982.

Slater, M.J., "The relative sizes of fixed bed and continuous countercurrent flow ion exchange quipment." Trans. I. Chem. E. 54-58 Vol.60 1982.

DESIGN METHODS FOR ION EXCHANGE EQUIPMENT

J.Loureiro, C.Costa, M.Dias, J.Lopes and A.Rodrigues

Department of Chemical Engineering, Unversity of Porto, Portugal

ABSTRACT

The methods for designing ion exchange equipment, namely batch, continuous perfectly mixed and fixed beds, are discussed.

For each type of equipment we get, either analytically or numerically, the solution concentration (and resin concentration) as a function of time by using three different models:

- equilibrium model
- film diffusion model
- homogeneous particle diffusion model.

In all cases the ion exchange equilibrium isotherm is nonlinear.

The numerical solutions were obtained using the method of lines with a finite element collocation technique for the spatial coordinate.

INTRODUCTION

Design of ion exchange equipment is often made in an empirical or semi-empirical way, using the concept of mass transfer zone.

Computer aided design is being more and more used by chemical engineers; then modelling and simulation of ion exchange processes will enable us to understand the process dynamics, by learning about the influence of each parameter and as a consequence we get an economy of experiments.

Development of numerical methods such as collocation made possible the solution of complex models which arise from the consideration of complete particle diffusion equations.

In the following, the fixed bed is simulated by a cascade of J perfectly mixed sorbers.

BATCH ION EXCHANGE

Equilibrium Model

Assuming instantaneous equilibrium between fluid and solid phases (equilibrium model), and complete mixing within the reactor, the model equations are:

- Unsteady state material balance; the mass balance for an ionic species over the batch reactor is

$$v_s \frac{dc}{dt} + v_r \frac{dq}{dt} = 0 \qquad |1|$$

- Equilibrium law at the interface; in order to analyse the influence of the nonlinearity of the equilibrium isotherm, we assume a constant separation factor $K = q(c_o - c^*)/\{(Q - q)\, c^*\}$, which can also be written as

$$q = \frac{QKc^*}{c_o + (K-1)c^*} \qquad |2|$$

- Kinetic law; in this case (equilibrium model), there are no mass transfer resistances to the transport of an ionic species from the bulk of fluid to the sites on the solid phase, then

$$c = c^* \qquad |3|$$

- Initial condition; we assume that the solid is completely regenerated and that the concentration of ions in the fluid phase is c_o at $t = 0$, then

$$t = 0 \qquad c = c_o \quad \text{and} \quad q = 0 \qquad |4|$$

Introducing the dimensionless variables

$$x = \frac{c}{c_o}\ , \ y = \frac{q}{Q} = \frac{Kx}{1+(K-1)x}$$

equation |1| becomes

$$\frac{dx}{dt} + \xi \frac{dy}{dt} = 0 \qquad |5|$$

where $\xi = \frac{1-\varepsilon}{\varepsilon}\frac{Q}{c_o}$ is the capacity parameter and ε is the porosity or the fraction of the reactor occupied by the fluid.

Upon integration with the initial condition, equation |4|, equation |5| conduces to the final equilibrium concentration

$$x_f = \frac{-\{2+K(\xi-1)\}+\sqrt{K^2(\xi-1)^2+4\xi K}}{2(K-1)} \qquad |6|$$

In other words: the initial point (1,0) moves on the operating line $y = (1-x)/\xi$ to the equilibrium point (x_f, y_f) as shown in Figure 1.

The equilibrium concentration, x_f, depends on two parameters: ξ, the capacity factor, which takes into account the initial concentration, ion exchange capacity of the resin, and void fraction; K, the equilibrium constant, accounts for the nonlinearity of the isotherm. Figure 2 shows the importance of K on x_f, with ξ as a parameter.

Film Diffusion Model

Introducing now one resistance to mass transfer (the external mass transfer resistance) due to a stagnant film of fluid around the particles, the model equations are still the mass balance, Eq. |1|, the equilibrium isotherm, Eq. |2|, and the initial condition, Eq. |4|. The kinetic law of transport is now

$$\frac{dq}{dt} = K_f a_p (c - c^*) \qquad |7|$$

Upon the introduction of the dimensionless variables x and y, the system of equations |1|, |2| and |7| can be written

$$\frac{dx}{dt} + \xi\frac{dy}{dt} = 0 \qquad |8a|$$

$$\frac{dy}{dt} = \frac{1}{\xi\tau_f}\{x - \frac{y}{K+(1-K)y}\} \qquad |8b|$$

where $\tau_f = \varepsilon/\{(1-\varepsilon)K_f a_p\}$ is a measure of the mass transfer resistance by film diffusion.

Integrating, with the initial condition, Eq. |4|, this system conduces to the solution

$$\frac{t}{\tau_f} = \frac{K(\xi-1)}{2\sqrt{K^2(\xi-1)^2+4\xi K}} \ell n \frac{\dfrac{1}{K(\xi-1)-\sqrt{K^2(\xi-1)^2+4\xi K}} - \dfrac{(K-1)x}{K(\xi+1)+\sqrt{K^2(\xi-1)^2+4\xi K}}}{\dfrac{1}{K(\xi-1)+\sqrt{K^2(\xi-1)^2+4\xi K}} - \dfrac{(K-1)x}{K(\xi+1)-\sqrt{K^2(\xi-1)^2+4\xi K}}} +$$

$$+ \frac{1}{2} \ell n \frac{\xi K}{(K-1)x^2+\{2+K(\xi-1)\}x-1} = \qquad |9|$$

$$= \frac{K(\xi-1)}{2\sqrt{K^2(\xi-1)^2+4\xi K}} \ell n \frac{1+\dfrac{2\xi(1-K)}{K(\xi+1)+\sqrt{K^2(\xi-1)^2+4\xi K}}y}{1+\dfrac{2\xi(1-K)}{K(\xi+1)-\sqrt{K^2(\xi-1)^2+4\xi K}}y} -$$

$$- \frac{1}{2} \ell n\{\xi \frac{K-1}{K} y^2 - (\xi+1)y + 1\}$$

The evolution of the concentrations in the fluid and solid phases, x and y, respectively, are functions of the model parameters, ξ, K, and τ_f, the film diffusion time constant.

As can be seen from Figure 3, it is possible to determine the value of τ_f (and then, $K_f a_p$) from one batch experiment. Of course, the concentrations tend to the equilibrium values (which depend only on ξ and K).

Homogeneous Particle Diffusion Model

If, instead of an external mass transfer resistance, we introduce an intraparticle diffusion resistance, the model equations become now, in terms of the dimensionless concentrations x and y:

- Mass balance

$$\frac{dx}{dt} + \xi \frac{d\bar{y}}{dt} = 0 \qquad |10|$$

- Equilibrium law at the interface

$$y(1,t) = \frac{Kx}{1+(K-1)x} \qquad |11|$$

- Kinetic law

$$\frac{\partial y}{\partial t}(r,t) = \frac{D_{eff}}{r^2 R_o^2} \frac{\partial}{\partial r} \left(r^2 \frac{\partial y}{\partial r}\right) \qquad |12|$$

- Initial and boundary conditions

$$t = 0 \quad x = 1$$

$$r < 1 \quad y = 0 \qquad |13a|$$

$$r = 1 \quad y = 1$$

$$r = 0 \quad \frac{\partial y}{\partial r} = 0 \quad \forall t \qquad |13b|$$

$$r = 1 \quad \frac{\partial y}{\partial t} = -3\xi \frac{D_{eff}}{R_o^2} \frac{\{K+(1-K)y(1,t)\}^2}{K} \left.\frac{\partial y}{\partial r}\right|_{r=1} \quad \forall t \qquad |13c|$$

In these equations, $\bar{y} = \frac{3}{R_o^3} \int_0^{R_o} y r_1^2 \, dr_1$, for spherical particles of radius R_o, $r = r_1/R_o$, r_1 is the radial position in the particle, and D_{eff} is the effective internal diffusivity.

If we introduce the independent radial variable $u = r^2$ and the dimensionless time variable $\theta = t\,D_{eff}/R_o^2 = t/\tau_d$, we obtain the nonlinear parabolic partial differential equation

$$\frac{\partial y}{\partial \theta} = 4u \frac{\partial^2 y}{\partial u^2} + 6u \frac{\partial y}{\partial u} \qquad |14|$$

subject to the initial and boundary conditions

$$\theta = 0 \quad u = 1 \quad y(1,0) = 1$$

$$u < 1 \quad y(u,0) = 0 \qquad |15a|$$

$$u = 0 \quad \frac{\partial y}{\partial u} = 0 \quad \forall \theta \qquad |15b|$$

$$u = 1 \quad \frac{\partial y}{\partial \theta} = -6\xi \frac{\{K+(1-K)y(1,\theta)\}^2}{K} \left.\frac{\partial y}{\partial u}\right|_{u=1} \quad \forall \theta \qquad |15c|$$

The boundary value problem (spatial variable) was solved using an orthogonal collocation on finite elements technique. In brief, it consists on representing the solution by a series expansion of polynomials of degree N, |1|

$$y_N(u,\theta) = \sum_{i=1}^{N+1} a_i(\theta)\, H_i(u) \qquad |16|$$

where the $H_i(u)$ are known functions and the coefficients $a_i(\theta)$ are to be determined in order to have null residuals at N+1 points of the domain.

In this particular case, in which the initial profiles in the particle are steep, and so, difficult to represent, the spatial integration interval ($u \in |0,1|$) was partitioned into subintervals (three, normally) and the u variable inside each of them was normalized in order to vary between 0 and 1 (domain of the chosen polynomials for the series expansion). For the subinterval k:

$$v_k = \frac{u-u(k)}{h(k)} \quad , \; h(k) = u(k+1) - u(k) \qquad |17|$$

and we have, for this subinterval

$$y(v_j) = \sum_{i=1}^{N+1} H_{ji}\, a_i \qquad |18a|$$

$$\frac{dy}{dv}(v_j) = \sum_{i=1}^{N+1} A_{ji}\, a_i \qquad |18b|$$

$$\frac{d^2y}{dv^2}(v_j) = \sum_{i=1}^{N+1} B_{ji}\, a_i \qquad |18c|$$

where the matrices H, A and B are known. Substituting the equations |18a,b,c| on Eq.|14|, equating the residuals to zero and introducing $\ell = N(k-1)/2$, after putting $N+1 = 4$ (used in the solution) we obtain

$$\sum_{i=1}^{4} H_{ji} \frac{da_{i+\ell}}{d\theta} = \frac{4\{u(k)+h(k)v_j\}}{h_k^2} \sum_{i=1}^{4} B_{ji}\, a_{i+\ell} + \frac{6}{h_k} \sum_{i=1}^{4} A_{ji}\, a_{i+\ell} \qquad |19|$$

$$\text{for} \quad k = 1,\cdots,SN \quad \text{and} \quad j = 2,3$$

Applying the same treatment to the boundary conditions, Eqs. |15b,c| we get, for the first subinterval

$$u = 0 \qquad \sum_{i=1}^{4} A_{1i}\, a_i = 0 \qquad |20a|$$

which, for the Hermite polynomials used, is equivalent to

$$a_2 = 0 \qquad |20b|$$

and, for the last subinterval $(k = SN)$

$$u = 1 \quad \sum_{i=1}^{4} H_{4i} \frac{da_{i+\ell}}{d\theta} = -6\xi \frac{\{K+(1-K)\sum_{i=1}^{4} H_{4i}\, a_{i+\ell}\}^2}{K} \frac{1}{h_k} \sum_{i=1}^{4} A_{4i}\, a_{i+\ell} \qquad |21a|$$

which can be reduced to

$$\frac{da_{3+\ell}}{d\theta} = -6\,\xi \frac{\{K+(1-K)a_{3+\ell}\}^2}{K} a_{4+\ell} \qquad |21b|$$

Eliminating a_2, we then obtain a system of $2SN+1$ nonlinear ordinary differential equations with the initial conditions (recall Eqs.|15a|)

$$0 < u < 1 \qquad \sum_{i=1}^{4} H_{ji}\, a_{i+\ell} = 0 \qquad j = 2,3;\ k = 1,\cdots,SN \qquad |22a|$$

$$u = 1 \qquad \sum_{i=1}^{4} H_{4i}\, a_{i+\ell} = 1 \qquad k = SN \qquad |22b|$$

This initial value problem, generally stiff, is solved using the GEARIB package,|2|, included in the more general PDECOL package, |3|.

Numerical simulations were made in order to analyse the influence of the model parameters (ξ, K and $\tau_d = R_o^2/D_{eff}$, the particle diffusion time constant, which takes into account the influence of the intraparticle mass transfer resistance) on the evolution of the fluid concentration, x, and of the concentration profile, y, inside the particle.

These results were plotted and, as well as for the film diffusion model, we can determine the value of τ_d (and then of D_{eff}, known R_o) from one batch experiment, as can be seen from Figure 4.

Figure 5 shows the evolution of the concentration y inside the particle with time ($\theta = t/\tau_d$) for given values of ξ and K. It evidences the steepness of the initial profiles referred to before.

ION EXCHANGE IN A CONTINUOUS SYSTEM (CSTR or COLUMN)

As stated in the introduction, the column will be simulated by a series of J perfectly mixed cells.

Equilibrium Model

The model equations are

- Material balance over the ith cell

$$U c_{i-1} = U c_i + \frac{v_s}{J}\frac{dc_i}{dt} + \frac{v_r}{J}\frac{dq_i}{dt} \qquad |23|$$

- Equilibrium law

$$q_i = \frac{QKc_i^*}{c_o+(K-1)c_i^*} \qquad |24|$$

- Kinetic law

$$c_i = c_i^* \qquad |25|$$

- Initial and boundary conditions. Assuming the solid completely regenerated and a step input of concentration $c_o = c^o H(t)$, where $H(t)$ is the Heaviside function

$$t = 0 \qquad c_i = q_i = 0 \qquad |26a|$$

$$t > 0 \qquad c_o = c^o H(t) \qquad |26b|$$

The solution for this case and the influence of the model parameters ξ, K and J, the number of perfectly mixed cells, which accounts for the axial dispersion, is known and is published elsewhere, |4|.

Film Diffusion Model

The system of equations (mass balance, kinetic law and equilibrium law) in dimensionless form, is reduced to

$$\frac{dx_i}{d\theta^*} = J(x_{i-1}-x_i) - N_f\{x_i - \frac{y_i}{y_i+K(1-y_i)}\} \qquad |27|$$

$$\frac{dy_i}{d\theta^*} = \frac{N_f}{\xi}\{x_i - \frac{y_i}{y_i+K(1-y_i)}\} \qquad |28|$$

where $\theta^* = t/\tau$, $\tau = v_s/U$ (space time)

$N_f = \tau/\tau_f$, $\tau_f = \varepsilon/\{(1-\varepsilon) K_f a_p\}$

with the initial and boundary conditions

$$\theta^* = 0 \qquad x_i = y_i = 0 \tag{29a}$$

$$\theta^* > 0 \qquad x_o = 1 \tag{29b}$$

The system of Eqs. |27| and |28| is a stiff system and its solution was accomplished using either Michelsen code, |5|, or DIFSUB routine, |6|. Some results have been published before, |4,7,8|.

For a CSTR (J = 1), Figure 6 shows the influence of N_f, the number of external mass transfer units, on the effluent concentration x, for fixed values of ξ and K. As it was to be expected, θ^*_{Bp} (the breakthrough time, time at which the concentration reaches a certain value, say $0.05\,c_o$), increases with N_f. In Figure 7 we can see the histories of concentration for various values of K (ξ and N_f fixed). Again, θ^*_{Bp} increases with K, that is to say, with the nonlinearity of the isotherm.

For a column, Figure 8 shows the influence of the axial dispersion (represented by J) on the concentration history, for given values of ξ, K and N_f; whereas Figure 9 displays the effect of N_f (ξ, K and J fixed). Identical conclusions can be drawn from these figures: the greater J (or N_f), the greater θ^*_{Bp}.

Homogeneous Particle Diffusion Model

After some mathematical manipulation, the model equations reduce to

$$\frac{1}{N_d}\frac{\partial y_i}{\partial \theta^*} = 4u\,\frac{\partial^2 y_i}{\partial u^2} + 6\,\frac{\partial y_i}{\partial u} \tag{30}$$

where $N_d = \tau/\tau_d$, $u = r^2$, with the initial and boundary conditions

$$\theta^* = 0 \qquad y_i(u,0) = 0 \tag{31a}$$

$$u = 0 \qquad \frac{\partial y_i}{\partial u} = 0 \tag{31b}$$

$$u = 1 \qquad \frac{\partial y_i}{\partial \theta^*} = \frac{\{K+(1-K)y_i(1,\theta^*)\}^2}{K}\,\{J(x_{i-1}-x_i)-6\xi N_d\,\frac{\partial y_i}{\partial u}\} \tag{31c}$$

with $x_o = 1$

We have now a system of J nonlinear parabolic partial differential equations. The methodology and technique used in its solution were the same explained above, when treating this very model in batch systems.

Computer results, which show the importance of the model parameters K, ξ, J and N_d (particle transfer units) were plotted.

We can see the influence of N_d on the concentration histories, for $J = 10$, $\xi = 1$, $K = 10$, in Figure 10, while Figure 11 shows the profiles inside the particle in a CSTR ($J = 1$) for the same values of ξ and K ($N_d = 0.01$).

Figures 12 and 13 show the influence of K and ξ, with $J = 10$ and $N_d = 0.1$, for fixed values of ξ ($\xi = 1$) and K ($K = 10$), respectively.

Finally, Figure 14 shows the importance of the axial dispersion, measured by J, for given values of the other parameters ($\xi = 1$, $K = 10$, $N_d = 0.1$).

CONCLUSIONS

Mass transfer resistances (film diffusion and internal diffusion) were considered in modelling ion exchange processes both in batch and continuous operations, for nonlinear equilibrium situations.

We used a collocation on finite elements method to solve the nonlinear parabolic PDEs which arise in model equations.

For the problem treated in this paper, three intervals and cubic Hermite polynomials were sufficient to achieve the solution.

For low axial dispersion (high J) the staged approach seems not to be very interesting due to the high number (8xJ) of nonlinear ODEs (in the case of internal diffusion) which requires high CPU capacity.

NOMENCLATURE

a_p specific area of particle

c fluid concentration

c_o feed concentration or initial concentration in batch systems

c^* fluid equilibrium concentration

D_{eff} effective diffusivity
K constant separation factor
k_f film mass transfer coefficient
J number of cells
N_f, N_d number of transfer units for film and particle diffusion models
q solid concentration
Q ion exchanger capacity
r radial particle coordinate
R_o particle radius
t time
U flowrate
v_r volume of resin in the reactor
v_s volume of solution in the reactor
x reduced fluid concentration
y reduced solid concentration
$\bar{y}$ average reduced solid concentration

Greek letters

ε porosity of the system
ζ capacity factor
τ_d time constant for particle diffusion
τ_f time constant for film diffusion
τ space time
θ, θ^* dimensionless time

REFERENCES

1. Finlayson,B.A., "Nonlinear Analysis in Chemical Engineering", McGraw-Hill, New York (1980)

2. Hindmarsh,A.C., "Preliminary Documentation of GEARIB: solution of implicit systems of ordinary differential equations with banded Jacobian", Rep. UCID-30130, Lawrence Livermore Lab., Livermore, Calif., Feb. 1976

3. Madsen,N.K., and R.F. Sincovec, "PDECOL: General Collocation Software for Partial Differential Equations", ACM Trans. Math. Software 5 (1979) 326

4. Rodrigues,A.E. and E.C. Beira, "Staged Approach of Percolation Processes", AIChE J. 25 (1979) 416

5. Michelsen,M., AIChE J. 22 (1976) 594

6. Gear,C., ACM Comm. 14 (1971) 185

7. Rodrigues,A.E., Dias,M.Q., Lopes,J.C., and D. Tondeur, "Design of Perfectly Mixed Sorbers", CHEMPOR'78, Braga (Portugal)

8. Rodrigues,A.E., Lopes,J.C., Dias,M.Q., and M.R. Costa, "Computational Aspects of the Dynamics of Sorption Operations", CACE 79, Montreux

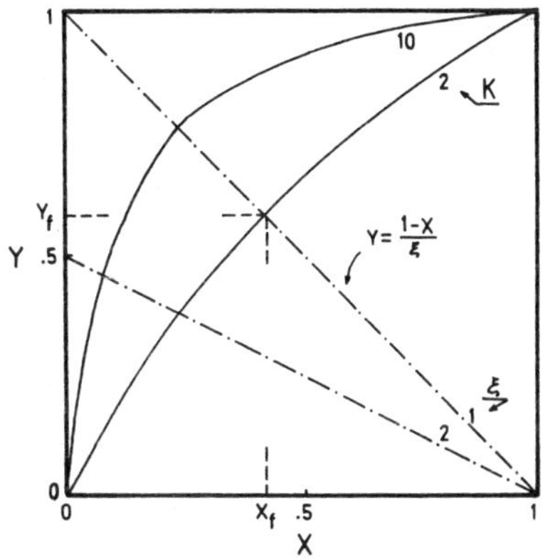

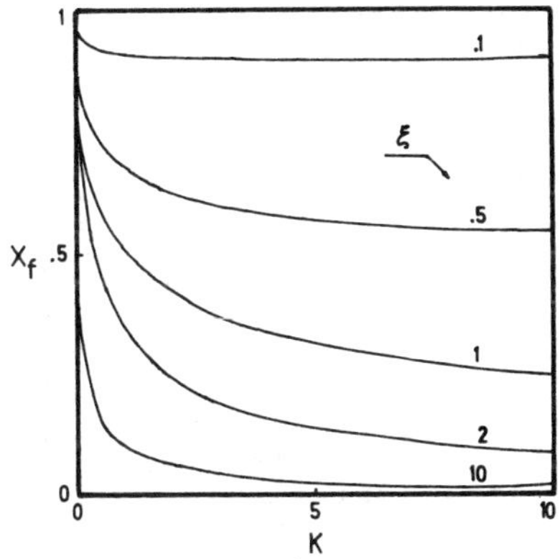

Figure 1 - Batch ion exchange (equilibrium model). (——) isotherm; (—·—) operating line; (x_f,y_f) equilibrium point.

Figure 2 - Batch ion exchange (equilibrium model). Influence of the nonlinearity of the isotherm on the final equilibrium fluid concentration.

Figure 3 - Batch ion exchange (film diffusion model). Evolution of the fluid (x) and solid (y) concentrations with time (reduced by the film diffusion time constant, τ_f).

Figure 4 - Batch ion exchange (homogeneous particle diffusion model). Evolution of the fluid (x) and average solid ($\bar{y}$) concentrations with time (reduced by the particle diffusion time constant, τ_d).

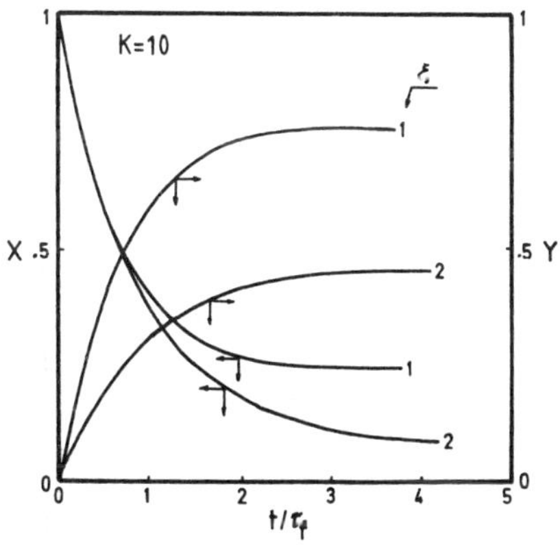

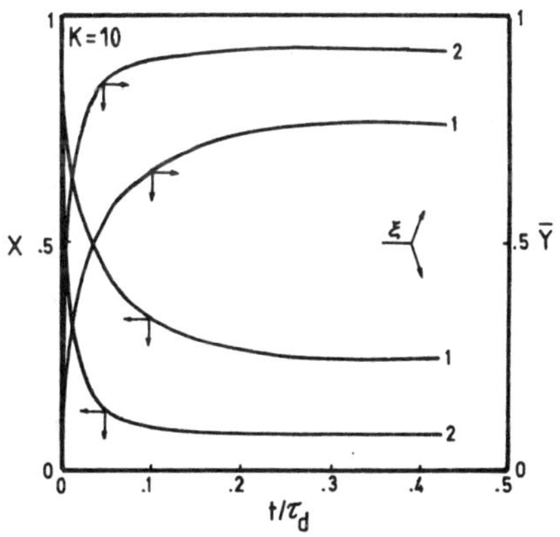

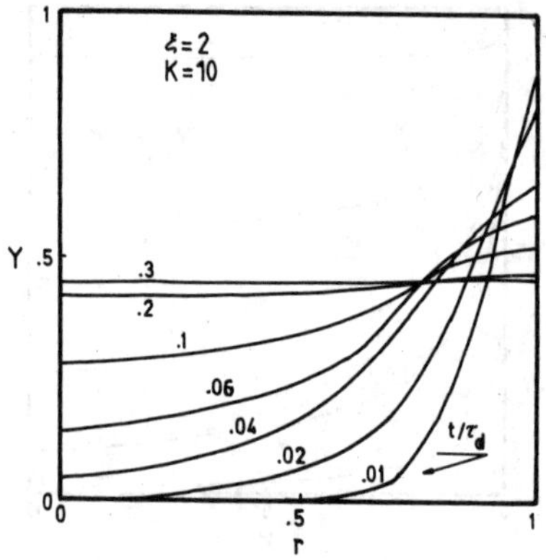

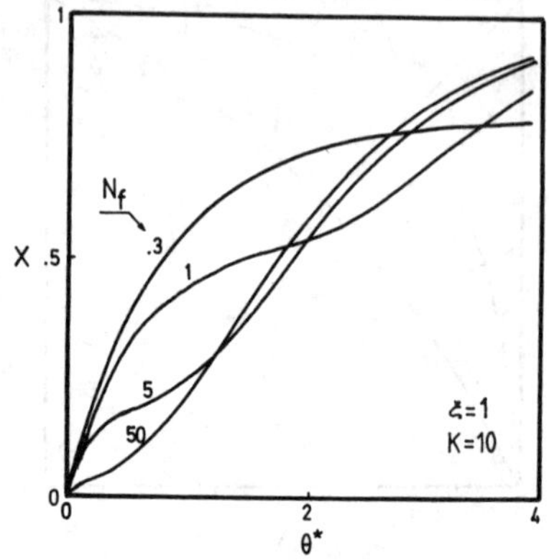

Figure 5 - Batch ion exchange (homogeneous particle diffusion model). Concentration profiles inside the particle for different values of t/τ_d.

Figure 6 - Ion exchange in a CSTR (film diffusion model). Effluent concentration,x as a function of time for different values of N_f (number of film mass transfer units).

Figure 7 - Ion exchange in a CSTR (film diffusion model). Effluent concentration, x as a function of time with K as a parameter.

Figure 8 - Ion exchange in a column (film diffusion model). Influence of the axial dispersion (measured by J) on the history of concentrations.

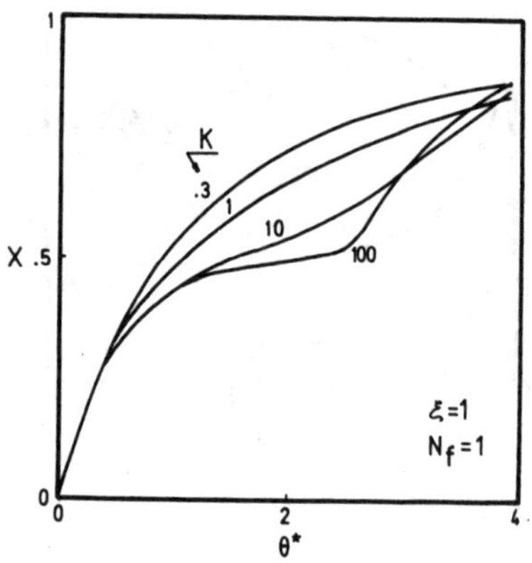

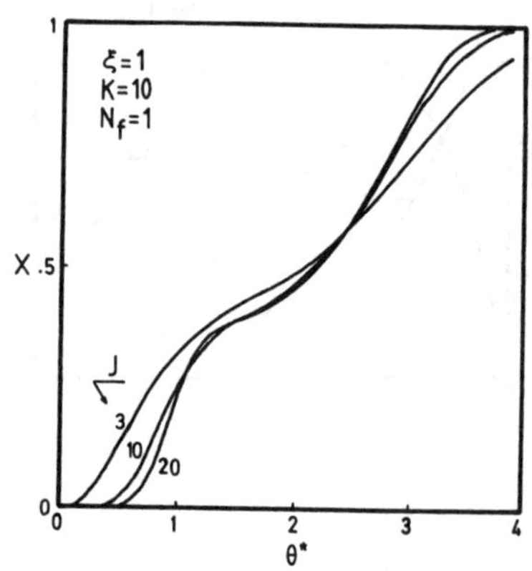

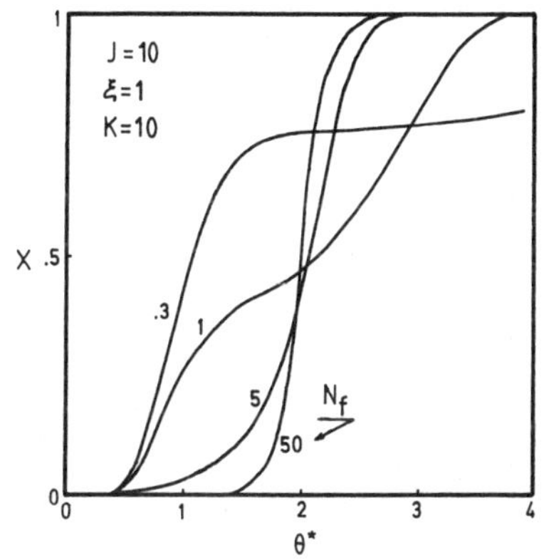

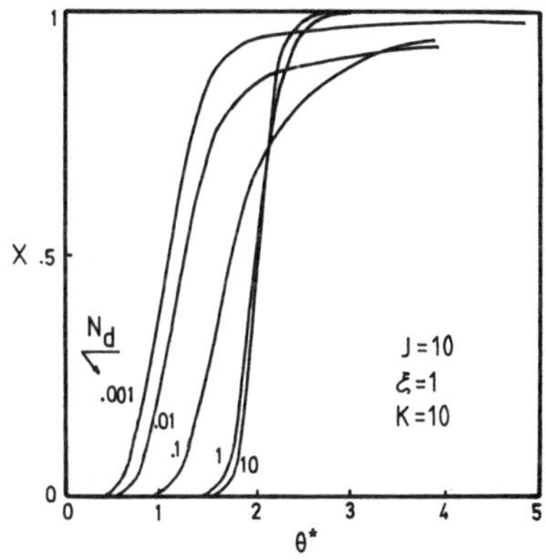

Figure 9 - Ion exchange in a column (film diffusion model). Influence of the number of film transfer units, N_f on the history of concentrations.

Figure 10 - Ion exchange in a column (homogeneous particle diffusion model). Influence of the number of particle transfer units on the history of concentrations.

Figure 11 - Ion exchange in a CSTR (homogeneous particle diffusion model). Concentration profiles inside the particle at different times.

Figure 12 - Ion exchange in a column (homogeneous particle diffusion model). Influence of the nonlinearity of the isotherm, K on the history of concentrations.

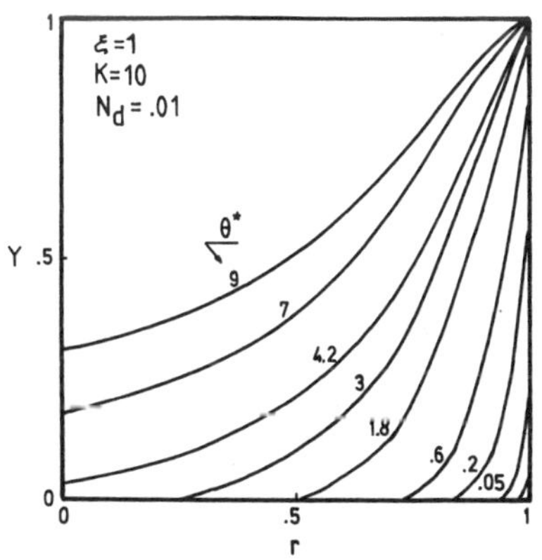

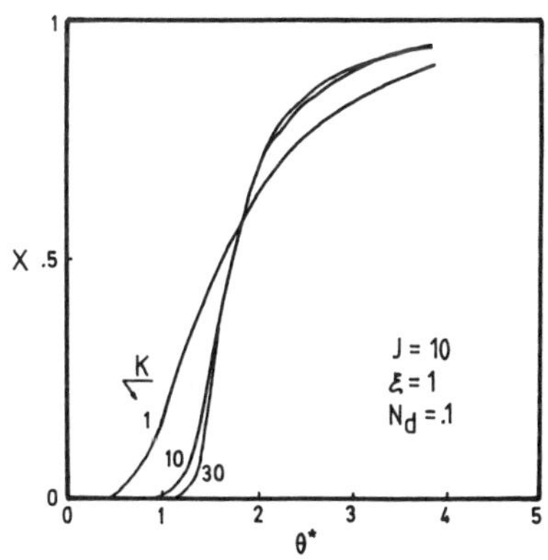

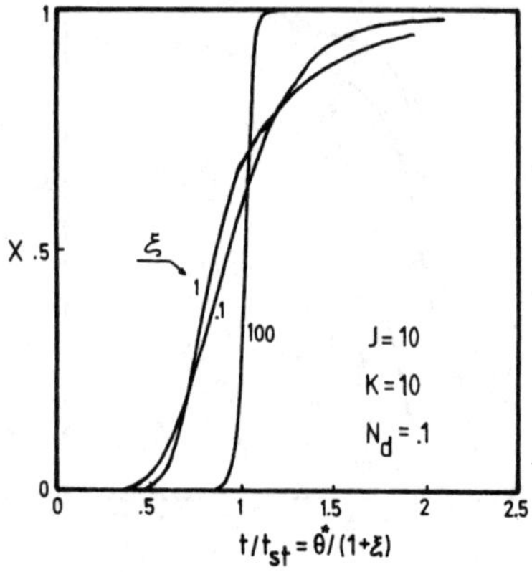

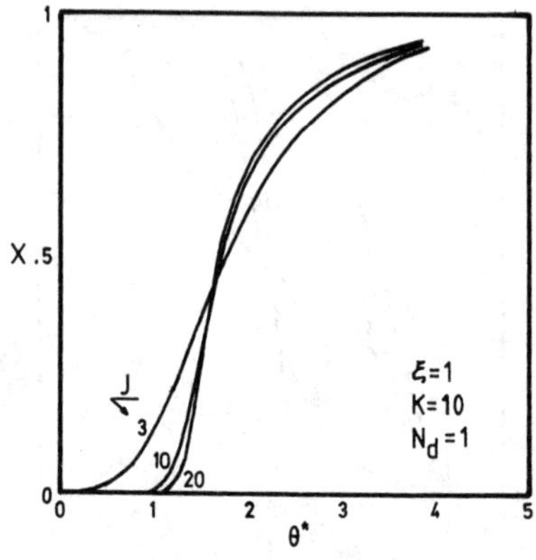

Figure 13 - Ion exchange in a column (homogeneous particle diffusion model). Influence of the capacity parameter, ξ on the history of concentrations.

Figure 14 - Ion exchange in a column (homogeneous particle diffusion model). Influence of the axial dispersion (measured by J) on the history of concentrations.

FUTURE CRITERIA FOR THE DESIGN AND MANUFACTURING OF WATER TREATMENT PLANTS.

Pier Giorgio BELLOTTI

S.p.A. TERMOMECCANICA ITALIANA
LA SPEZIA - ITALY

Nowadays, it is no longer a problem for Manufacturing Companies of water treatment plants to supply plants which absolutely assure good operating conditions and which guarantee maximum resistance from both a mechanical and chemical point of view.

Therefore, it is not here where these companies are to make their major efforts in the forthcoming years, although, of course, they still are to keep a look out for any innovations which the market could offer regarding the construction material, automatic controls and monitoring.

On the contrary, the engineering and manufacturing companies are to concentrate most of their efforts in designing plants which, with the same reliability, have lower plant costs and meet with the Customer-User requirements who continuously ask for better quality of the product water, obtained with lower working costs and with processes which do not create pollution problems.

This may be said for all the plants, in particular for plants using ion exchange resins for water treatment.

The collaboration between the Customer-User, who with his demands and observations is the basic point of reference, the resin manufacturer and the designer-constructor of the water treatment plant, is fundamental for achieving these objectives.

The problem in the next few years, as it has been in the past recent years, is that of increasing the efficiency of the regeneration in order to improve the quality of the product water and to reduce the regenerant consumption, hence, meeting with the user's demands which, either for the boiler feed or for the various production processes (electronic, pharmaceutical, chemical, food processing, etc.) require purer water and following the key-indications of our times, and which we have placed at the base of our meeting, which are energy saving (hence, the chemical products too for the regeneration of the resins, the production of which requires energy consumptions) and the reduction of the use of polluting products in order to help fight against water pollution.

We must remember that raw water also has its cost and that in many areas is quite precious, therefore, the raw water consumption/product water ratio is also to be kept more and more in consideration.

Without predicting which programmes the ion exchange resin producers have in store for us (will we reach the stage when resins will be regenerated without the use of chemical products?), even from a plant design point of view a lot has been done and a lot has still to be done.

Once, in order to have high quality demineralized water, with low leakage, the only method used was that of increasing the regenerative level of the resin. The engineering innovations which were introduced gave the same results but with lower regenerant consumptions; I refer to the counter-current regenerations, to the fluidized bed systems, to the dual layer units and so forth.

In other cases, with the same purpose of reaching the best possible results, optimizing the quality/cost ratio, plants have been designed which combine the ion exchange resin process with the other processes, for example, membrane or evaporative processes; this proved particularly interesting when the salinity of the raw water is high.

And what about the future?

If no revolutionary discoveries are made, which are not predictable at the moment, and the exchange resin will continue to be the heart of the water demineralizing plants, we shall continue in the same direction in the next years.

It is therefore worthwhile, even if briefly, to go into more detail, to see how we have proceeded, what results have been obtained, and what can be expected, giving also some figures which will underline these concepts.

I would like to mention some experiences about the demineralization of the raw waters and the condensate polishing, not to exhaust the argument, but simply to give an example.

Demineralization

It is always possible to have different solutions for the design of a demineralization plant. But which one, with all things considered, is really the most convenient for the User?
Let us see the example given below:

1) Water analysis ($CaCO_3$) -

Ca	90	HCO_3	110
Mg	30	SO_4	30
Na	80	Cl	60
	200		200

SiO_2 12 as SiO_2

2) Design parameter - Demineralization TDS 2ppm as $CaCO_3$
SiO_2 0.1 ppm as SiO_2

Flow 45 m3/h; 12 hour run

Cation Exchanger

Base solution : Strong acid resin (H+ cycle)

124 ft^3 strong acid resin
Regeneration level 6.5 lbs/ft^3 H_2SO_4
Chemical cost for regeneration $ 0.35/1000 galls.

Alternative solution : Dual layer (weak acid/strong acid resin)

28 ft^3 of weak acid resin

72 ft^3 of strong acid resin

Regenerative level 5.2 lbs/ft^3 H_2SO_4

Chemical cost for regeneration: $ 0.25/1000 galls.

Anion Exchanger

Base solution : Strong base type I resin

124 ft^3 of strong anion resin
Regener. level 6 lbs/ft^3 NaOH
Chemical cost: $ 1.10/1000 galls.

Alternative solution a) : Strong/weak anion resin dual-layer resin.

47 ft^3 weak anion resin
76 ft^3 strong anion resin
Regener. level 4.35 lbs/ft NaOH
Chemical cost: $ 0.80/1000 galls.

Alternative solution b) : Strong base type 2

127 ft^3 of strong base resin
Regener. level 3.7 lbs/ft^3 NaOH

Chemical cost: \$ 0.70/1000 gallons.

Alternative solution c) : Decarbonator + strong base resin type 1

74 ft^3 of strong base resin
Regener. level 5.7 lbs/ft^3 NaOH
Chemical cost: \$ 0.62/1000 gallons.

The lowest chemical cost is obtained by use of dual layer cation, decarbonator and strong base type I (Cation Alt. a) and Anion Alt. c).

Based upon stated chemical consumption and 24 hours operation/day, this solution together with the use of the decarbonator, will reduce chemical costs by \$ 46,000.00/year in comparison with chemical consumption of equipment offered in base solution.

Cost of selected plant is \$ 256,000 while the base solution plant is valued at \$ 239,000.00.-

The difference in cost between the two plants is recuperated in less than six months due to lower chemical consumption.

Another interesting example is to compare the results from the demineralization plant with the co-countercurrent regeneration (CCCR**), with the plants which have the conventional downflow regeneration. Although the data focus on a cation exchanger (effect on Na leakage), the same philosophy applies to anion exchangers (effect on SiO_2 leakage).

The co-countercurrent regeneration of the cation exchanger lowers both the Na leakage and acid consumption.

** Cochrane-Termomeccanica Process

The benefits of lower Na leakage are not limited to lower anion effluent conductivity; the level of the silica leakage from the anion exchangers is also effected by the amount of cation Na leakage (lower Na leakage = lower SiO2 leakage).

This system can provide substantial reduction in operating chemical costs and some reduction in capital equipment expenditures when compared against co-current (downflow) regeneration on waters with the following characteristics:-

1. High sodium content (50% to 80%) water
2. High dissolved solids water (400-800 TDS as $CaCO_3$)
3. Low alkalinity water
4. Combination of high sodium and high dissolved solids

In these applications the CCCR system may be the most cost effective means of producing quality effluent (less than 2, and in some cases even less than 1 ppm TDS) from a 2-bed system.

The co-counter current regeneration provides simultaneous chemical introduction to the exhausted resin bed from the bottom and the top of the bed. The regeneration is collected at an interface collector located so that one third of the bed is above the collector.

Regenerant dilution control equipment is pre-set to deliver one third of the total amount of regenerant to the top of the bed at a lower concentration than that applied to the bottom of the bed. This procedure avoids calcium sulfate precipitation in the upper regions of the bed.

The application of regenerant from both the top and bottom of the bed helps maintain the bed in a non-fluidized condition necessary for proper regeneration and corresponding low sodium leakage during service.

Because the majority of particulate matter is removed in the upper portion of the bed, a mini-backwash is utilized to clean the upper portion of the bed without disturbing the lower bed.

A full bed backwash is utilized every 10-20 regeneration cycles. Following this backwash procedure a regeneration must include 2-3 times normal acid dosage.

Whereas cation leakage in downflow regeneration is normally figured at minimum 1% of total influent cations at maximum chemical dosage, cation leakage from CCCR units may be generally figured at 0.75 - 0.5% of FMA (not as a percent of total cations) and at lower chemical dosages.

The following two examples may be considered fairly significant:

1. Water analysis -

Ca	75	HCO_3	100
Mg	25	SO_4	100
Na	300	Cl	200
	400		400
		SiO_2	12

Project data:

- flow : 45 m^3/h - 8 hour run
- Demineralized water : TDS 4 ppm as $CaCO_3$
  SiO_2 0.1 ppm as SiO_2

Downflow regeneration -

152 ft^3 of resin, regenerative level 5 lbs/ft^3 of H_2SO_4
leakage 4 ppm
Chemical cost for regeneration: $ 0.49/1000 galls.
Total cost fo plant: $ 361,000.00.-

CCCR -

123 ft^3 of resin regeneration level 5 lbs/ft^3 of H_2SO_4
leakage 2 ppm

Chemical cost for regeneration: \$ 0.40/1000 gallons
Total cost of the plant : \$ 347,000.00.-

2\. Water analysis -

Ca	175	HCO_3	275
Mg	100	SO_4	100
Na	275	Cl	175
	550		550
		SiO_2	12

Project data:

- flow : 45 m^3/h - 8 hour run

- demineralized water: TDS 6 ppm as $CaCO_3$
  SiO_2 0,1 ppm as SiO_2

Downflow regeneration -

184 ft^3 of resin, regeneration level 8 lbs/ft^3 of H_2SO_4
leakage 5-6 ppm
Chemical cost for regeneration: \$ 0.95/1000 gallons
Total cost of the plant: \$ 411,000.00.-

CCCR -

181 ft^3 regeneration level 5 lbs/ft^3 of H_2SO_4
leakage 2 ppm
Chemical cost for regeneration: \$ 0.58/1000 gallons
Total cost of plant: \$ 409,000.00.-

In both cases, the CCCR system offers a slight cost reduction of the plant and a considerable reduction in the chemical consumption, therefore, not only does it offer economical benefits but also a decrease in the pollution.

What do all these figures mean?

Simple, the design of the plants is under constant and continuous progress and that the designer-manufacturer has his own part to play.

Condensate Polishing

The condensate water recoveries have acquired great importance in energy saving in the past recent years and will continue to do so in the years to come.

There is a trend to recuperate not only the condensate from the utility power plants, but also the industrial process condensate, which creates various problems as it is more polluted and because it would often be convenient to treat it at higher temperatures.

The use of ion exchange resins is fundamental in this case too, either alone or together with other well known processes (carbon filters, pre-coat filters) or newer ones (magnetic filters).

The main purpose is that of always obtaining purer waters (leakage below 1 ppb are often required in the high pressure fossil or nuclear boilers) with limited chemical consumptions and reduced waste volume, especially in the nucelear power plants.

As an example, I would like to illustrate the advantages which, from this point of view, may be attained by using mixed beds with the AMMONEX process**.

What is AMMONEX?

Ammonex is basically a regeneration tecnique which adopts the use of condensate polishing deep mixed beds in the ammonia cycle.

** Cochrane-Termomeccanica Process

The Ammonex process is applicable to condensate polishing system in power plants where the operating pH is 9.0 or above.

This procedure permits lower sodium leakages by removal of sodium contamination from:-

. Cation resin fines in the anion resin section.

. Cation resin inadvertantly carried over during a poor resin separation.

. Cation resin fines in the inert layer, if present.

. Surplus sodium from the anion resin itself (due to the soaking from the caustic soda regenerant).

The following regeneration sequence summarizes the procedure and the salient features:

a) Scrubbing: The resins are air scrubbed and backwashed to remove crud.

b) Separation: The resins are hydraulically separated by a final backwash.

c) Anion Resin Transfer: The anion resin is removed by hydraulically transferring it to the anion tank. The cation resin layer is left behind on the cation/separation tank. In doing so, traces of cation resin are inadvertently carried over with the anion resin.

d) Cation Regeneration: Sulphuric (or hydrochloric) acid is used to regenerate the cation resin.

e) Anion Regeneration: Sodium hydroxide is used to regenerate the anion resin. During this caustic regeneration phase, the cationic resin fines that were carried over are converted to the sodium form.

f) Normal Anion Rinse: This rinse removes predominant amount of caustic soda from the anion bed, but sodium still remains on the cation fines and on the anion resin itself.

g) Ammonia Rinse of the Anion Resin: (The Ammonex Procedure.
Dilute 0.3 - 0.5% ammonium hydroxide is applied to the anion bed to effect the following two main items:

1. Exchange ammonium for the sodium on the cation fines.
2. Purge the sodium hydroxide residuals from the anion resin itself. The purge is necessary if low ppb sodium is to be expected during the service cycle.

h) Final Displacement and Rinse from the Anion Resins

i) Cation and Anion Resins Remix for Service

In this way the sodium leakage is lower in the ammonium mixed beds than in the normal H cycle when the sodium has not been scavanged by the ammonia, as carried out in the above mentioned procedure.

Moreover, the operation run is much longer, as it continues even when the ammonia in the condensate has exhausted the resins.

In the utility power plants, the high pressure fossil and nuclear boilers require a treatment system for the entire condensate flow (and also a good make-up demineralization plant) in order to protect the boilers and the turbine.

The polishing plant with only the mixed bed still offers the most practical system with low equipment and operating costs.

With properly selected resin size and ratio, a possible introduction of a layer of inert resin and a good ammoniation procedure, will give long extended runs with lower leakages.

In conclusion, on having examined these examples, we can say that the design-manufacturing is to face two apparently contradictory requirements; the need to have always better quality product water and always lower operating and plant costs.

"DO MORE WITH LESS" could be our philosophical motto for the future.

******

WATER SOFTENING FOR FOOD PROCESSING

Albano Malaguti

Culligan Italiana S.p.A.

The aim of this brief presentation is to emphasize the important role that soft water-or any water treated by ion-exchange-plays in the field of food processing and in many other fields in which soft water can be employed.

Furthermore, it is meant to stress the fact that this important issue has not received adequate attention by the EEC law regulation No. 80/778 concerning the characteristics and requirements for drinking water. The EEC regulation which set limits to the softening of water for drinking purposes, are based on assumptions and supported by arguments that are both contradictory and erroneous.

Water softening--or other ion-exchange water conditioning techniques--in the field of food processing are in practice used in the following applications:

1) Water for food industry, pharmaceutical chemistry and industrial chemistry.

   Processes:
   - Boiling and steam cooking of food stuffs to be processed and canned
   - Diluting syrups for soft drinks
   - Beer brewing
   - Distillation/Alcoholic liquors processing
   - Dairy farming and the cheese industry
   - Pharmaceutical and cosmetic industry (these industries cannot be included in the field of food processing. Yet, whether directly or not, their products are made for human consumption).

2) Household Economy

   Most foods, such as vegetables and legumes, improve in taste and flavour if cooked in soft water.

Also kitchen utensils are more easily cleaned when food has been cooked in soft water because no built-up scale will be found in pots and pans and so more hygienic conditions will be achieved.
Despite these facts, the EEC water policy in terms of water conditioning, by limiting the practice of water softening does limit the benefits resulting from the use of soft water; on the other hand no attempts have been made to solve the problem represented by the very poor quality of water distributed by most city mains.
The two EEC arguments against water softening-which heavily jeopardize such treatment--are the following:
a) Soft water has a higher sodium concentration in comparison with the same amount of raw water and sodium is bad for the health of human beings.
b) Unlike sodium, calcium is not bad for the human body.
As opposed to this, we would point out that:
- The EEC regulations do not provide any information with regard to the minimum quantity of calcium that must be present in water, so that, it can be inferred that waters without any content of calcium are perfectly acceptable providing that the absence of calcium is not due to conditioning processes but is a natural characteristic of the water.
- The upper limit of sodium present in soft water obtained with ion-exchange resin treatment can be maintained in strict accordance with desired allowance values set by the EEC; moreover some conditioning processes can even reduce or, at least, not alter the concentration of sodium that was originally present in the raw water source.
- It is possible to obtain water having good quality standards--as good as those of the best mineral waters--by treating (softening or other water conditioning) water even if it has the worst possible chemical properties.
- There are other elements (such as cadmium, zinc, ecc.) that can be held responsible for cardiovascular diseases for which too often calcium is blamed; these elements may cause serious problems and are sometimes overlooked and underrated.
- Statistical researches on sodium and calcium have never been carried out on softened water (that is water originally hard, which has been conditioned by ion-exchange) but rather on naturally soft water, which is chemically and physiologically important.

  Regulations such as those reported here do not exist in the U.S.A. where, in a more reasonable way, since current numerous studies on sodium are not considered exhaustive, the federal authorities, while awaiting the results of a forthcoming monitoring campaign, have temporarily set the value limits of the sodium content in aqueous solutions in conformity with the European standard values.

According to the U.S. federal recommendations, the sodium quantity ingested in drinking should not exceed 5 - 10% maximum of the total sodium consumption in a day's diet.
This sounds very reasonable considering that a healthy person normally consumes 10 to 12 g. of sodium chloride per day, corresponding to 3900 up to 4700 mg. of sodium element; if this person drinks 1.5 litres of water, the hardness of which was originally 30°fr (which corresponds to the usual Italian conditions), this individual will ingest 210 mg. of sodium equal to 4-5% of a total daily intake of sodium. This shows the unimportant role that sodium, resulting from the softening of water, plays in the total daily intake of sodium per person.
As far as ion-exchange and food processing are concerned all domestic and community laws recommend the use of potable water, yet in practice such laws are imperfect and do not lead to the observation of the requirements for safe drinking water since in the law texts "potable water" is a definition that is only referred to in connection with microbiological and toxic parameters.
Derogations such as these show the contradictory ambivalence of such regulations; on one hand food industries (namely soft drinks divisions) have their work facilitated, on the other it should be noted how incongruent these laws and regulations are, above all the already mentioned EEC regulations, with particular reference to art. 2, 3, 7, and 17, a summary of which is given below:

Art. 2 According to the present regulations "waters suitable for human consumption" are any natural or treated waters, whatever their origin might be, namely

- either tap water for human consumption,
- or waters,
- for use in the food industry for the manufacturing, processing, preserving, or marketing of food or substances suitable for human consumption
- that may affect the wholesomeness of the final products.

Art. 3 As for the waters, that are described in art. 2 point 2, the State members apply the values of the toxic and microbiological parameters, as listed in table D and E, respectively, of schedule I, in addition to the values of other domestic parameters--supplied by the competent national authorities--that may prejudice the wholesomeness of the final food products.

Art. 7 1) With regard to the parameters listed in schedule I, the State members thereby indicate the values applicable to waters suitable for human consumption.

2) Regarding the parameters for which no values are indicated in schedule I, the State members, if they so wish, are entitled to postpone their decision to set values for the application of the first paragraph, until these have been fixed by the Council.

3) With regard to the parameters in table A,B,C, D, and E of schedule I:
   - the values to be fixed by the State members should be lower than or equal to those indicated in the column "Maximum Allowed Concentration"
4) With regard to the parameters in table F, schedule I, the values that will be fixed by the State members must be higher than or equal to those indicated in the Column "Minimum Required Concentration" related to waters in art. 2 point 1, that have undergone softening treatment.
5) The values that are listed in schedule I, must be interpreted taking the "Observations" into account.
6) The State members will take the necessary steps to ensure that the waters destined for human consumption conform at least to the requirements specified in schedule I.

Art.17 The State members may adopt particular regulations for wording on packages labels and advertising slogans in terms of the the suitability of waters for infants.Such provisions of the law may also concern the properties of the water for use as advertised.
The State members that intend to adopt such measures must inform the other State members and the Commission, in advance.
The State members may forbid, or subordinate to particular requirements,the mentioning on packages, labels and also advertisements, of any specific reference to the appropriate characteristics of water intended for the preparation of infants' feeding.

From the articles reported above, one could deduce that:

1) Water used in soft drinks may be "non-potable", at least in relation to the concept of potability accepted for mains water.
2) Since limitations set for water softening are clearly indicated with reference ONLY to mains water--namely tap water--, and in consideration of the fact that the EEC laws regulating this matter define as "edible food stuff" products and substances that are suitable for human consumption, including drinks, and finally, since art. 17 clearly states that the bottling of "any" kind of water is allowed providing that said water is safe from a microbiological point of view and suitable for drinking as a consequence of all this, it can be inferred that properly conditioned tap water--distributed by water mains--is to be considered "non potable" if it is drunk directly from a tap, while the very same water becomes potable after bottling and labelling.

   I want to conclude my presentation with the affirmation that soft water obtained by ion-exchange resin treatment is specific

water impossible to replace in food processing.
This strong belief comes from the experience gained in connection with thousands of working softeners installed by us in Italy and the hundreds of thousands installed in the rest of the world.

ION EXCHANGE IN THE FOOD INDUSTRY

Lucio Galeazzi

Ausind, Montedison Group

ABSTRACT

The paper briefly reviews main applications of ion exchange resins in food processing and the legislation in some EEC countries and in USA.
The most common exchangers and processes are mentioned.
National legislations concerning ion exchangers in food processing in Italy, France, Germany and USA are examined as well as exchanger grades and related application fields.
National authorizations requested and methods governing the use of exchangers are listed.
Main national laws and regulations are mentioned.

INTRODUCTION

Ion exchange technology, which is widely used in many industrial processes, and in the food industry as well, involves the utilization of a broad range of modern synthetic ion exchangers.
The food industry requires ion exchangers that
- simplify production processes and reduce production costs
- do not deteriorate foods from the hygiene and health stand-points
- do not alter food basic equilibria

Consequently, suitable ion exchangers are needed to meet the specific requirements of the food industry.
Ion exchangers must not contaminate processed foods, and must be used in compliance with correct manufacturing practice norms so as

to prevent any food alteration.
Therefore, these principles must be applied in the light of duly studied and enforced regulations.
The following is a summary of the main ion exchanger applications in the food industry and of the main legislations governing their utilization.

APPLICATIONS

The description of the main applications worldwide follows hereby irrespective of, more or less limiting legislations enforced in some countries.

Potable water

Water softening by means of ion exchangers is a widely used process in industry. It has also been adopted as a part of central potable-water treatment systems or domestic systems for many years in a large number of countries.
As to potable water, a topical problem is that of nitrates. Many types of water that are potentially suitable for use in the food industry, have an excessively high nitrate content.
The reduction of nitrates to nitrites may involve the danger that blood hemoglobin be converted into metahemoglobin which cannot transfer oxygen to organic tissues. Moreover, nitrites may react with many amines contained in foods, thus forming nitrosamines, some of which are cancerogenic.
For these reasons, some legislations indicate very strict limits of admitted concentrations of nitrates in potable water.
Ion exchange removes nitrates from water, and gives an effective contribution to the solution of the problem. Ion exchange is also used for the adsorption of organic substances and for the removal of heavy metals from water.

Water used in the food industry

Water may be used as a basic ingredient in food in high percentages (e.g., beer, soft drinks) or as washing water for meat, vegetables, containers, etc. In all these cases, water must have optimal characteristics for the specific application.
In general, water used in the food industry should be neither scale forming nor corrosive and, at the same time, have all the

necessary potability and chemical-physical organoleptic properties. For instance, a correct degree of hardness, alkalinity and salinity of water is essential in the manufacturing process of beer and other drinks.
An optimal calcium concentration is required in the water used to treat vegetables before canning.
The use of softened and deionized water to wash bottles and drink containers, prevents scale problem.
Softening processes, decarbonation of deionization by means of ion exchangers make it possible to obtain water suitable for various applications.
Ion exchange is also used in the production processes of a number of foods or products utilized in the food industry making it possible to

- simplify production processes
- obtain high quality products
- improve yields

as illustrated below.

Sucrose

Ion exchange has been a well established technology in sugar production for already some decades with major qualitative and technical-economic advantages.
Sucrose or common sugar is the pure organic compound manufactured in the largest quantity. It is obtained by extraction and purification both from beets and canes.
Ion exchange is used in various stages of the sucrose production process. Table 1 lists the most widely used processes.
Through these processes high-purity crystalline or liquid sugar is obtained.

Glucose and sugar syrups from hydrolyzed starch compounds

Starch, mainly obtained from maize, is submitted to acid or enzymic hydrolysis.
Obtained syrups undergo a purification process including deionization by means of ion exchangers.
Glucose as crystals and syrup is obtained from the process.
In recent years, the use of high-fructose content sugar syrups has spread, especially in the USA.
Because of their sweetening power, these syrups are increasingly used instead of sucrose in beverages and in the food industry in

general.
High-fructose content syrups are obtained by means of glucose enzymic isomerization.

Wine

In wine production, the ion exchange process replaces other techniques with advantage, and makes it possible to obtain high-quality products.
The most widely used processes include :
- control of K content in wine so as to prevent K acid tartrate precipitation, thus avoiding the long and costly wine refrigeration which is normally used to this end
- control of Fe and Cu content so as to prevent wine clouding. This eliminates the need for potassium ferrocyanide
- correction of must and wine acidity as an alternative to the addition of organic acids or inorganic bases and salts that may give wine an unpleasant taste
- production of concentrated deionized musts with a very high sugar and very low mineral salt content.

These musts are used, for example, in the preparation of special wines.

Milk and related product industry

An increasing use of ion exchange is expected in the milk and related product industry, especially for the recovery, purification and utilization of by-products.
Whey, the watery part of milk discarded during cheese and casein production, contains lactose and proteins as well as mineral substances. Considerable quantities of whey are left over following cheese production. Whey is mainly used as feed.
After treatment with membrane techniques and deionization by means of ion exchangers, whey may be dried, thus originating a high lactose content powder containing proteins, that may be used as an ingredient in foods for children, powdered milk and other foods.
By means of acid or enzymic hydrolysis, lactose may be transformed into glucose and galactose, and used as sweetener.
Proteins may be industrially separated from whey and fractioned by means of special ion exchangers.

Other uses

Ion exchangers are also used in extraction and purification processes of many other products for the food industry such as :
- amino acids from sugar juices or effluents from slaughtering
- organic acids or their salts : tartaric, lactic and citric acids and sodium glutamate
- natural sweeteners and sweetening derivatives : xylithol, sorbitol, mannite
- anthocyanines from vegetables to be used as natural dyes.

LEGISLATION

The legislation concerning the use of ion exchangers in the food industry differs in different countries. In fact, not all countries envisage the use of the same types of exchangers, the same applications, the same controls.
In order to simplify matters, we are going to take into consideration legislations in the USA and in three major EEC countries (Italy, Federal Republic of Germany and France).
Even in the EEC, national legislations are quite different.
In general, the fundamental principles national legislations are based on, include :
- health protection
- preservation of foods quality
- fraud repression

Ion exchange resins are generally considered as technological auxiliary substances. In other words, ion exchange resins are not considered as food additives (as, for example, preserving agents).
During a specific time interval, resins may be in contact with water used for food processing, with food itself or with one or more ingredients.
During such time, and although to a very limited extent, matter transfer from the exchanger to food may occur.
Hence, health protection requires the use of exchangers with a known composition, and designed so as to avoid any harmful effect on man.
Moreover, the quantity of matter transferred from exchangers to foods, must be rigorously limited.
The safeguard of foods quality implies the definition of precise methods of exchanger utilization.
The following is a comparative analysis of legislation as to :

- types of exchangers
- applications
- authorizations and methods of exchanger utilization

Types of exchangers

A list is provided of the exchangers specifically mentioned by national legislations, irrespective of the applications they are designed for.
Table 2 lists a small but significant number of exchangers that are indicated by almost all legislations. These exchangers are among the most widely known and used also in fields other than the food industry.
Table 3 summarize the chemical structure of the resins mentioned in Table 2.
European legislations do not explicitly mention other types of exchangers. The only exception is the West German legislation which also admits the use of :
- weakly acid resins with carboxylic exchange groups consisting of phenoxyacetic acid and formaldehyde condensation products
- anion resins resulting from epichlorhydrin and aliphatic amines condensation products

In addition to resins listed in Table 2, the USA legislation also admits the use of many other exchangers.
In order to present a comprehensive picture, Table 4 provides the full list of resins admitted by the USA legislation.
They are subdivided by type of functional group or polymeric matrix.
Some of the approved exchangers are subject to application limits.
In the frame of "Codex Alimentarius" - a joint WHO-FAO programme in which over 90 countries partecipate - the approval of several types of exchangers was recently proposed following the guidelines of the USA legislation.

Applications

USA. The USA legislation admits the use of approved resins for the treatment of potable water and of food in general (1).
Moreover, it provides specific regulations for the treatment of wines (4).
Federal Republic of Germany. In Germany the Government may (9) enforce regulations concerning production, marketing and purity of ion exchangers to be used in the food industry.
Moreover, the Government has the power to prohibit or limit the

use of certain exchangers in the production of foods.

Until now, the Government has never used these powers. As to ion exchangers, the Kunstoffe-Kommission of the Federal Health Bureau issued a recommendation (2).

This recommendation, which is currently being updated, lists some regulations concerning the utilization of ion exchangers in the treatment of potable water to be used in the production of foods and not in central distribution systems.

France. The French legislation envisages the use of cation resins for the treatment of potable water as well as of styrene type strong anion resins to remove colour from sugar solutions (3).

Applications in other fields and with other resin grades are allowed for upon authorization of the Ministry of Agricolture in compliance with the circular letter dated August 8, 1980, (7) concerning additives and technological auxiliary substances used in food production.

Italy. The legislation admits the use of sulphonic resins in the treatment of base wines for aromatic wines (5) and the use of sulphonic and anionic resins in the process of concentrated and rectified grapes must (6).

EEC. The EEC regulation Nr. 2310/80, dated September 3, 1980 (8) envisages the use of styrene sulphonic and anion exchange resins in the process of concentrated and rectified grapes must (for the 1980-81 and 1981-82 viticultural seasons).

Consequently, this regulation is in force in all EEC countries. Table 5 summarizes applications envisaged by the various legislations.

Authorizations and methods governing the use of exchangers

USA : before being used (1), resins must be preconditioned according to manufacturer's instructions.

The organic extract must not exceed 1 ppm in the extracting solution when performing extraction tests after preconditioning, at the highest process temperature.

These tests are carried out using distilled water, 15% V/V alcohol in water and 5% m/m acetic acid in water 1)

Resins must be used for the functional applications they are supplied and officially intended for.

---

1) Sulphonic resins are not submitted to the extraction test with acetic acid.

Federal Republic of Germany : the water outflow from the exchanger must have the same permanganate consumption of the water inflow (2). Before being used, ion exchangers must be repeatedly exhausted and regenerated until reaching compliance with this regulation. The exchanger-treated water must contain neither formaldehyde nor amines.

Treated water must be odourless and with no taste.

France : as to the treatment of potable water (3), an authorization of the Health Ministry is required. In order to obtain such an authorization, the resin manufacturer has to submit a documentation specifying :

- resin manufacturing method
- results of extraction and toxicity tests carried out by experts authorized by the Health Ministry.

The central treatment of potable water has to be approved by the Public Health National Board. As to the treatment of water used in the processing cycle of foods or food solutions (7), an authorization of the Ministry of Agriculture -Repression des Fraudes- is required. To this end a documentation is to be submitted specifying :

- exchanger type and characteristics
- utilization field and method; technical and other reasons for the use thereof
- toxicity and transfer test data
- references as to authorization for use in the EEC and other countries.

Italy : as to the treatment of base wines for aromatic wines (5), an authorization of the Ministry of Agriculture is required. Before use, the (sulphonic) resin must undergo two exhaustion and regeneration cycles following resin manufacturer's instructions, and using food reagents in compliance with the Official Pharmacopeia. The ready-for-use resin must not loose more than 1 mg/l in extraction tests performed with distilled water and 15% V/V ethyl alcohol solution in water.

As to the treatment of must (6), in compliance with EEC regulations (8), resins must not loose more than 1 mg/l in the extracttion test with distilled water, 15% ethanol in water and 5% m/m acetic acid in water 1).

---

1) The general regulation envisages and extraction test with an acetic acid solution. However, in consideration of the well-known analysis difficulties involved by sulphonic resins, an improvement of the legislation would be advisable similar to that enforced in relation to base wine for aromatic wines.(5).

The comparative analysis of the legislations mentioned herein outlines the substantial differences as to types and number of admitted exchangers, application fields and procedures required to obtain the authorization for use by competent authorities.
We conclude this review of the legislation concerning ion exchange in the food industry by expressing the wish that

- a gradual integration and harmonization of the various national legislations implying an effective action of the competent international organizations such as the EEC and Codex Alimentarius
- the extensive knowledge acquired on a large number of exchangers as to their effect on man (many exchangers used by the food industry also find an application in the pharmaceutical industry)
- the continuous improvement of exchanger production processes and preservice regulations (cutting the transfer of material to foods to very low values)
- exchanger utilization methods by users according to sound industrial practice
- the development of applications and the improvement of exchanger application technology

provide an increasingly valuable contribution to the food industry in view of obtaining high-quality foods by means of simplified and less costly production processes while ensuring the best safeguard of human health and food basic characteristics.

Acknowledgements

We wish to thank Mr. Tiziano Garlanda and Mr. Lucio Ciccardini of Montedison SpA for their advice and co-operation as to legislation on ion exchange.

REFERENCE

(1) Code of Federal Regulation Title 21 Food and Drugs § 173.25 (revised as of April 1. 1980)
(2) 36. Mitteilung : Bundesgesundheitsblatt 10.120 (1967)
(3) Circulaire du 3.5.63 et 1.10.63; arrêté du 12.1.80
(4) Code of Federal Regulations 27 Alcohol Tobacco Products and Firearms § 240.1051 (revised as of April 1.1980)
(5) Italian Ministerial Decree as of March 22.1978
(6) Italian Ministerial Decree as of November 28.1980
(7) Circulaire du 8.8.1980 Journal Officel du 25.9.80

(8) EEC Regulation Nr. 2310/80 as of September 3.1980
(9) Lebensmittel und Bedarfsgegenständegestz 20 August 1974

Table 1
Typical Applications of Ion Exchange in Food Processing

Application Field	Type of Exchanger	Ions or Substances to remove from the solution	Process
Potable water	SA	hardness	softening
	WA	bicarbonate hardness	decarbonation
and	SB	bicarbonates	neutral exchange
	SB	nitrates	neutral exchange
water for the	SB	organic substances	adsorption
food industry	SA/WB	salts	deionization
Sucrose	SA	hardness	softening
	SA/WB	non-sugars	deionization-decolourization
	SB	coloured-organic substances	decolourization
	SA	K + Na	Quentin
	SA	hardness (K + Na)	Grillus
	SA	-	inversion
	SB	anions	alkalization
	SA	-	ion exclusion
	SB/WA	salts	deionization-decolourization
Glucose	SA/WB	salts	deionization
Fructose Syrups	SA/WB	salts	deionization
	SA	-	chromatographic separation
Date Juice	SA/WB	salts-non sugars	deionization-decolourization
Wine and Must	SA	K	neutral exchange
	SA	Fe and Cu	neutral exchange
	SA	hardness	softening
	SA	cations	decationization
	WB	acids	deacidification
	SA/WB	salts	deionization
	WB	SO_2	desulphitization
Milk	SA	hardness	softening
Whey	SA/WB	salts	deionization
Lactose	SA	-	hydrolysis
	SA/WB	salts	deionization

SA = strong acid cation exchanger - WA = weak acid cation exchanger
WB = weak basic anion exchanger - SB = strong basic anion exchanger

Table 2
Main Grades of Ion-Exchange Resins Included in Legislation Specifications 1)

Resin Grade	Italy (6)	France (3) and (8)	Germany F.R. (2) and (9)	USA (1)
- S-DVB 2) sulphonated copolymer	yes	yes 3)	yes	yes
- Resin with acrylic ac.-DVB or methacrylic ac.-DVB copolymer structure	no	yes 5)	yes 4)	yes
- S-DVB copolymer chloromethylated and aminated with TMA 6)	yes 9)	yes	yes 8)	yes 7)
- S-DVB copolymer chloromethylated and aminated with DMA 10)	yes 9)	yes 9)	yes 8)	yes 11)

1) When the structure of the resin included in the table does not fully correspond to the definition in the legislation the definition used in the legislation is mentioned in a special note
2) S = styrene, DVB = divinylbenzene
3) "Cation exchange resins with sulphonic groups"
4) "Copolymers of acrylic acid, or derivatives thereof and DVB, with carboxylic groups
5) "Cation exchange resins with carboxylic groups"
6) TMA = trimethylamine
7) "Cross-linked polystyrene chloromethylated and aminated with TMA"
8) "S-DVB copolymers with active groups containing aliphatic nitrogen basic groups"
9) "S-DVB copolymers containing ammonium (8) groups"
10) DMA = dimethylamine
11) "Cross-linked polystyrene chloromethylated and aminated with DMA"

Table 3
Chemical Structure of Resin Grades mentioned in Table 2

- Sulphonated Styrene-Divinylbenzene Copolymer

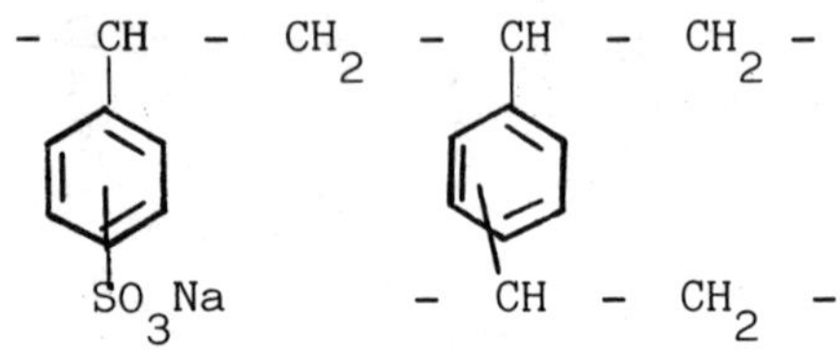

- Acrylic grade carboxylic resins

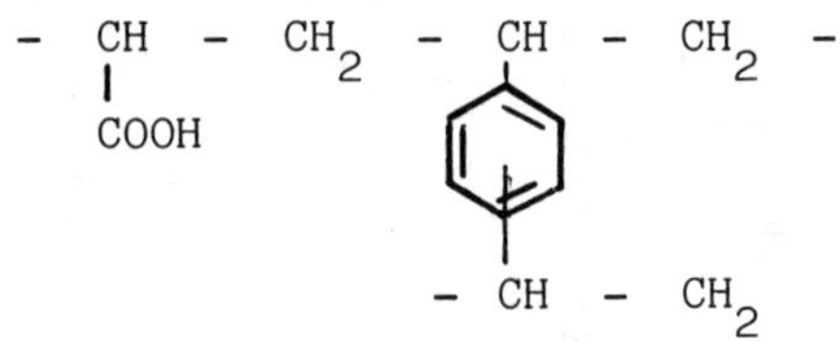

- Methacrylic grade carboxylic resins

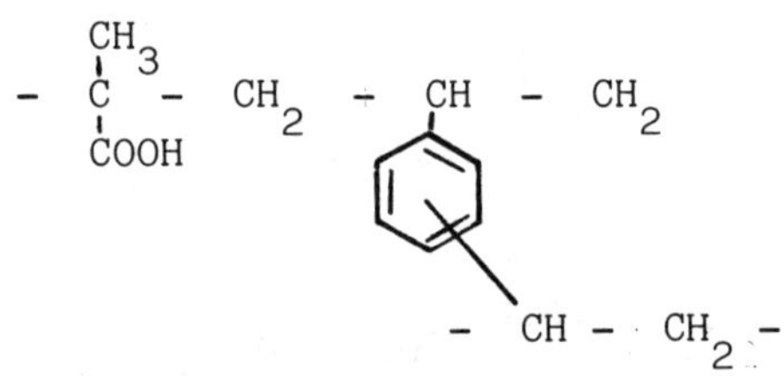

- Styrene-Divinylbenzene copolymer chloromethylated and aminated with trimethylamine

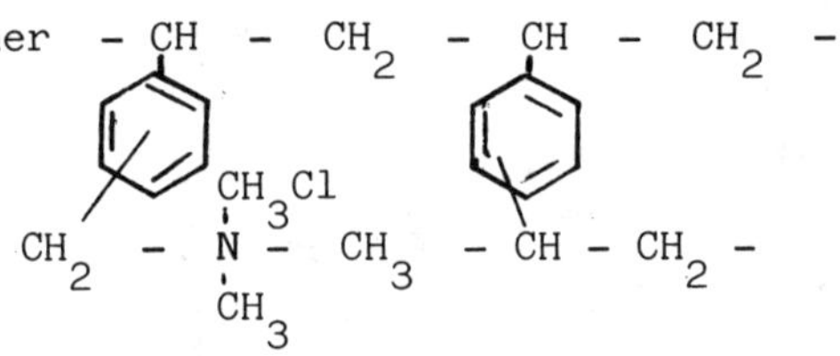

- Styrene-Divinylbenzene copolymer chloromethylated and aminated with dimethylamine

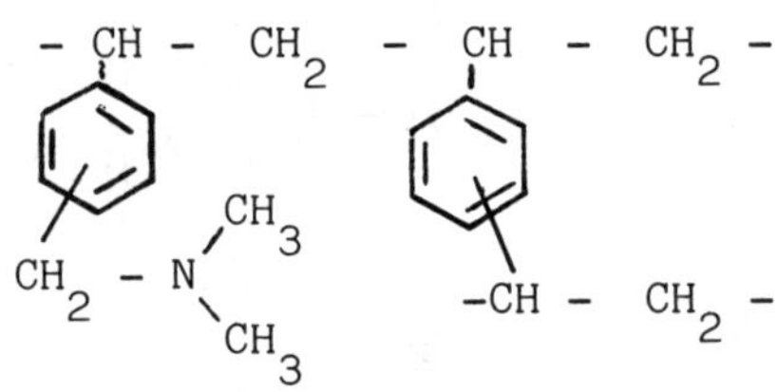

Table 4
Ion-exchange resins for use in the food industry according to the USA legislation (1)

- Cation resins with sulphonic groups
  . Sulphonated S-DVB 1) copolymer
  . Sulphonated anthracite coal 2)
  . Sulfite-modified cross-linked phenol-formaldehyde, with modification resulting in sulfonic acid groups on the side chains
  . Sulphonated S-DVB-AN 3) or AM 4) terpolymers
  . Sulphonated S-DVB-AN-AM 3) 4) tetrapolymer. The tetrapolymer is to be derived from a monomer blend with 2% maximum content of AM and AN
- Cation resins with carboxylic groups
  . Methacrylic acid-DVB copolymers
  . Completely hydrolyzed AM-DVB copolymers
  . Completely hydrolyzed AM-DVB-AN terpolymers
- Styrene grade anion resins with groups with different basic strength
  . Cross-linked polystyrene first chloromethylated then aminated with TMA and DMA or DETA or DMEA 5)
  . S-DVB copolymer first chloromethylated then aminated with DMA, and oxidized with H_2O_2 6)
- Anion resins with groups with different basic strength derived from polycondensations
  . DETA, TETA 7) or TEPA 7) cross-linked with epichlorohydrin
  . Phenol-formaldehyde cross-linked and actived with one or both of TETA and/or TEPA
  . Formaldehyde, acetone and TEPA reaction resin
  . Epichlorohydrin cross-linked with ammonia
- Acrylic grade anion resins with groups with different basic strength
  . AM-DVB copolymer containing not less than 2% by weight of DVB aminolyzed with DMAPA 8) 9)
  . AM-DVB copolymer containing not less than 3.5% by weight of DVB aminolyzed with DMPA 9)
  . AM-DVB-DEGDVE 10) terpolymer containing not less than 3.5% by weight of DVB and no more than 0.6% by weight of DEGDVE, aminolyzed with DMAPA 9)
  . AM-DVB-DEGDVE terpolymer containing not less than 7% by weight of DVB and not more than 2.3% by weight of DEGDVE, aminolyzed with DMAPA and quaternarized with CH_3Cl 9)

Table 4

1) S = styrene, DVB = divinylbenzene
2) The product must comply with ASTM D 388-38 Class 1 Group 2
3) AN = acrylonitrile
4) AM = methylacrylate
5) TMA = trimethylamine, DMA = dimethylamine,
   DETA = diethylene-triamine, DMEA = dimethylethanolamine
6) As to applications, the resin is subject to limitations and must undergo control tests different from those relating to other products
7) TETA = triethylenetetramine, TEPA = tetraethylenepentamine
8) DMAPA = dimethylaminopropylamine
9) As to applications, the resin is subject to limitations
10) DEGDVE = diethyleneglycoldivinylether

Table 5
Application fields of ion exchange resins envisaged by legislation

	Italy (5) and (6)	France (3) and (7)	Germany F.R. (2) and (9)	EEC (8)	USA (1)
- Water used in food production processes	NR	SA	GA	NR	GA
- Sugar juices	NR	SA	GA	NR	GA
- Wines	SA	SA	GA	NR	GA
- Musts	GA	GA	GA	GA	GA
- Other sectors	NR	SA	GA	NR	GA

GA = General authorization
SA = Special authorization
NR = No regulation

ION EXCHANGE FOR THE RECOVERY OF CONCENTRATED AMMONIUM SULPHATE FROM THE PROCESS CONDENSATE OF AN AMMONIA-UREA FACTORY

G. Campolo, F. Ciallié – ITALBA S.p.A. Milano
G. Vergerio – ANIC S.p.A. Milano

- ABSTRACT

Many efforts are made nowadays to limit the concentration of nitrogen compounds both in the waste waters and in the waters reclaimed for human consumption. Several process are under advanced study but still not widely applied.
The industries producing nitrogen compounds (Ammonia, Fertilizers, Explosives, Caprolactam etc.) have been formely burdened with this problem. Therefore it is just in the industrial field that the most significant appli ances have been accomplished.
This report outlines the main features and the performances of an ion exchange plant installed in an Ammonia-Urea factory and designed to achieve three main targets:
- removal of ammonia from the process condensate and its recovery in the existing ammonium sulphate production line
- relief of the nitrogen load entering the oxidation section of the existing waste water biological treatment plant (W.W.B.T.P.)
- exploitation of the methanol present in the condensate (catalitic reforming by product) as a valuable carbonaceous substrate indispensable for the final denitrification section of the W.W.B.T.P.

1 - INTRODUCTION

In the process condensate coming out from the production cycles of Ammonia-Urea factories can be found large amounts of NH_3 and CO_2 associated with important quantities of CH_3OH as reaction by-product.

A typical composition is given in Table 1.
The most commonly applied treatment consists of a steam stripping tower which removes the most part of the pollutants.
A further treatment with ion exchange resins is then required in the case of recovery of the condensate as boiler make-up.
This system takes its main advantage from recycling the relatively good quality condensate into the boiler cycle, nevertheless it shows the following limits:
- the stripping requires large quantities of steam: the order of magnitude is ranging around 70 kg of steam per m3 of recovered condensate.
Therefore this system shows an economical significance particularly when the cost of raw water demineralization is high because of high TDS.
- a certain amount of methanol and some organics still remains in the condensate after stripping. The heavier organics can represent a severe threat to the anionic resins while the methanol content is not affected by the ion exchange. Therefore the recovery into the boiler will be allowed only when the dilution ratio between the treated condensate and the freshly demineralized water is such to maintain the methanol concentration within the range acceptable for the boiler.
Attempts have been made to remove organics and methanol by means of activated carbon periodically regenerated with steam.
Nevertheless such system drastically increases both installation and operation cost giving as result the complete removal of the heaviest organics but poor efficiencies as regards the methanol (see table 1).
- the ammonia released in the stripper together with carbon dioxide and steam when becomes in contact with the fresh atmosphere. tends to form of ammonium bicarbonate which easily drop down to the earth.
By this way a consistent portion of nitrogen-salt will be conveyed through the sewer system to the final W.W.B.T.P. to which also the eluates of the ion exchange resins (containing the remaining portion of ammonia) will be fed.
A different way of facing the problem has been recently adopted in an Ammonia-Urea factory of ANIC S.p.A. (ENI group) where the problem was focused in the following terms:
- the high pressure boiler cannot accept but small amounts of organics
- the existing W.W.B.T.P. (two stages activated sludge with total nitrification and final anaerobic denitrification) is already

operating with a large excess of nitrogen in the oxidation section.
- the anaerobic denitrification requires consistent quantities of methanol to perform the reduction of the nitrates
- in the factory it does already exist a crystallizer for the product of the ammonium sulphate (obtained as by product of the caprolactam line).

On the basis of the above consideration the technical department of ANIC S.p.A. evaluated the feasibility and then assigned the construction of a new installation for the treatment of the process condensate with the purpose to:
- remove the ammonia down to the limits imposed by the public authority
- obtain an ammonium sulphate solution at the maximum allowable concentration to reduce to minimum the operating cost of crystallization.
- keep the sulphuric acid consumption as close as possible to the stoichiometric requirements (the ammonium sulphate is a poor product)
- preserve the methanol in the treated stream so to use it as carbonaceous substrate in the denitrification stage.

2 - FEASIBILITY STUDIES

Since years sulphonic type ion exchange resins are available suitable to fix ammonium ions and to release them during regeneration with highly concentrated sulphuric acid (up to 30% by weight).

The ammonium sulphate concentration in the eluates depends from several factors like void volume in the coloumn, displacement and rinse volume required etc. Concentration may be further increased applying the fractional regeneration systems but the limiting factor is the exchange capacity of the sulphonic resin itself ranging around 1 - 1,2 eq/l.

As in the present instance the most part of the ammonia was bound to carbon dioxide (see table 2) it has been envisaged the opportunity to employ a carboxylic resin ensuring capacities higher than 2 eq/l_R thus allowing higher concentrations.

Feasibility tests were formely performed by the resin manufacturer by using a two coloumns system (carboxylic sulphonic) regenerated in series (sulphonic   carboxylic) with countercurrent system on sulphonic resin and then repeated together with our Company to cover the precise range of feed concentrations and to closely define the operating parameters. The table 3 summarizes the most significant figures obtained during tests.

3 - PROCESS DESCRIPTION

3.1 -Plant arrangement

The installation operating in full automatic mode includes three ion exchange lines each one consisting of a series of two coloumns (see fig. 1): the first one filled with a macroporous polyacrylic resin (Duolite C 464), the second one filled with a macroporous polystyrene sulphonic resin (Duolite C 26).

An atmospheric degasser removes the carbon dioxide from the treated condensate before conveying it to the W.W.B.T.P.

3.2 -Exhaustion cycle

The main reactions occuring during the exhaustion cycle are represented by the following equations:

1) $R - COOH + NH_4OH \longrightarrow R - COONH_4 + H_2O$
2) $R - COOH + (NH_4)_2CO_3 \rightleftharpoons R - COONH_4 + NH_4HCO_3$
3) $R - COOH + NH_4HCO_3 \rightleftharpoons R - COONH_4 + CO_2 + H_2O$
4) $R - H + (NH_4)_2CO_3 \rightleftharpoons R - NH_4 + NH_4HCO_3$
5) $R - H + NH_4HCO_3 \rightleftharpoons R - NH_4 + CO_2+H_2O$
6) $R - H + (NH_4)_2SO_4 \rightleftharpoons R - NH_4 + NH_4HSO_4$
7) $R - H + NH_4HSO_4 \rightleftharpoons R - NH_4 + H_2SO_4$

written in the order of de creasing efficiencies.

Conductivity and pH are continuously monitored.

The pH of the treated effluent is the most significant parameter of the resins exhaustion and has been taken as end cycle criterion. Figure 2 shows a typical exhaustion curve and suggests some considerations:

- the series arrangement allows the highest exhaustion of the leading carboxylic column
- the pH evolution curve gives a comprehensive picture of the cycle progress. pH is very sensitive to even minimal leakages of bicarbonates from the strong cationic exchanger thus allowing a timely shut-off of the cycle
- taking into consideration the equations 1) 2) 3) the operating capacity of the weak resin is a function of the free ammonia content because reaction 1) is not subject to an equilibrium being completely shifted to right while reactions 2) and 3) are subject to the relevant equilibrium constants.

On this last subject it is worth to be noted that because of a temporary reduction of the production of the ammonia, the composition of the raw condensate during the start-up was particularly unfavourable as evidenced in the last line of Table 2. Nevertheless the actual performances obtained (last column of Table 3) are more satisfactory than expected from the laboratory tests.

3.3 -Regeneration

The equilibria reactions governing the release of ammonium sulphate are the following:

8) $R - NH_4 + H_2SO_4 \rightleftharpoons R - H + NH_4HSO_4$

9) $R - COONH_4 + NH_4HSO_4 \rightleftharpoons R - COOH + (NH_4)_2SO_4$

9a) $R - COONH_4 + NH_4HSO_4H_2SO_4 \rightleftharpoons R - COOH + 2NH_4HSO_4$

10) $2(R\text{-}COONH_4) + H_2SO_4 \rightleftharpoons 2R - COOH + (NH_4)_2SO_4$

As the minimum ammonia leakage and maximum regeneration efficiency are required the sulphonic resin is regenerated by countercurrent with a fixed bed system with upper inhert mass filling which, beyond ensuring an almost perfect stability of the bed reduces the void volume of the column, limiting the dilution of both regenerants and eluates and minimizing the rinse requirements.

The attainable concentration of the recovered ammonium sulphate is primarly due to the peculiar properties of the carbocylic resin:

- it offers an extremely high operating capacity (3 eq/l approx in proper conditions)
- it is suitable to withstand high sulphuric acid concentration (up to 30%)
- swelling of the resin in exhausted form, notwithstanding the extreme operating conditions, does not exceed 1,5 ÷ 1,6 times the regenerated form
- it can be regenerated even by an acid weak as bisulphate thus ensuring the maximum conversion of the eluted ammonium ion into the neutral sulphate form (which is not possible with a conventional strong sulphonic resin).

The last property is evidenced by fig. 3 and 4 showing the elution curves for the case of a couple of strong sulphonic exchangers and for a couple of weak and strong exchangers.

The comparison shows that the carboxylic resin effect is to postpone the elution of the excess acid thus allowing to recovery the most part of it.

Moreover it has been evidenced that in the first case an higher free acidity excess has to be neutralized to allow the recovery of the product while with the carboxylix resin the neutralization consumption is kept to a minimum.

Because of the high concentration of the acid, the total volume of sulphuric acid to be fed during regeneration is less than 1 BV and the total rinse requirement do not exceed 2 BV.

Fig. 5 shows the sequence of the regeneration and the splitting up of the eluates:

- the first fraction (1 BV) containing only traces of $(NH_4)_2SO_4$ is recycled to the raw condensate tank
- the second fraction (0.7 BV) is conveyed to a product tank, neutralized with gaseous ammonia and then delivered to the crystallization
- the third fraction (0,5 BV) is collected to a recovery tank where H_2SO_4 is reintegrated up to 250 g/l approx
- the fourth fraction (0,8 BV) containing only traces of free acidity is conveyed directly to waste

3.4 -Automatic criteria

All the above operations are performed in a full automatic mode based on prefixed volumes and times (exception made for product neutralization which is automatically pH controlled).

4 - CONCLUSIONS

The performances of the industrial plant after some months of service are well above the granted figures and in some cases even more satisfactory than the ones obtained in the pilot tests. The plant has demonstrated the possibility to use carboxylic type resin with extremely high acid concentration thus enhancing exchange capacities and recovered product concentration.
The leakage of ammonia also with high loads in the feed is practically absent.
The excess of sulphuric acid is kept within the predicted limits and it is such for the purpose to provide an extremely simple installation: a further fractionation of the eluate could easily lead toward the stoichiometric figures.

5 - ACKNOWLEDGEMENTS

The plant has been designed, fabricated, erected and commissioned by ITALBA S.p.A. during the 1981.
The datas contained in the present report have been obtained by courtesy of:
- ANIC - Technical Dept/
- DUOLITE INTERNATIONAL s.r.l.

Table 1

	Units	Raw condensate min. max.	Stripper effluent min. max.	Activated carbon effluent min. max.	I.E. effluent min. max.
NH_3	ppm	400 - 1500	20 - 30	20 - 30	nil
CO_2	ppm	1100 - 1500	40 - 60	40 - 60	nil
CH_3OH	ppm	400 - 1500	60 - 80	30 - 50	30 - 50

Table 2 - Ammonia process condensate
Comparison between the design, test and actual composition

	Units	Design Figures min. max.	Pilot influent 1st test	Pilot influent 2nd test	Actual Con-densate	Expected composition range min. max
Temp.	°C	40	60	40	40	40
NH_3	mg/l	500-2200	580	2200	370	400 - 700
CO_2	mg/l	500-1500	750	1930	960	900 -1500
CH_3OH	mg/l	600-1500	-	-	400	400 - 700
Excess NH_3(1)	mg/l	300-1200	290	1470	0	50 - 120

(1) calculated as quantity of ammonia in excess to the NH_4HCO_3 salt

Table 3 - Process performances
Comparison of the performances obtained in laboratory tests and in the plant

	Units	Pilot 1st test	Pilot 2nd test min. max.
Resins Operating Capacity	eq/l_R	2.1	2.1 - 2.2
$(NH_4)_2SO_4$ concentr/			
- before neutr.	g/l	130	178 - 196
- after neutr.	g.l	149	215 - 234
Specific acid consumption	$\frac{kgH_2SO_4}{kg\ NH_3}$	4.22	3.15-3.60
Stoichiometric excess	%	146	110 - 125
NH_3 leakage	mg/l		0.1 - 0.5
NH_3 breakthrough	mg/l		10 - 40

	Units	Required guarantees	Plant perform. min. max.
Resins Operating Capacity	eq/l_R	1.9	1.9 - 2.0
$(NH_4)_2SO_4$ concentr.			
- before neutr.	g/l	-	155 - 165
- after neutr.	g/l	137	170 - 190
Yield (1)	%	95	99.7
Specific acid consumption	$\frac{kg H_2SO_4}{kg NH_3}$	3.65	3.52
Stechiometric excess	%	127	122
NH_3 leakage	mg/l	15	0.1 - 0.5
NH_3 breakthrough	mg/l		5 - 10

(1) calculated as quantity of ammonia recovered in the product (before neutralization) and total quantity of ammonia entering the plant.

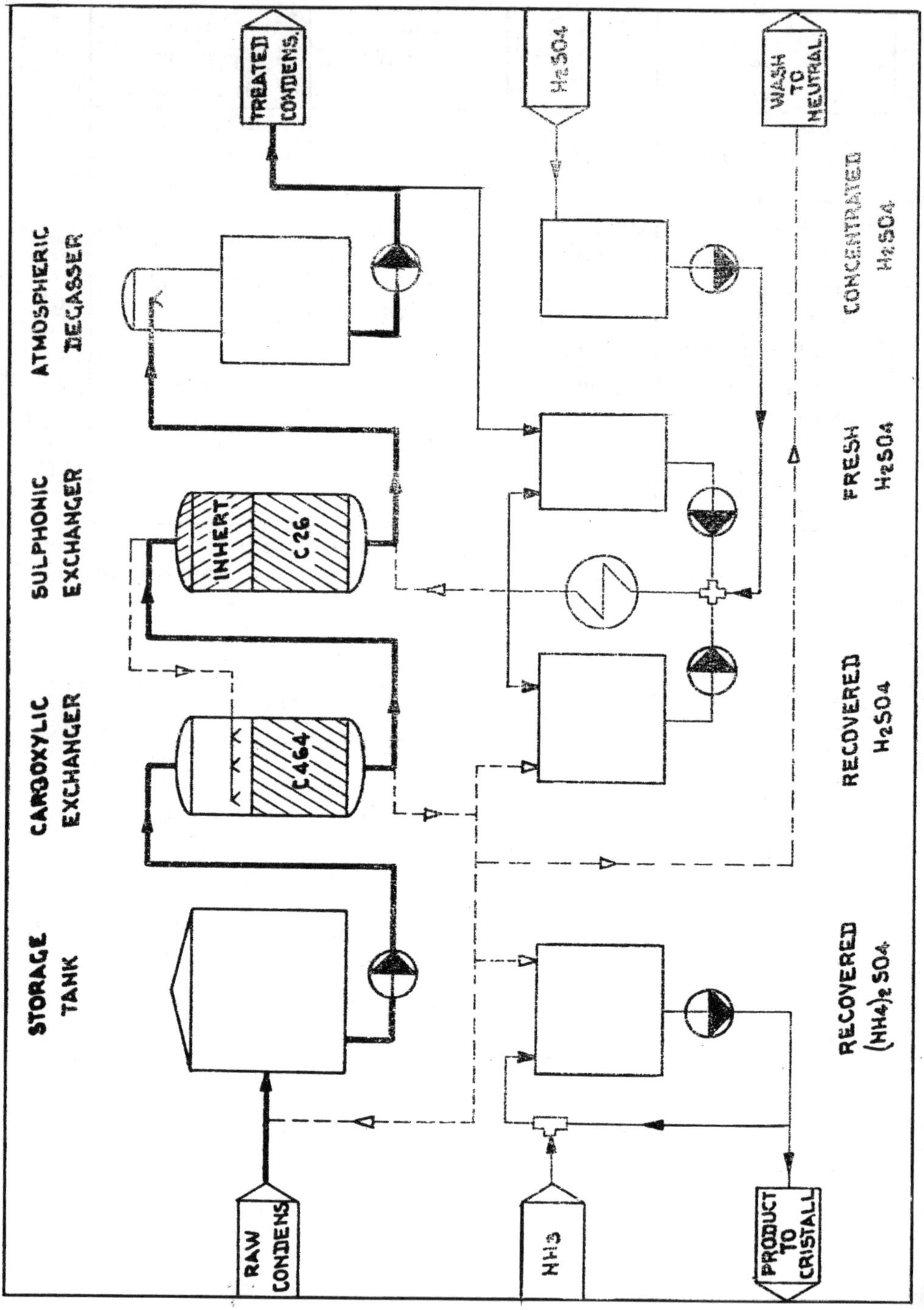
STORAGE TANK
CARBOXYLIC EXCHANGER
SULPHONIC EXCHANGER
ATMOSPHERIC DEGASSER
C464
INHERT
C 26
TREATED CONDENS.
H_2SO_4
WASH TO NEUTRAL
RAW CONDENS
NH_3
PRODUCT TO CRISTALL
RECOVERED $(NH_4)_2SO_4$
RECOVERED H_2SO_4
FRESH H_2SO_4
CONCENTRATED H_2SO_4

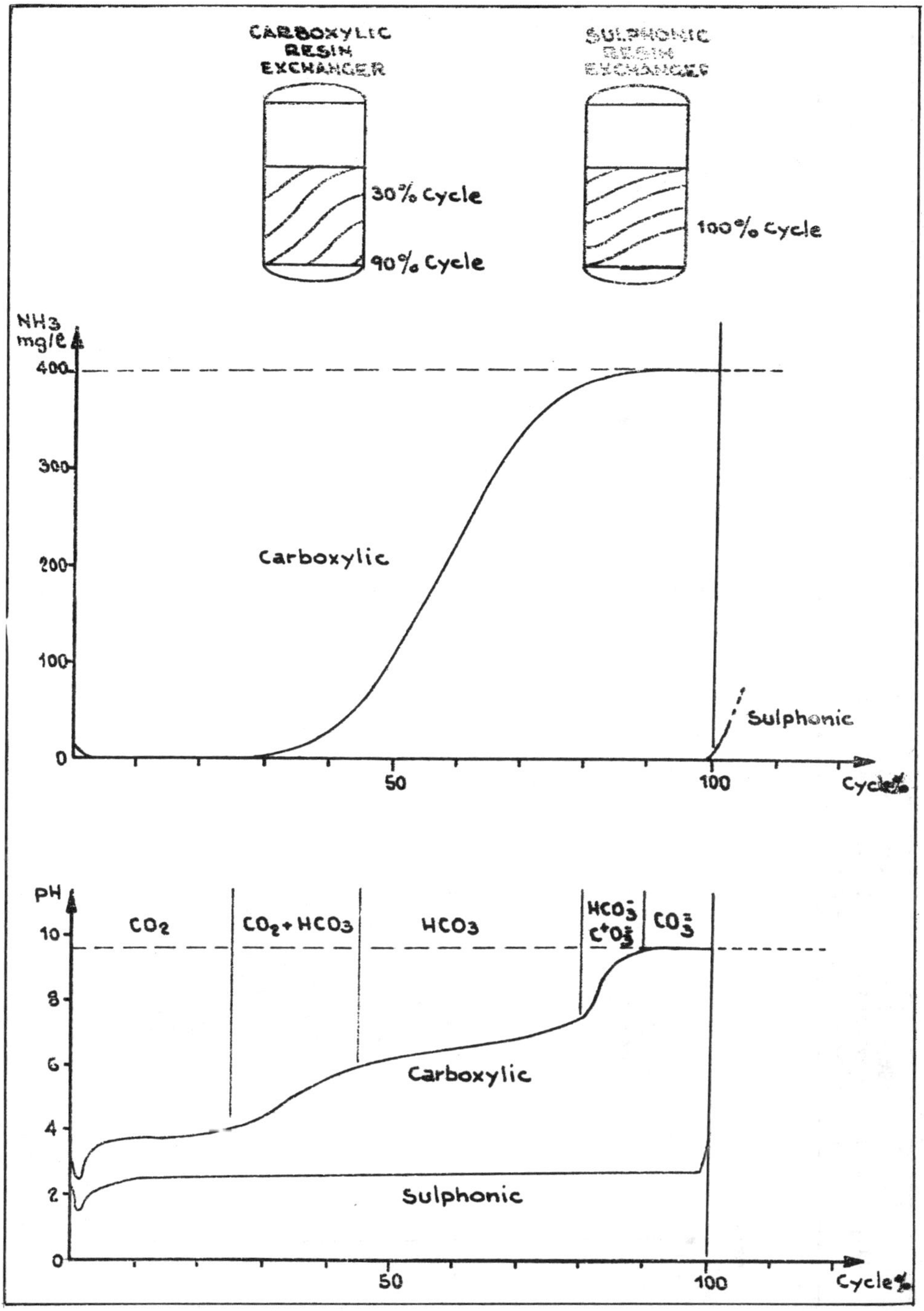

Fig. 2- Typical exhaustion curve

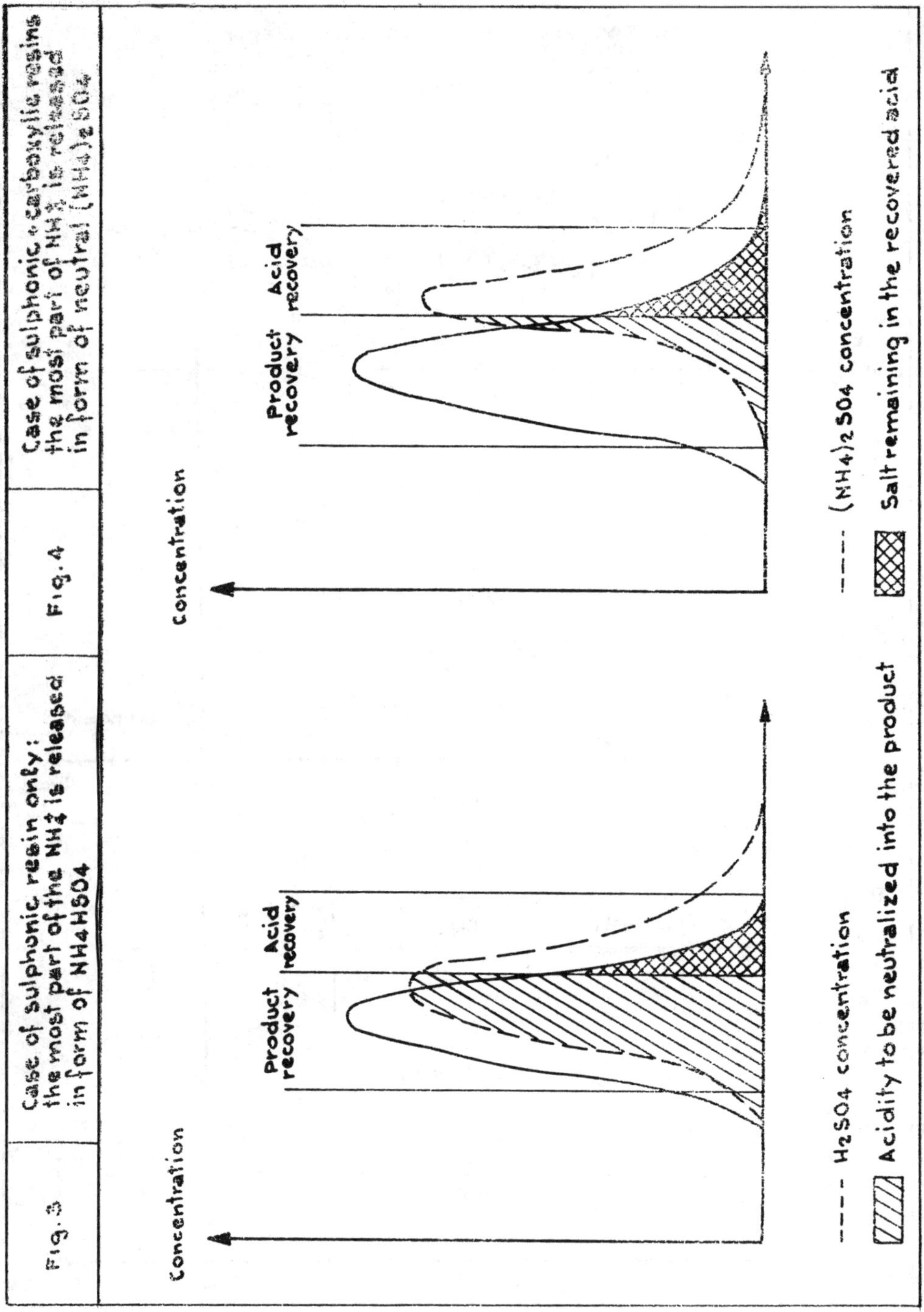

Fig.3 & 4 - Elution curves

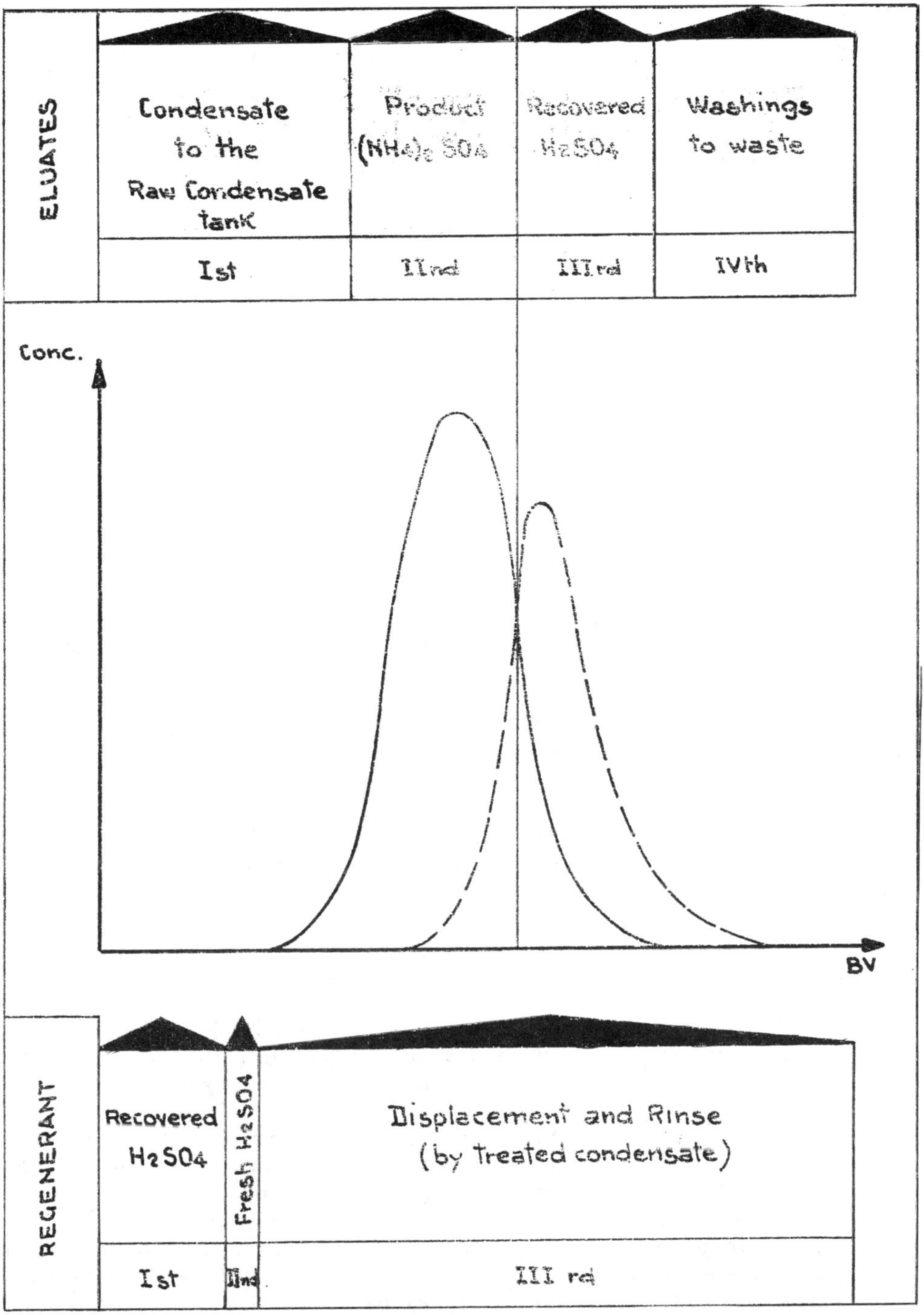

Fig. 5 - Regeneration sequence and eluates fractionation

THE MECHANISM OF ION EXCHANGE ON MICROCRYSTALS OF INORGANIC OXIDE-HYDROXIDES.

by Russell Paterson & Habibur Rahman,

Department of Chemistry,
University of Glasgow,
Glasgow,G128QQ
Scotland, U.K.

Insoluble hydrous oxides of the trivalent and tetravalent metals have been of interest as ion exchangers since the mid 1950s (1). Early researches showed that most hydrous oxides had pseudo-amphoteric exchange properties, with pH variable capacities, anionic in acid, cationic in alkali. Experimental curves relating capacity to pH for these materials have been widely reported (2,3), but because they were gels and difficult to prepare reproducibly, these curves showed wide variations, even between samples prepared batch-wise in the same laboratory (4,5).

Possibly because of this irreproducibility, the mechanism of ion exchange on these oxides has not been fully investigated, till now. In this research thermodynamic and mechanistic aspects of exchange have been investigated using crystalline analogues of the hydrous gels: the corresponding oxide-hydroxides. Being crystalline they were expected to be completely reproducible in their properties and to show well-defined selectivity patterns unattainable in gels; even ion sieving properties (6).

Three oxide-hydroxides were chosen for study; αFeOOH, βFeOOH, and monoclinic zirconia. In each case the maximum dimension of the crystals was 2000 Å. Consequently, although intracrystalline diffusio might be slow, the small crystal sizes ensured rapid equilibrations and in all cases, titration procedures were employed. αFeOOH (7) and monoclinic zirconia (8) were both found to have pseudo-amphoteric exchange properties, as shown in Figs 1 and 2, respectively. In solutions of pH less than 8.5 they were anion exchangers, in alkaline solutions, above pH 8.5, cation exchangers. For convenience, in these figures, cation capacities (for sodium) are represented as positive and anion capacities(for chloride) as negative. For both these crystals it is obvious that capacity was not a single valued function of pH. Series of such curves were obtained by adding increasing amounts of the common salt (sodium chloride in this case) as shown in Figs 1,2. The lowest capacities were obtained at each pH when no added salt was present. Salt additions caused the equilibrium pH to rise in acidic solutions and to fall in alkaline ones. The results for ßFeOOH (9) which had only anionic capacity in acid (and no cationic branch in alkali) were similar, Fig 3. In the titration procedures used to generate these capacity data, changes of one pH unit were common on addition of as little as one millimole of salt to the titration vessel (~30ml).

Analyses showed that increase (or decrease) in chloride capacity of the oxide-hydroxide crystals were always accompanied by loss (or gain) of hydrochloric acid from the solution phase. Similarly cation exchange capacity was due to uptake of sodium hydroxide. In both cases the processes were reversible.

The observed salt and pH dependence of these capacity curves may be explained by a pore mechanism.

In acidic solutions a positively charged matrix is created by 'protonation' of the crystal pore walls, eqn(1), which is a process which cannot be sustained without a Donnan uptake of acid from solution, eqn(2). In these equations barred species are those specifically within the pore structure and away from the bulk solution phase.

$$\overline{H}^{+} + \overline{MeOOH} \rightleftharpoons \overline{MeOOH_2^{+}} \qquad (1)$$

$$H^{+} + Cl^{-} \rightleftharpoons \overline{H}^{+} + \overline{Cl}^{-} \qquad (2)$$

This Donnan uptake provides both the protons necessary (assigned here provisionally to the conversion of surface hydroxyls into bound water molecules) and the accompanying chloride (counterion) to preserve electroneutrality. The overall reaction is given by eqn(3).

$$H^{+} + Cl^{-} + \overline{MeOOH} \rightleftharpoons \overline{MeOOH_2^{+}} + \overline{Cl}^{-} \qquad (3)$$

It is noteworthy that pore protonation could not proceed significantly by a reaction such as ;

$$H^{+} + \overline{MeOOH} \rightleftharpoons \overline{MeOOH_2^{+}}$$

because of the very large Donnan potential which would develop between the solution and the pore.

In this pore model, chloride activity in solution would be expected to have an equal role to that of the proton in determining the capacity of the exchanger for chloride. It is obvious from eqn(3) that increase in chloride concentration would drive the equilibrium to the right, causing increased capacity for chloride on the exchanger and a rise in the solution pH. A quantitative test of these predictions was applied to the data of Figs 1,2,3. In each case capacities were plotted as functions of pA = pH + pCl.

In accordance with eqn(3), it was observed that the chloride capacities for αFeOOH, monoclinic zirconia and βFeOOH were single-valued functions of pA, Figs 4,5,6. The divergence of capacities observed in the cationic branches of αFeOOH and zirconia merely indicates that the mechanism of acid uptake(eqn(3)) no longer applies under alkaline conditions. Later work on βFeOOH showed that perchlorate ion was sterically excluded from the hollandite pores of its structure and it has been possible to distinguish surface sites which, having no Donnan limitations,are purely pH dependent, (6).

Sodium capacities in αFeOOH and zirconia are due to a similar mechanism. In these cases a 'hydroxylation' of the crystal pore wall accounts for the anionic matrix charge. Once again, such a charging process cannot be sustained within the pore, except by a Donnan uptake of sodium hydroxide. In place of eqn(3) we have an analogous mechanism involving sodium hydroxide, eqn(4).

$$Na^+ + OH^- + \overline{MeOOH} \rightleftharpoons \overline{Na}^+ + \overline{MeOO}^- + \overline{H_2O} \qquad (4)$$

(once more pore species are indicated by superscript bars)

The addition of sodium chloride to the equilibrium , eqn(4), would raise sodium activities and move the equilibrium to the right. The result would be increased cation capacity and a fall in pH. These predictions are in accord with experimental observations. As a quantitative test the data of Figs 1,2 were plotted as functions of pB = pOH + pNa. In both cases the result was a single curve for cationic capacities, indicating that eqn(4) was obeyed and the relative proportion of sites on the matrix pore walls which were hydroxylated was indeed a single-valued function of the product of the activities of sodium and hydroxide ions in the equlibrium solution. Anionic capacities, which were shown above to obey eqn(3) remain divergent, Figs 7,8.

The Donnan pore model has therefore been shown to apply to all the data collected so far for these three microcrystalline oxide-hydroxides. The only exception has been for perchlorate ion on βFeOOH (6) and in that case the perchlorate was excluded by its size from entering the pore structure and confined to surface exchange sites, which are,as expected,purely pH dependent. The crystals of the oxide-hydroxides provide us with a new genre of crystalline ion exchangers of great practical and theoretical interest.

References

1. Kraus,K.A., Phillips,H.O., Carlson.T.A.,& Johnson,J.S., Proc. Second Int.Conference on "Peaceful Uses of Atomic Energy", United Nations,Geneva,28,3, (1958).
2. Amphlett,C.B.,"inorganic Exchangers" Elsevier,Amsterdam,(1964).

3. Clearfield,A., Nancollas,G.H., and Blessing,R.H., in "Ion Exchange and Solvent Extraction", Chapter 1,vol 5,(1973) Editors; J.A. Marinsky and Y. Marcus: Marcel Dekker,N.Y.
4. Nancollas,G.H. and Paterson,R., J.Inorg.Nuclear Chem.,22,259,(1961).
5. Nancollas,G.H. and Paterson,R., J.Inorg.Nuclear Chem.,29,565,(1967).
6. Paterson,R. and Rahman,H.,J.Colloid and Interface Science; in press.
7. Paterson,R. and Rahman,H.,J.Colloid and Interface Science; in press.
8. Paterson,R. and Rahman,H.,J.Colloid and Interface Science; in prep.
9. Paterson,R. and Rahman,H.,J Colloid and Interface Science; 94,60,1983.

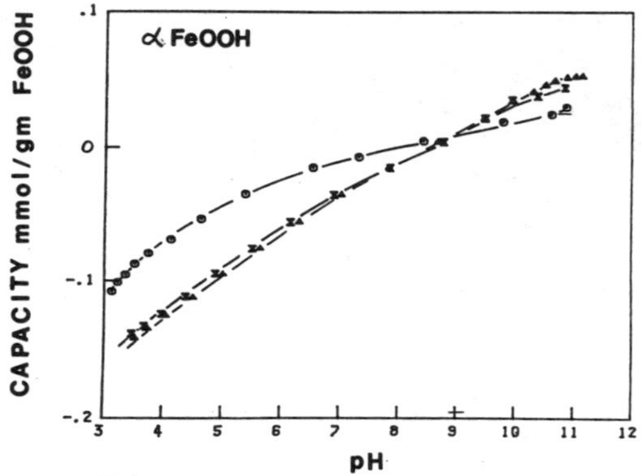

FIGURE 1

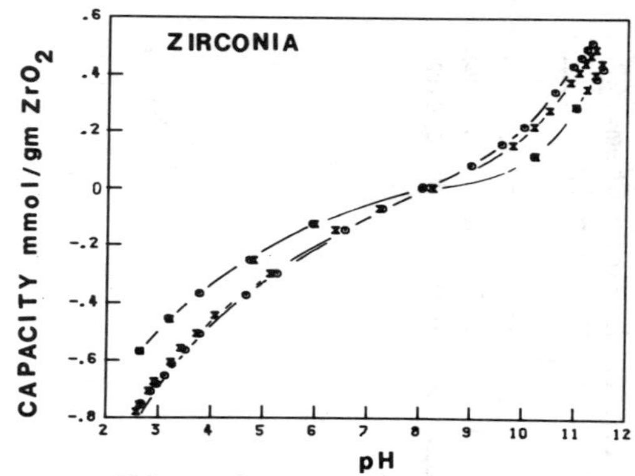

FIGURE 2

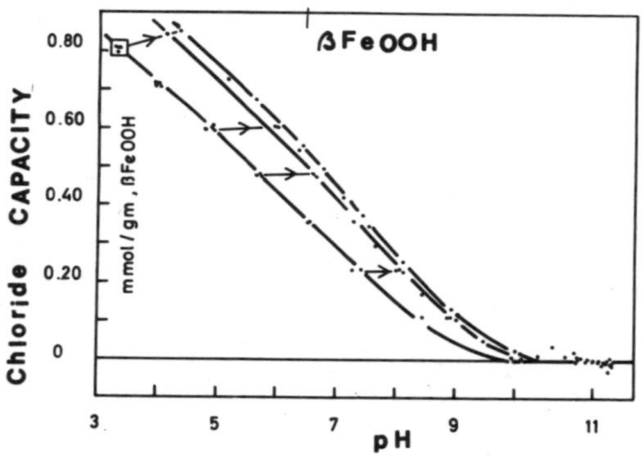

FIGURE 3

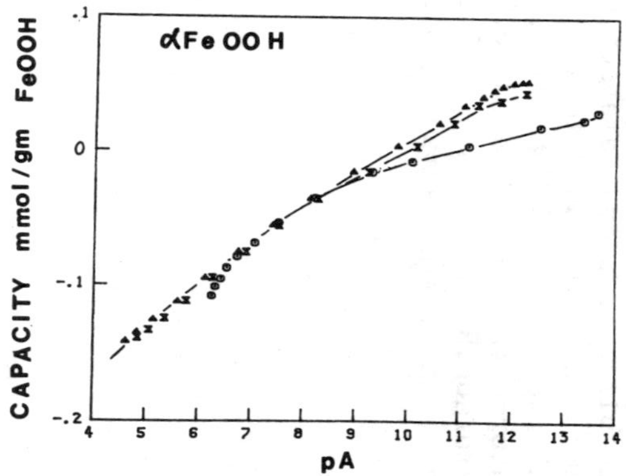

FIGURE 4

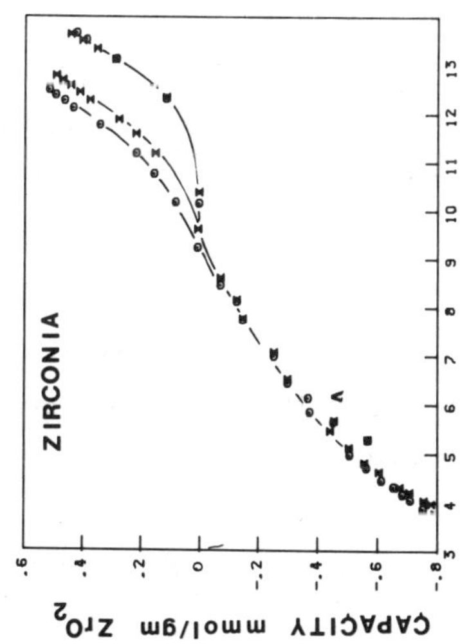

FIGURE 5

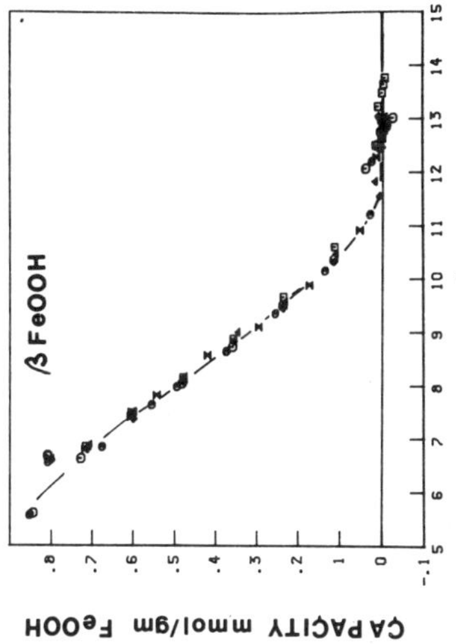

FIGURE 6

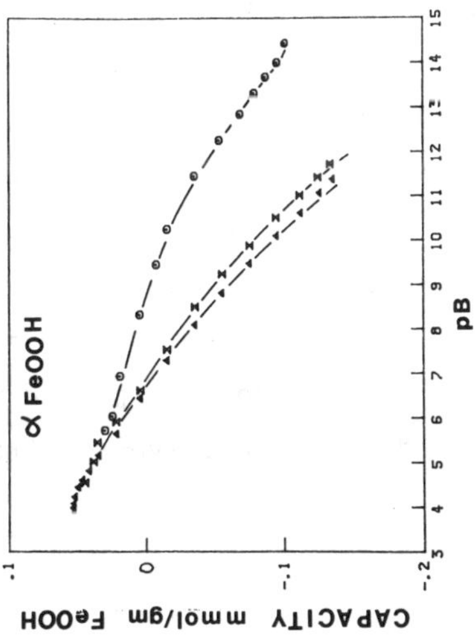

FIGURE 7

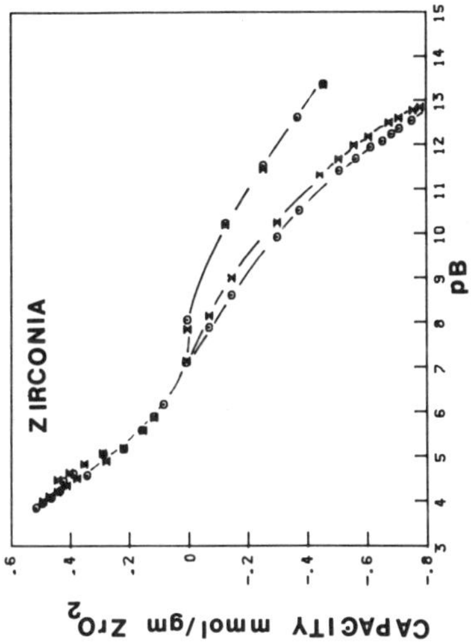

FIGURE 8

A Useful Method to Summarize Data in Ion Exchange
2. Influence of Temperature and Ionic Size

E Högfeldt[1], A V Mikulich[2] and V S Soldatov[2]

[1]Department of Inorganic Chemistry, The Royal Institute of Technology, S-100 44 Stockholm, Sweden

[2]Institute of General and Inorganic Chemistry, Academy of Sciences BSSR, 220603 Minsk, The Soviet Union

Abstract

The reaction: $(CH_3)_4N^+ + HR \rightleftharpoons (CH_3)_4NR + H^+$

has been studied at 0°C, 25°C and 63°C on the Russian cation exchanger KRS-8. The plots $\log\kappa(x)$ are linear implying that the system behaves as a regular solution. κ is the equilibrium quotient of the reaction above, x=mole fraction of $(CH_3)_4NR$.

The water uptake during ion exchange has been studied at 25°C on the same resin for the pairs: $CH_3NH_3^+-H^+$, $(C_4H_9)_2NH_2^+-H^+$ and $(C_2H_5)_4N^+-H^+$. In all cases negative deviations from linearity are found in the plots W(x). W=number of H_2O per sulfonate group.

The good fit to the curves: $W=w(1)x + w(0)(1-x) + Bx(1-x)$ (B=const) shows the applicability of a simple model proposed using mole fraction (x) as independent variable (mole fraction statistics).

No improvement can be expected from using volume fraction (ϕ) as independent variable (volume fraction statistics).

Introduction

Recently a simple approach for summarizing thermodynamic data on micelles of the liquid cation exchanger dinonylnaphtalene sulfonic acid (HD) was suggestcd[1,2]. In the present paper its application to solid resins[3] is continued.

The Model

Consider a number of available sites in an ion exchanger. For a statistical distribution of ions over the available sites, the probability that component A has another A as nearest neighbor is equal to x_A^2 and similarly for component B. For A having a B as nearest neighbor the probability is equal to $2x_Ax_B$. It is now assumed that for any property Y to have the same value as in pure A (y_A) the probability is the same, i e

$$Y = y_Ax_A + y_Bx_B + 2y_mx_Ax_B \tag{1}$$

y_m is the value of Y in the mixtures.
This homogeneous equation is related to

$$Y = y_Ax_A + y_Bx_B + Bx_Ax_B \tag{2}$$

by $$y_m = \tfrac{1}{2}(y_A + y_B + B) \tag{3}$$

B is a constant taking care of deviations from linearity (additivity) in the plot Y(x).

Applications

The model outlined above applies to ion exchange equilibria. For the exchange between the cations Am^+ and H^+ on the resin R^-

$$Am^+ + HR \rightleftarrows AmR + H^+ \tag{4}$$

the equilibrium quotient (κ) of reaction (4) is defined by

$$\kappa = \frac{[AmR][H^+]}{[HR][Am^+]} = \frac{x_1c_2}{x_2c_1} = \frac{xc_2}{(1-x)c_1} \tag{5}$$

x=mole fraction of AmR

Concentrations in the resin phase are expressed by mole fractions and in the aqueous phase by molarities.
In the following it will be assumed that the activity coefficient ratio y_{H^+}/y_{Am^+} stays practically constant and can be included in κ, and that all variation in $\kappa(x)$ can be attributed to nonideal behavior of the resin phase.
The water uptake during exchange (W), can conveniently be defined by

$$W = [H_2O]_{res}/S_0 \quad (S_0\text{=capacity of resin}) \tag{6}$$

W is thus the number of water molecules per sulfonate group.

1. <u>The system $(CH_3)_4N^+$-H^+ on KRS-8 at 0-63°C</u>

Mikulich[4,5] studied the exchange between various organic ammonium ions and hydrogen ion on the Russian cation exchanger KRS-8, similar to Dowex x-8) and the ionic strength I=0.100 M in the aqueous phase. Most studies were carried out at 25°C but for some ions also 0°C and 63°C were investigated. In this paper the system: $(CH_3)_4N^+$-H^+ will be used to illustrate use of the model to data at different temperatures. In this as well as all other systems studied the plots $\log\kappa(x)$ could to a good approximation be fitted with straight lines, i e B in eqn (2) is equal to zero. Thus

$$\log\kappa = \log\kappa(1)x + \log\kappa(0)(1-x) \tag{7}$$

From eqn (3) with B=0

$$\log\kappa_m = \tfrac{1}{2}[\log\kappa(0) + \log\kappa(1)] \tag{8}$$

As seen from eqn (7) only two parameters, the limiting values for the two pure ionic forms, are needed to describe the system.
From plots $\log\kappa(1/T)$ the following expressions were obtained by linear regression:

$$\begin{aligned}\log\kappa(0) &= -0.727 + 582/T \\ \log\kappa(1) &= -0.366 - 204/T\end{aligned} \tag{9a,b}$$

In Table 1 experimental $\log\kappa$-values are compared with those computed from eqn (7) with $\log\kappa(0)$ and $\log\kappa(1)$ computed from eqn (9).

Table 1

Comparison of experimental and computed logκ-values for the reaction: $(CH_3)_4N^+ + HR \rightleftharpoons (CH_3)_4NR + H^+$ at 0°C, 25°C, 63°C

t=0°C			t=25°C			t=63°C		
x_{AmR}	logκ exp	logκ calc	x_{AmR}	logκ exp	logκ calc	x_{AmR}	logκ exp	logκ calc
0	-	1.405	0	-	1.226	0	-	1.005
0.189	0.977	0.929	0.181	0.798	0.814	0.172	0.698	0.665
0.315	0.571	0.612	0.301	0.533	0.541	0.299	0.415	0.414
0.418	0.350	0.352	0.403	0.313	0.308	0.391	0.224	0.232
0.499	0.164	0.149	0.489	0.143	0.113	0.467	0.090	0.081
0.538	0.049	0.050	0.528	0.017	0.024	0.519	-0.047	-0.022
0.603	-0.146	-0.113	0.598	-0.151	-0.136	0.583	-0.180	-0.148
0.668	-0.264	-0.277	0.665	-0.270	-0.288	0.634	-0.245	-0.249
0.701	-0.390	-0.360	0.696	-0.383	-0.359	0.703	-0.411	-0.386
0.801	-0.586	-0.612	0.796	-0.551	-0.586	0.801	-0.536	-0.579
1	-	-1.113	1	-	-1.051	1	-	-0.973
	σ(logκ)	±0.030		σ(logκ)	±0.022		σ(logκ)	±0.026

As a measure of the goodness of fit the quantity σ(Y) has been used.

$$\sigma(Y) = \pm\sqrt{\Sigma(Y_{calc} - Y_{exp})^2/(n-1)} \tag{10}$$

n = number of experimental points

In Table 1 Y=logκ. As seen a fit to within ±0.03 in logκ is obtained in the range 0-63°C using the four parameters given in eqns (9a,b).

It has been shown elsewhere[3] that for logκ(x) given by eqn (7) the activity coefficients on the mole fraction scale can be expressed by

$$\log f_{AmR} = b(1-x)^2 \qquad \log f_{HR} = bx^2 \tag{11a,b}$$

where

$$b = \tfrac{1}{2}[\log\kappa(1) - \log\kappa(0)] \tag{12}$$

From eqns (9a,b)

$$b = 0.181 - 393/T \tag{13}$$

Eqns (11a,b) imply that the system can be formally treated as a regular solution.

The thermodynamic equilibrium constant for reaction (4) can be computed from[3,6-8]

$$\log K = \int_0^1 \log\kappa\, dx = \tfrac{1}{2}[\log\kappa(0)+\log\kappa(1)] = \log\kappa_m \qquad (14)$$

From eqns (9a,b)

$$\log K = \log\kappa_m = -0.547 + 189/T \qquad (15)$$

In Table 2 $K(=\kappa_m)$ and the b-values are listed for the three temperatures studied. As seen the b-values are negative making the activity coefficients smaller than unity; the system shows negative deviations from ideal behavior.

Table 2

Thermodynamic equilibrium constants and b-values for the reaction: $(CH_3)_4N^+ + HR \rightleftharpoons (CH_3)_4NR + H^+$ at 0°C,25°C and 63°C

t°C	logκ(0)	logκ(1)	logκ$_m$	κ$_m$=K	b
0	1.405	-1.113	0.146	1.40	-1.26
25	1.226	-1.051	0.088	1.22	-1.14
63	1.005	-0.973	0.016	1.04	-0.99

2. The water uptake for reaction (4) and various amines at 25°C

Mikulich[4,5] also studied the water uptake during exchange. Table 3 gives results for $Am^+ = CH_3NH_3^+$, $(C_4H_9)_2NH_2^+$ and $(C_2H_5)_4N^+$. The expressions used are

$$\begin{aligned} &Am^+=CH_3NH_3^+ && W=8.18x + 12.79(1-x) - 0.95x(1-x) && w_m=10.00 \\ &Am^+=(C_4H_9)_2NH_2^+ && W=1.84x + 12.82(1-x) - 7.07x(1-x) && w_m=3.80 \\ &Am^+=(C_2H_5)_4N^+ && W=5.38x + 12.71(1-x) -10.52x(1-x) && w_m=3.79 \end{aligned} \qquad (16a\text{-}c)$$

Observe that the secondary and quarternary ammonium ions show large deviations from additivity, while the free energies are additive as demonstrated for $(CH_3)_4N^+$ above. Moreover, the fit is equally good for the larger cations as for $CH_3NH_3^+$, indicating that size has little influence on the randomization of the ions over the available sites. No improvement in fit can be expected by using volume fractions instead of mole fractions as independent variables. Table 3 also shows the excellent fit to the model in spite of the large negative deviations from additivity for the two larger cations. In these cases the value of w_m is small indicating the incompatibility of the ammonium ion with H^+ as far as the arrangement of water molecules around the ions in the resin phase. This seems to be a general principle, the deviation being larger the greater the incompatibility of the two ions. For similar ions on the other hand like Cs^+ and Na^+, the water uptake to a good approximation is linear with composition[9].

Table 3

Water uptake at 25°C on KRS-8 and some organic ammonium ions

$Am^+ = CH_3NH_3^+$			$Am^+ = (C_4H_9)_2NH_2^+$			$Am^+ = (C_2H_5)_4N^+$		
x_{AmR}	W_{exp}	W_{calc}	x_{AmR}	W_{exp}	W_{calc}	x_{AmR}	W_{exp}	W_{calc}
0	12.76	12.79	0	12.76	12.82	0	12.76	12.71
0.161	11.86	11.92	0.190	9.72	9.65	0.175	9.90	9.91
0.287	11.38	11.27	0.382	7.05	6.96	0.331	7.91	7.95
0.414	10.69	10.64	0.532	5.15	5.22	0.403	7.13	7.22
0.526	10.24	10.12	0.667	3.90	3.93	0.476	6.57	6.60
0.577	9.77	9.89	0.734	3.30	3.38	0.490	6.44	6.49
0.615	9.64	9.72	0.771	3.13	3.11	0.518	6.24	6.29
0.686	9.31	9.41	0.818	2.80	2.79	0.550	6.12	6.07
0.725	9.24	9.24	0.846	2.66	2.61	0.577	6.04	5.91
0.810	8.93	8.89	0.868	2.34	2.48	0.587	5.87	5.86
0.829	8.82	8.82	0.918	2.33	2.21	0.633	5.75	5.63
1	8.19	8.16	1	-	1.84	1	5.33	5.38
	$\sigma(W)$	±0.07		$\sigma(W)$	±0.08		$\sigma(W)$	±0.07

Acknowledgement

The work in the present study is carried out within the agreement between the Swedish and Russian Academics of Science. Their support is gratefully acknowledged.

References

1. Högfeldt, E, Acta Chem Scand *A33* (1979) 557
2. Högfeldt, E and Soldatov, V S, J Inorg Nuclear Chem *41* (1979) 575
3. Högfeldt, E, Reactive Polymers, In press
4. Mikulich, A V, Thesis, Minsk, 1982
5. Soldatov, V S, Mikulich, A V and Högfeldt, E, In preparation
6. Ekedahl, E, Högfeldt, E and Sillén, L G, Acta Chem Scand *4* (1950) 556
7. Argersinger, W J, Davidson, A W, and Bonner O D, Trans Kansas Acad Sci *53* (1950) 404
8. Högfeldt, E, Arkiv Kemi *5* (1952) 147
9. Soldano, B, Larson, Q V and Myers, G E, J Amer Chem Soc *77* (1955) 1339

REVERSIBILITY AND PERFORMANCES IN PRODUCTIVE ION-EXCHANGE CHROMATOGRAPHY

M.BAILLY and D.TONDEUR

Laboratoire des Sciences du Genie Chimique

CNRS-ENSIC 1, rue Grandville NANCY FRANCE

I INTRODUCTION

In chromatographic separations, the dilution of the products by the eluent is a direct measure of the work of separation, which includes the minimum thermodynamic work and the work needed to compensate the irreversibilities. Therefore, reducing the irreversibilities has direct incidence on the eluent consumption and on the dilution of the products.

The purpose of the present paper is twofold:
First, to show that one of the important irreversibilities is generated by the non-linearity of the equilibrium, and is related to the propagation of compressive fronts (shocks) in the chromatographic column.
Second, to illustrate some non-conventionnal modes of chromatography, which reduce this particular source of irreversibility, and correlatively diminish the chemical work (eluent consumption) required to reach a given product purity.
Four operating modes will be analysed from the point of view of shock propagation, and correlative entropy production on one hand, and from the point of view of eluent consumption and immobilized capacity on the other hand. The experimental process used for illustration is the ion-exchange separation of K^+ and Na^+ cations with H^+ as eluent, in aqueous chloride solution, on a sulfonic cation exchanger.

II IRREVERSIBILITY DUE TO SHOCK PROPAGATION

In the present approach, we do not give attention to irreversiblilities due to thermal or mechanical effects, such as irreversible heat effects, expansion work, friction, which can be considered as negligible in ion exchange. We shall focus our attention on irreversible mass transfer related to departure from equilibrium between phases. To introduce the pertinent notions, let us first consider some ideal chromatographic processes.

The ideal non-linear pulse

Figure 1 shows schematically the propagation of a pulse of solute under conditions of non-linear sorption equilibrium, described by an isotherm concave toward the mobile-phase concentrations (see Figure 2).

It is known that non-linear equilibrium implies that different concentrations travel at different velocities. More precisely, in a continuous concentration distribution, and when equilibrium prevails between the two phases, the velocity v_c of a given concentration c is obtained from a differential material balance as:

$$V_c = \left(\frac{\partial z}{\partial t}\right)_c = \frac{u}{1 + \frac{1-\varepsilon}{\varepsilon}\frac{dq}{dc}} \qquad (1)$$

where u is the interstitial fluid velocity, ε the packing porosity, and dq/dc is the slope of the sorption isotherm at concentration c. Equation (1), the basic equation of chromatography, has been well established by De Vault (1) and many later authors.

With a sorption isotherm such as that of Figure 2, dq/dc will be high at small concentrations, and small at high concentrations. Thus high concentrations move faster than low concentrations: the leading front of the pulse of Figure 1 will therefore tend to sharpen (compressive, or self-sharpening behaviour), whereas the rear front will on the contrary spread out (dispersive, or spreading behaviour).

The spreading behaviour is in no way incompatible with the assumption of local equilibrium between phases, as a limiting behavior. We shall see that such is not the case for the sharpening behaviour.

Clearly, the sharpening tendency is limited by the fact that an overlapping (multivalued) concentration distribution is prohibited. The limiting shape attained by the front is then simply a discontinuity (in the ideal case).

In fact, the presence of diffusion, hydrodynamic dispersion, and the discrete nature of the packing will give the front a finite thickness. When this limiting (asymptotic) shape is attained, the front is a so-called "shock wave", or briefly a shock, by analogy with the aerodynamic phenomenon. From then on, it will travel through the column conserving its shape (constant pattern behaviour).

The existence of these shocks in chromatography is well known (De Vault:1, Rhee: 2). A fundamental property is that their "operating line" (that is the line plotting the actual fixed phase concentation versus the mobile phase concentration in the shock region), is a straight line. This property can be shown from a simple material balance between say the inlet and any point in the shock region; it is formally similar to the property of straight operating lines in conventionnal counter-current operations at steady-state.

The important consequence is that, since the concentrations existing in a shock lie on a straight line, they cannot conceivably approach the curved isotherm (Figure 2): the shock is intrinsically a non-equilibrium distribution. As it moves, mass is transferred (from the fluid to the solid in the present case) across a finite driving force, and this causes entropy production. Clearly, this production is largest when the driving force is largest, in other words, when the distance between operating line and equilibrium curve is largest. This distance in turn is seen (Fig 2) to depend on the concentration difference between the end points of the shock, in other words, on its "amplitude".

A further observation is that after the shock is formed the spreading of the rear front is no longer "compensated" by the sharpening of the leading front, and there results an overall spreading of the pulse, accompanied by a dilution of the solute in the carrier solvent. This dilution is the thermodynamic price to pay for the irreversibility due to shock propagation. It is largest when the distance travelled by the shock is largest.

Note that in the process considered here, no separation between solutes is performed, and all irreversibilities are neglected, except that due to non-linearity in the shock; thus no thermodynamic work is needed other than that mentionned above. The case of an ideal separation is considered below.

The ideal linear binary pulse

Figure 3 shows the ideal chromatographic separation of two solutes A and B adsorbed independently and following linear isotherms of different slopes (thus travelling at different velocities). The difference with the single solute case is that each molecule of B taken out of the A pulse is replaced by a molecule of the solvent (assuming all these mixtures are ideal, say), and reciprocally. Thus the A and B peaks, as defined by their distribution of concentration c_A and c_B respectively, are unaltered, but the accompanying species has been in each replaced by solvent. The net result is that a total amount n_A+n_B moles of solvent have been "consumed" to dilute the A and B products. The corresponding work of mixing exactly compensates the work of separation of A and B.
The case of the ideal non linear binary pulse will be treated in some detail in the next section.

Provisional conclusions

We may summarize the situation as follows.

1-In a chromatographic separation, the "basic" source of energy is the energy of mixing of the solvent with the solutes. The dilution of the products, that is the solvent "consumption", is thus a direct measure of the work of separation required, and may be subdivided into a thermodynamic minimum, and a part required to compensate the irreversibility.

2-Among the sources of entropy, an important one is intrinsic to non-linear sorbent-solute interactions, and results from the propagation of concentration shocks in the chromatographic column. The entropy production is an increasing function of the amount of material transferred within shocks, in other words, to the distance travelled by the shocks, and to the (integrated) deviation from equilibrium within the shock.

3-Separation procedures which minimize the amplitude of the shocks and the distance they travel, are likely to bring about sizeable savings in solvent consumption and products dilution.

It is the purpose of the next section to illustrate a few such procedures, and compare them with a "classical" chromatographic scheme.

III COMPARISON OF CHROMATOGRAPHIC MODES

For a specified product purity, we shall consider four different chromatographic modes, from two different points of view: the eluent consumption on one hand, the shock propagation on the other hand. Our purpose is to show qualitatively, that there is a correlation between the entropy produced by the shocks and the eluent consumption.

No attempt is made here to use explicit expressions for the entropy production, thus to establish a direct quantitative relationship.

This will be the subject of a forthcoming paper. For our present purpose, we need only to admit, as mentionned earlier, that the entropy produced by a shock is an increasing function of the distance travelled (for a given shock amplitude and column cross section), and of the integrated deviation from equilibrium; this deviation is itself a non-linear function of the amplitude, measured for example by the concentration jump across the shock.

Description of the four chromatographic modes

The four chromatographic modes considered are: classical chromatography, two-way chromatography, simple recycle chromatography, and segmented recycle chromatography. The first three modes have been analysed in detail, experimentally and theoretically, in (3) and (4) to which the reader is referred. Here, we shall recall only what is necessary for comprehension, and use these results from a somewhat different point of view. Figure 4 to 7 present the flow sheets and experimental effluent histories for the four modes, and we shall comment briefly on each.

The experimental system considered is the separation of Na^+ and K^+ cations, on a conventionnal sulfonic cation exchanger, with H^+ as eluent, and Cl^- as common anion. Working conditions are summarized in Table 1.

* "Classical" chromatography consists in injecting a (square) pulse of the mixture to be separated (here, Na^++K^+) into a column initially in regenerated form (H^+), and "push it through" with eluent (fig 4). The conditions here have been worked out so that the Na^+ and K^+ peaks are just separated at the outlet (Ref 3).

* Mixed recycle chromatography consists in having an incomplete separation of the peaks, collecting the unseparated fraction, and recycle it after mixing with fresh mixture (fig 5).

* Segmented recycle consists, in the preceding flowsheet, in collecting in distinct fractions the unseparated part of the effluent, and recycling these fractions separately, in reverse order of their exit (last fraction out is recycled first), following by fresh mixture, and then eluent (fig 6).

* Two-way chromatography, in its simplest version, involves introducing the mixture and the eluent at opposite ends of the column, and in opposite flow directions, and taking one product out as a sidestream, the other at the column end (fig 7).

Common features of chromatographic modes

Figure 8 shows different stages of the elution of a ternary square pulse of K^{+}+Na^{+}+H^{+} by H^{+} eluent. These profiles, or part of them, are generated in the four different modes considered. We notice that two shocks appear: a fast shock F1, separating pure H^{+} from Na^{+}+H^{+}, and a slow shock F2. These two shocks enclose the Na^{+} peak, and thus determine the concentration of the Na^{+} product. We are thus concerned by the amplitude and the distance travelled by these shocks in each mode.

Two dispersive fronts, F3 and F4, are also generated, which influence the K^{+} product, and in case of recycle, the composition of the recycled fractions. In our approach, these fronts correspond to a local equilibrium situation, and thus, do not contribute to the entropy production. Besides, we have shown (4) that, in the four modes, the K^{+} product has the same composition. The dispersive fronts thus play no role, neither in the entropy production, nor in the performances, and we need not discuss this aspect further.

Qualitative discussion of shock propagation

A complete analysis of the shock paths and the corresponding entropy production is beyond the scope of this text, and will be presented in further publications (5). We merely present here the main qualitative features of the four operating modes, which differ essentially by the behaviour of the fast shock F1.

In the classical mode, F1 has initially the maximum amplitude (between x_H=1 and x_H=0), which decreases only toward the end of the traject, through interference with F3 (after stage 4, in Figure 8).

In the mixed recycle mode, the initial amplitude of F1 is smaller than in the previous case (owing to the mixing of fresh feed with recycle containing already some H^{+}), and this amplitude remains constant, in the optimal case. In the segmented recycle mode, several shocks of small amplitude are generated, which eventually merge into a single one by the time they reach the column outlet, resulting in a peak of highly concentrated sodium. The departure from equilibrium (thus the entropy production) is thus small along most of the path, and increases only in the very last stages of the process.

In the two-way mode, shock F1 does not exist initially, and will only appear after "recompression", in backward flow, of the dispersed profile generated in forward flow. This is clearly desirable, since it allows a highly concentrated Na^+ peak.

It appears that in order to produce the least entropy and achieve the most concentrated Na^+ product, two conditions should be fulfilled. First, avoid having shocks of large amplitude travelling large distances in the column; second, build a shock of large amplitude only near the column exit, by merging of smaller shocks, or by recompression of a dispersed profile. Clearly, the two-way mode and the segmented recycle mode achieve this best, whereas in the classical mode, the amplitude of F1 is uselessly large at the beginning, and decreases toward the column exit.

A more complete analysis shows that shock F2 contributes much less than F1 to the entropy production, and that this contribution is of the same order of magnitude in the four modes. In addition, the distances travelled by the shocks in all four modes (when optimized) are nearly the same. The considerations on the amplitude of F1 developped above thus furnish a suitable criterion for comparing the operating modes.

We shall now substantiate these considerations by comparing the performances of the four operating modes, not on the basis of entropy production, but on the basis of criteria defined below.

IV PERFORMANCES OF THE CHROMATOGRAPHIC MODES

In references 3 and 4, we have introduced overall indices which, for specified purities of the products, characterize the performance of the process. These indices are:

SIC: The **S**pecific **I**mmobilized **C**apacity, defined as the adsorbent capacity necessary to treat one mole of feed mixture per cycle.

SEC: The **S**pecific **E**luent **C**onsumption, defined as the number of moles of eluent necessary to treat one mole of feed per cycle.

PRO: The Productivity, defined as the ratio of the time

of fresh feed injection to the time of eluent or recycle injection, in a cycle; PRO can also be viewed as an inverse cycle time.
Table 2 gives a summary of the values of these indices for the four chromatographic modes, including the experimental values which correspond to Figures 4 to 7, and theoretical values calculated from an equilibrium model with constant separation factors, as developped in (3) and (4). The average concentration of the Na^+ product is also given. The purity specification is the same.

V CONCLUSION

The conclusion is furnished by comparing the conclusions of the analysis of shock propagation and the results of Table 2. From the point of view of eluent consumption SEC and Na^+ product concentration, the four modes can be classified as

→ increasing product concentration →
Classical>Mixed recycle>Segmented recycle>Two-way
← increasing eluent consumption ←

The same order results from the comparison of entropy production by shock F1 as discussed above. Inasmuch as shock F1 is the main responsible for entropy production, it appears that operating modes which minimize the amplitude and/or distance travelled by this shock are less dissipative, and thus more economical in chemical work. Means for achieving this goal are segmentation of the shock <u>and</u> flow reversal (the reversal in recycle processes consists in recycling fractions in reverse order or their exit).

In a forthcoming publication, we shall present explicit calculations of entropy production in shocks, which will substentiate more rigourously the above conclusions, and furnish a quantitative basis for process analysis.

TABLE 1

experimental conditions

Ion exchanger:
DUOLITE C20 (8% DVB sulfonated polystyrene)
Granulometry:0.3-0.5mm
Column:
length 1.2m (except 3.2 for two-way chromatography)
diameter 2.5 cm
Flowrate: $5 cm^3 mn$
0.2N chloride solutions for all experiments

TABLE 2

Performance indices for the separation of K^+ and Na^+

Specified product purity: K < 0.5% in Na product
Na < 0.5% in K product

	Classical		Recycle Mixed		Recycle Segmented*	Two-way	
	exp	calc	exp	calc	exp (2 seg)	exp	calc
SIC	4.8	4.5	3.1	4.	4.	9.5	12.5
SEC	9.4	8.75	6.35	4.85	6.3	6.1	6.9
PRO	9.6	10.3	12.4	12.8	9.	8.	5.
x_{Na}	0.21		0.5		0.7	0.99	

* experiment not optimized, not truly cyclic

BIBLIOGRAPHY

1 DE VAULT D., The theory of chromatography, J.Am.Chem.Soc, 65 (1943) 532-540

2 RHEE H.K, AMUNDSON N.R, Shock layer in two solute chromatography : effect of axial dispersion and mass transfer Chem.Eng. Science, 29 (1974) 2049-2060

3 BAILLY M., TONDEUR D., Two-way chromatography, Chem.Eng. Science, 36 (1981) 455-469

4 BAILLY M., TONDEUR D., Recycle optimization in non-linear productive chromatography 1 Mixing recycle with fresh feed Chem.Eng. Science 37 (1982) 1199-1212

5 BAILLY M., TONDEUR D., Reversibility and performances in productive chromatography, submitted to Chemical Engineering Fundamentals

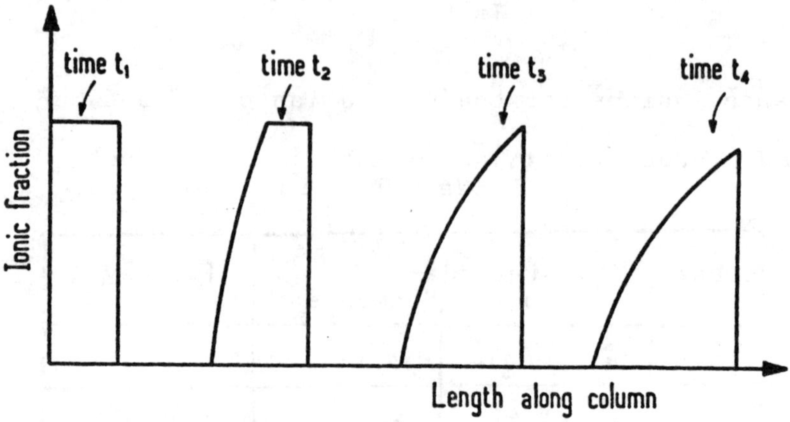

Fig1 The ideal non-linear pulse: Propagation of a finite pulse of solute under conditions of non-linear sorption and quasi-equilibrium (schematic)

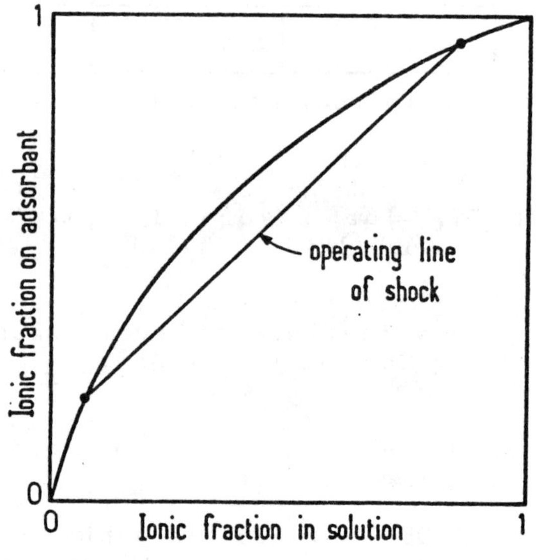

Fig2 Non-linear sorption isotherm, and operating line representing a steady shock.

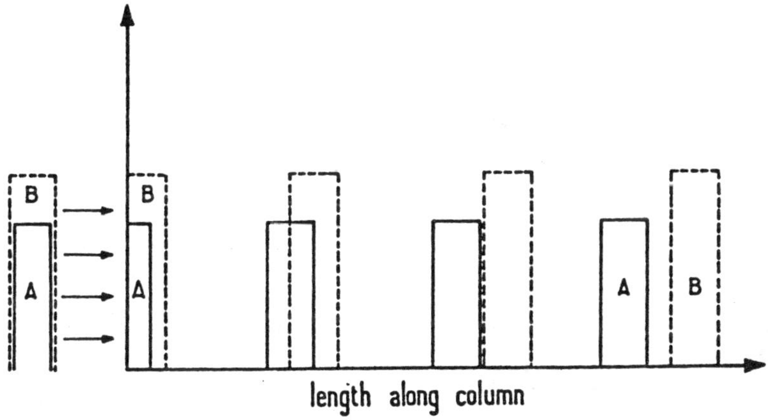

Fig3 The ideal linear binary pulse: Separation of two solutes A and B by a solvent S under conditions of linear (independent) sorption of A and B, and negligible irreversibilities. (schematic)

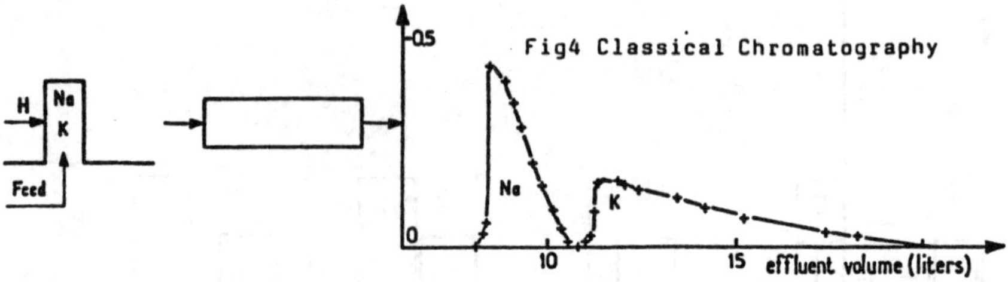

Fig4 Classical Chromatography

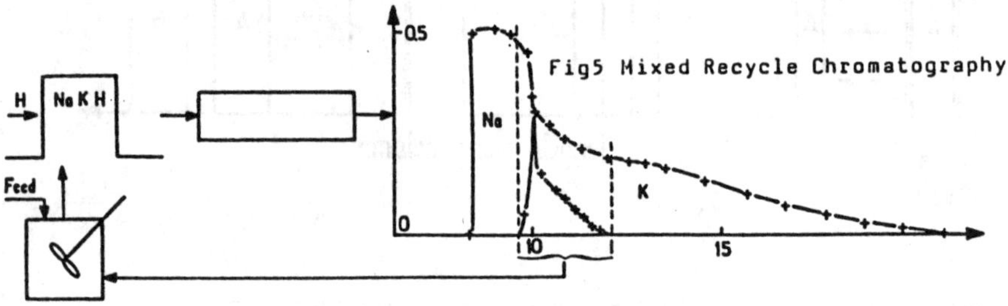

Fig5 Mixed Recycle Chromatography

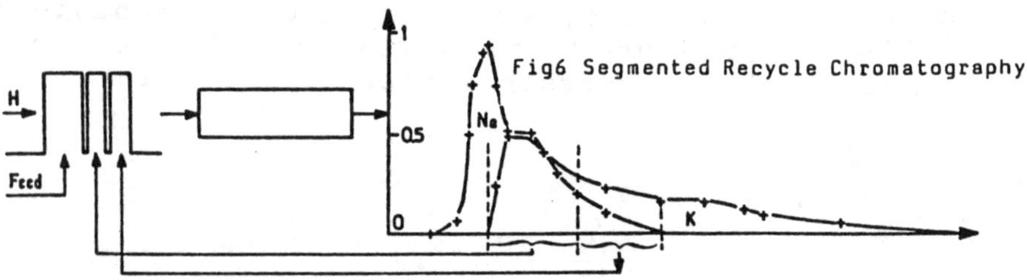

Fig6 Segmented Recycle Chromatography

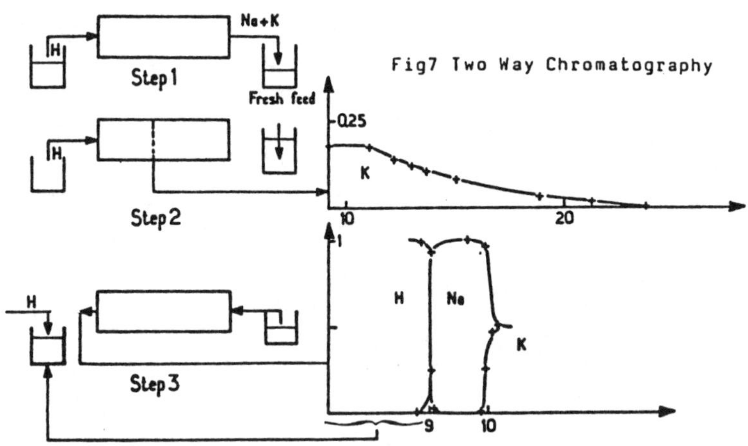

Fig7 Two Way Chromatography

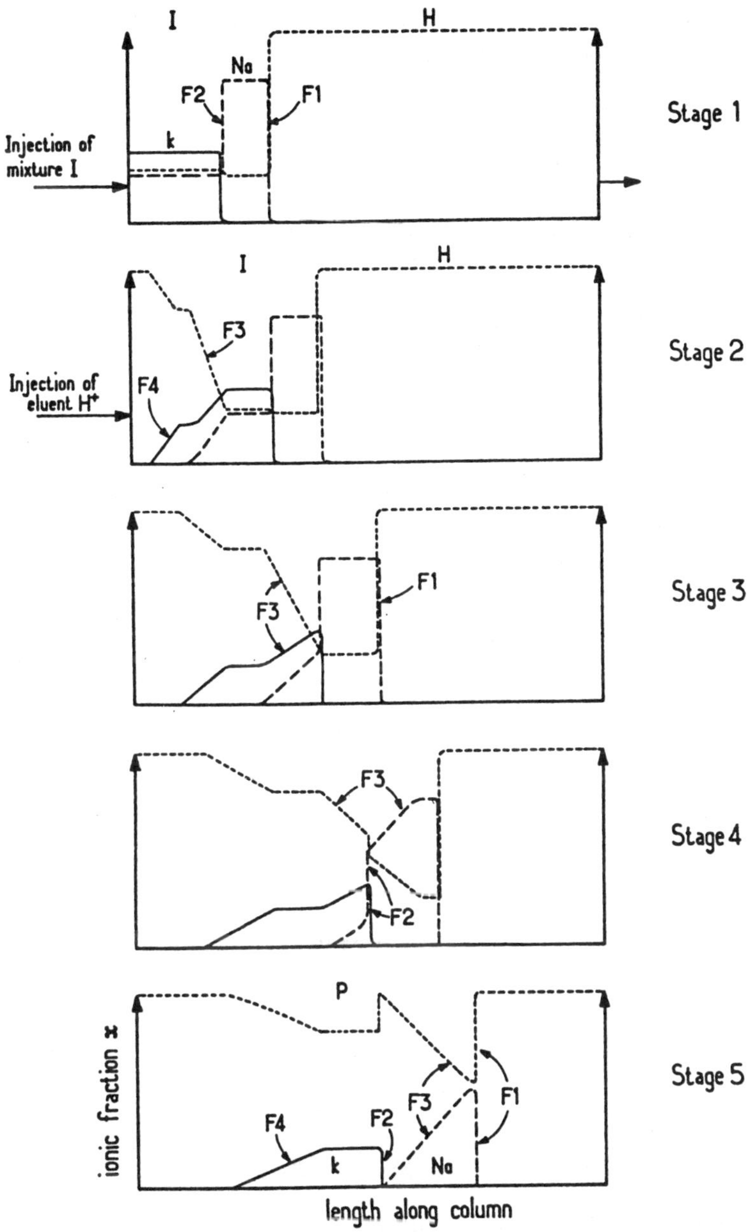

Fig8 Elution of a ternary pulse of K^+, Na^+, H^+ by H^+ eluent on an ion-exchange column. Successive stages of the process are shown.

MOSSBAUER AND ELECTRON MICROPROBE STUDIES OF PRECIPITATION PHENOMENA IN NAFION ION EXCHANGE MEMBRANES

A. Meagher[a,b], B. Rodmacq[a], J.M.D. Coey[b] and M. Pinéri[a]

(a) Centre d'Etudes Nucléaires de Grenoble, Département de Recherche Fondamentale, Section de Physique du Solide, Equipe Physico Chimie Moléculaire 85 X, 38041 Grenoble Cedex (France)

(b) Physical Laboratory, Trinity College, Dublin 2 (Ireland)

Introduction

Mössbauer spectroscopy [1, 2] is a useful probe of the microscopic environment of certain ions in ion exchange membranes and resins because it provides information both about the chemical environment of the exchanged ions and also about the rigidity of the lattice in which they are bound. This paper presents results obtained using ^{57}Fe Mössbauer spectroscopy applied to Nafion (R) perfluorinated cation exchange membranes in the Fe^{2+} and Fe^{3+} forms. Nafion has a polytetrafluoroethylene backbone with side groups of the form

$$- O - CF_2 - \underset{\displaystyle CF_3}{\overset{}{\underset{|}{CF}}} - O - CF_2 - CF_2 - SO_3^- \quad H^+$$

The Nafion used in our study was of equivalent weight 1200 corresponding to an ion exchange capacity of 0.83 meq/g. Nafion products exhibit exceptional chemical, thermal and mechanical stability and have good ionic conductivity and permselectiv properties. They have found applications as membrane separators in electrochemica cells [3, 4, 5]. A recent book [6] is mainly devoted tothis kind of material.

Mössbauer results on Nafion membranes exchanged with Fe^{2+} and Fe^{3+}

All Fe^{2+} samples examined by us and having more than 6 % water by weight only contain a single iron species identified as $[Fe\ (H_2O)_6]^{2+}$. Below 6 % water content we see evidence for a range of environments for the iron, reflecting the dehydration of the hexaquo complexes [7]. For each sample, the behaviour of the absorption area (f-factor) below 160 K is well fitted with a Debye model (θ_D = 140 K) ; the deviation from this model above 160 K is greater for more hydrated samples. This deviation can be attributed to a glass transition of the ionic phase [8, 9].

Ferric iron appears in a wide variety of structural forms in Nafion [8, 9]. A typical spectrum of a fully neutralized sample is shown in figure 1a top line. In figure 1b top line is shown the centre of this spectrum on an expanded scale. We have identified ferric iron in three different environments inside the Nafion. These are

(i) Fe^{3+} ions far from each other (> 12 Å) (sextuplet C of figure 1a)

(ii) Fe^{3+} dimers (doublet B of figure 1b)

(iii) Fe^{3+} ions clustered in small groups (doublet A of figure 1b)

These attributions are confirmed by the behaviour of the spectra in an applied magnetic field. The dimers we have found in Nafion give the same spectrum as those found in aqueous solution which Knudsen et al [10] assigned to the following species

$$[(H_2O)_5\ Fe - O - Fe\ (H_2O)_5]^{4+}$$

The amount of iron found in each of these three species varies with the total amount of iron present in the membrane. The data can be well fitted by a model of random filling by Fe^{3+} ions of sites distributed in unidimensional chains [11].

As in the case of Fe^{2+} Nafion, we see changes in all spectral parameters below 6 % water content. A Debye model with θ_D = 145 K gives a good fit for the Fe^{3+} Nafion absorption areas below 160 K. The Debye temperatures reported here are about 50 degrees lower than in frozen solutions [12, 13]. Infrared data [14] suggest that hydrogen bonding of water in the Na form of Nafion is weaker than in bulk water.

Precipitation phenomena

We have carried out experiments to study how iron is displaced from the membrane when it is reexchanged with another cation. Figure 1 shows the effect of soaking an Fe^{3+} Nafion in a solution of K^+. The first point to notice is that we still have a Mössbauer absorption – the iron has not been eliminated from the membrane. All of the iron is now contained in a single new phase which displays a magnetically split hyperfine pattern at 4.2 K (figure 1a bottom spectrum). The parameters of this spectrum (H_{hf} = 475 kOe, I.S. = .46 relative to iron, Q.S. = -.03) are quite similar to those reported by Coey and Readman [15] for amorphous ferric hydroxide. The magnetic sextuplet transforms with increasing temperature into a doublet (doublet D of figure 1b) in a manner characteristic of superparamagnetic small particles [16]. Magnetization measurements above the blocking temperature enable an estimate of the particle size to be given as 28 Å in diameter [11].

Our effort at reexchanging the iron Nafion has led to the precipitation of very small particles of ferric hydroxide inside the membrane. This occurrence is by no means restricted to reexchange with K^+. Table 1 presents results of reexchange of Fe^{2+} and Fe^{3+} Nafion samples using aqueous solutions of the chlorides of various cations (listed in order of increasing solution pH). The concentrations used were the same as for the original iron exchange (.2M for Fe^{3+}, .4M for Fe^{2+}). Solutions of NaOH, KOH, NH_4OH were also found to precipitate the iron in Fe^{3+} Nafion. We see that only solutions with a low pH have the effect of eliminating Fe^{3+} from the membrane. Fe^{2+} is more easily removed. Fe^{3+} is known to precipitate in aqueous solutions at pH $\gtrsim$ 2-3.

The temperature dependence of the Mössbauer absorption area is presented in figure 2 for a precipitated Fe^{3+} sample. The Debye temperature below 160 K has risen from 145 K to 250 K reflecting the increase in rigidity for the iron ions now bound in precipitated particles. However deviation from Debye behaviour still occurs near 160 K showing that the precipitated particles remain associated with the ionic phase [17].

A systematic study of the phenomenon of iron precipitation in Nafion has been carried out using both Mössbauer spectroscopy and electron microprobe ("Castaing probe" [18]). In the electron microprobe the sample is bombarded with a beam of electrons and the number of emitted x rays in a certain energy range (corresponding to the element we wish to examine) is counted. The number observed is proportional to the revelant elemental concentration. The volume of sample probed in this way is about $1\mu^3$.

Figure 3 shows the results obtained using the microprobe across the thickness ($\sim$ 250μ) of a Nafion Fe^{3+} membrane before and after reexchange with K^+. After reexchange we see that K^+ is neutralizing the sulphonate groups (it is evenly distributed across the membrane) while the precipitated iron hydroxide particles are concentrated in a layer in the membrane surface. The Mössbauer spectrum of a membrane which had its surface scraped off after reexchange gave the same spectral parameters (and blocking temperature) as the complete membrane. This shows that the concentration profile observed with the electron microprobe is due to a distribution in the average distance between the precipitated particles and not to a distribution of their sizes [11].

We interpret these results as follows : K^+ enters the membrane and replaces the iron by ion exchange. The iron tends to leave the membrane. However it meets near the surface regions of higher pH (due to inward diffusing electrolyte) and it precipitates.

Evidence concerning the evolution of precipitation with time is presented in figure 4. Initially identical Fe^{3+} samples were soaked for different amounts of time in .2M KOH. In the Mössbauer spectra we see the growth with soaking time of the magnetically split component due to precipitated particles. After 45 minutes more than 40 % of the iron has precipitated and after 120 minutes more than 70 %. The magnetic field characteristic of this species increases with soaking time indicating an increase in the size of the particles [11], [16]. The peaks in the profile at the membrane surfaces both broaden and heighten with soaking time. The precipitate first forms in the surface and later forms in regions further and further away from the surface as well as continuing to form in the surface.

Figure 5 shows the influence of the concentration of the reexchange solution on the precipitation phenomenon (soaking time = 40 hrs). 0.1 and 6M NaOH solutions have caused all the iron to precipitate. However none of the iron in the sample soaked in 12M NaOH has been exchanged or precipitated. The percentage of iron found in dimeric form has increased for this sample. A similar effect is observed when a normal Fe^{3+} Nafion is vacuum dehydrated [11, 8]. The profile is broader for the 6M case compared with the 0.1M.

In figure 6 we show the profiles obtained after successive precipitations. These were obtained by soaking the membrane first in $FeCl_3$ then in KOH solution and then repeating this procedure. Each precipitation takes place preferentially in the membrane surface. We see no sign of saturation in the surface.

Regarding the size of the precipitated particles these can be estimated from their blocking temperatures. We find that the size is generally independent of the chemical treatment used to cause precipitation [11].

Discussion

The results on the variation of the f-factor for all samples show that the iron is always associated with an aqueous phase which displays a glass transition [17] at about 160 K. This is true even for the precipitated particles.

The overall precipitation mechanism seems clear

$$3K^+ + \overline{\overline{Fe^{3+}}} \rightleftharpoons \overline{\overline{3K^+}} + \overline{Fe^{3+}}$$

$$\overline{Fe^{3+}} + \overline{3H_2O} \rightleftharpoons \overline{Fe(OH)_3} + 3H^+$$

We have here chosen K^+ as the reexchanging cation for illustrative purposes. Barred species are present in the membrane and double barred species also serve to neutralize the sulphonate groups. The precipitation releases H^+ ions into the external solution to balance the charge lost when K^+ ions enter the membrane. We see therefore why acidic external solutions (eg Zr^{4+} and H^+) favour the straightforward ion exchange reaction and the elimination of iron from the membrane. We might expect that the iron Nafion soaked in 12M NaOH would be completely precipitated. This is not the case (see figure 5). The Sodium/iron exchange has not even taken place. We explain this in terms of membrane dehydration being the dominant effect. Yeager et al [19] have observed that Na^+ self diffusion coefficients in Nafion drop sharply at high caustic strength.

Regarding the shape of the profile for samples completely precipitated (parameterized in terms of the width of the peaks of the concentration profiles) we have observed two effects :

(i) the peaks broaden somewhat with an increase in external solution molarity

(ii) the peak width is affected strongly by the cation used for reexchange [11]. In order of decreasing width we have

$$NH_4^+ \gg Na^+ \simeq K^+ > Ba^{2+} \simeq Al^{3+}$$

For the moment no theory has been found to explain these latter results. It is to be expected that such a theory would take into account simultaneously the tendency for iron to leave the membrane because of ion exchange and its tendency to precipitate in regions of higher pH due to electrolyte invasion. However, these preliminary results show that the utilization of techniques such as Mössbauer spectroscopy and electron microprobe can give useful information about the size and separation of precipitates in ion exchange membranes. Such studies could be usefully developed and applied to such systems as conductive oxide catalytic electrodes in solid polymer electrolyte technology [3].

Acknowledgements

We wish to thank J.P. Bossy (Service d'Etudes des Matériaux, Département de Métallurgie, CEN-G) who obtained the electron microprobe data. We thank Dr. R. Durand for a helpful discussion .

References

[1] N. GREENWOOD and T. GIBB, "Mössbauer Spectroscopy" Chapman and Hall, London, 1971

[2] V.I. GOLDANSKII and R.H. HERBER (EDS) "Chemical Applications of Mössbauer Spectroscopy", Academic Press, New York, 1968

[3] R.S. YEO, Chapter 18 of Reference 6

[4] P.W.T. LU and S. SRINIVASAN, J. Applied Electrochem., 9, 269, 1979

[5] A.B. LACONTI, E.N. BALKO, T.G. COKER, A.G. FAGALA in "Proceedings of the Symposium on Ion Exchange" R.S. YEO and R.P. BUCK (EDS), the Electrochemical Society, Pennington N.J., 1981

[6] A. EISENGERG and H.L. YEAGER (EDS) "Perfluorinated Ionomer Membranes" A.C.S. Symposium Series 180, 1982

[7] B. RODMACQ, M. PINERI, J.M.D. COEY, Revue Phys. Appl., 15, 1179, 1980

[8] B. RODMACQ, J.M.D. COEY, M. PINERI, Chapter 9 of reference 6

[9] B. RODMACQ, M. PINERI, J.M.D. COEY, A. MEAGHER, J. Polym. Sci. Polym. Phys. Ed., 20, 603, 1982

[10] J.M. KNUDSEN, E. LARSEN, J.E. MOREIRA, O. FAURSKOV NIELSON, Acta Chem. Scand., A29, 833, 1975

[11] B. RODMACQ, Thesis, Grenoble, 1982

[12] Our calculation based on graph given by S.L. RUBY in "Perspectives in Mössbauer Spectroscopy", S.G. COHEN and M. PASTERNAK (EDS), Plenum, New York, 1973

[13] S. MORUP, J.E. KNUDSEN, M.K. NIELSEN, G. TRUMPY, J. Chem. Phys. 65, 536, 1976

[14] M. FALK, Chapter 8 of reference 6

[15] J.M.D. COEY and P.W. READMAN, Earth Planetary Sci. Lett., 21, 45, 1973

[16] S. MORUP, J.A. DUMESIC, H. TOPSOE, Chapter 1 in "Applications of Mössbauer Spectroscopy vol. II" R.L. COHEN (ED), Academic Press, 1980

[17] N.G. BOYLE, J.M.D. COEY, V.J. McBRIERTY, Chem. Phys. Lett., 86(1), 16, 1982

[18] S.J.B. REED, "Electron Probe Microanalysis", Cambridge University Press, 1975

[19] H.L. YEAGER, B. KIPLING, R.L. DOTSON, J. Electrochem. Soc. 127, 303, 1980

	Fe^{3+} NAFION	Fe^{2+} NAFION
H^+	eliminated	eliminated
Zr^{4+}	eliminated	-
Al^{3+}	precipitated	eliminated
K^+	precipitated	eliminated
Na^+	precipitated	-
Mg^{2+}	precipitated	-
Li^+	95% precipitated	-
Ca^{2+}	precipitated	50% precipitated as ferric hydroxide 50% eliminated

TABLE 1 - Result of soaking an iron Nafion in various Chloride solutions.

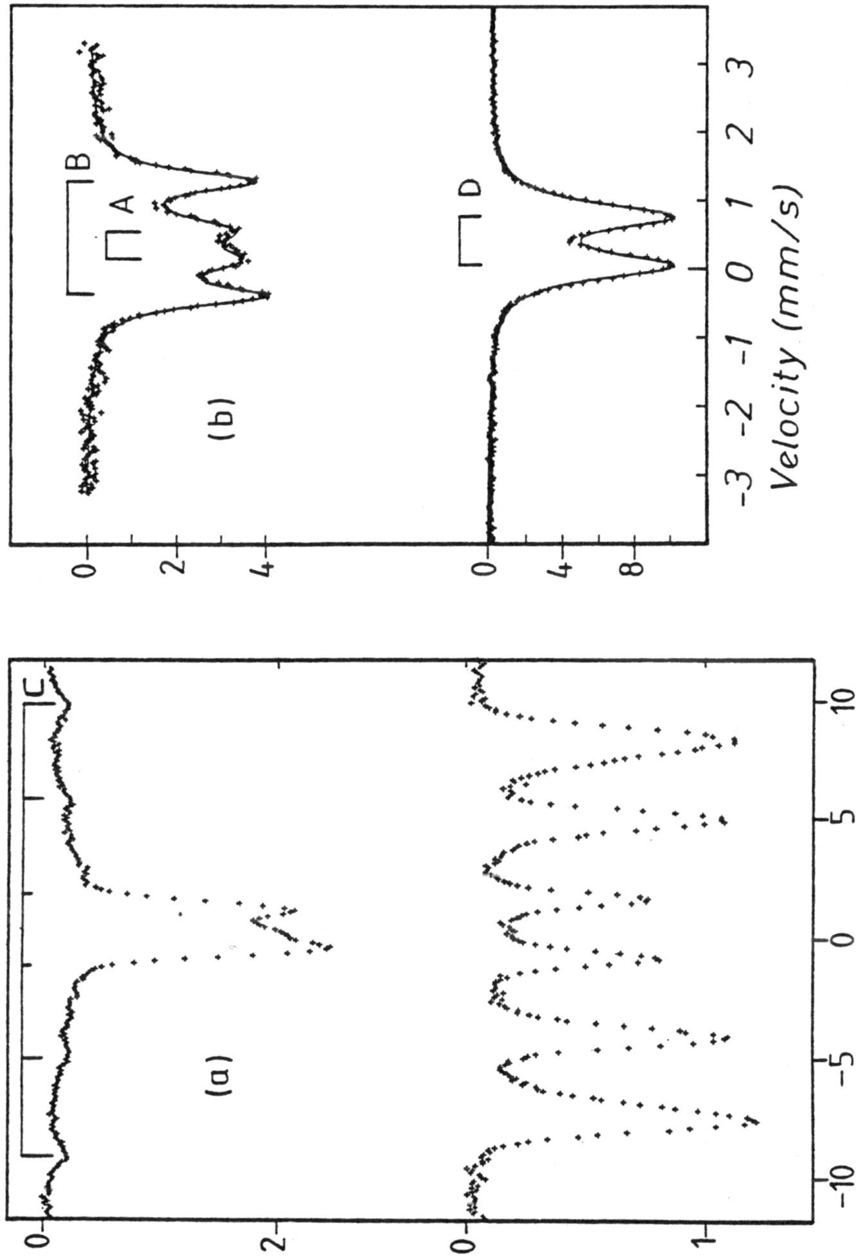

Figure 1 - Mössbauer spectra of a Fe^{3+} Nafion sample before (top spectrum) and after (bottom spectrum) reexchange with K^+.

(a) At 4.2 K

(b) At 80 K on expanded scale.

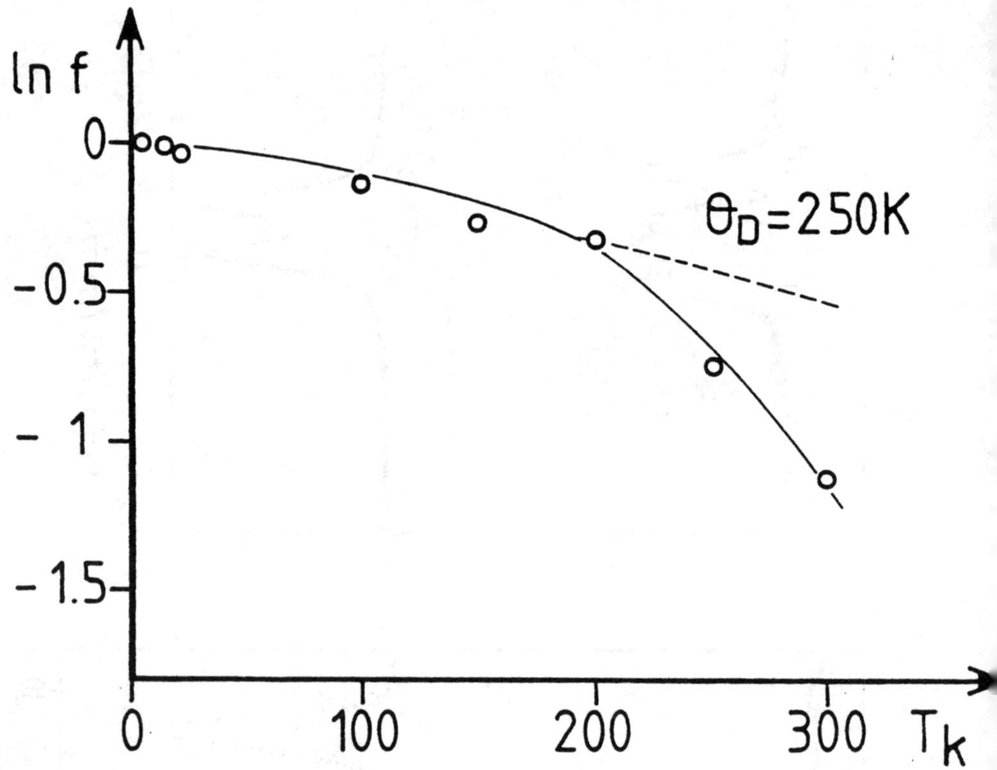

Figure 2 - Temperature dependence of the f-factor for precipitated iron in Nafion.

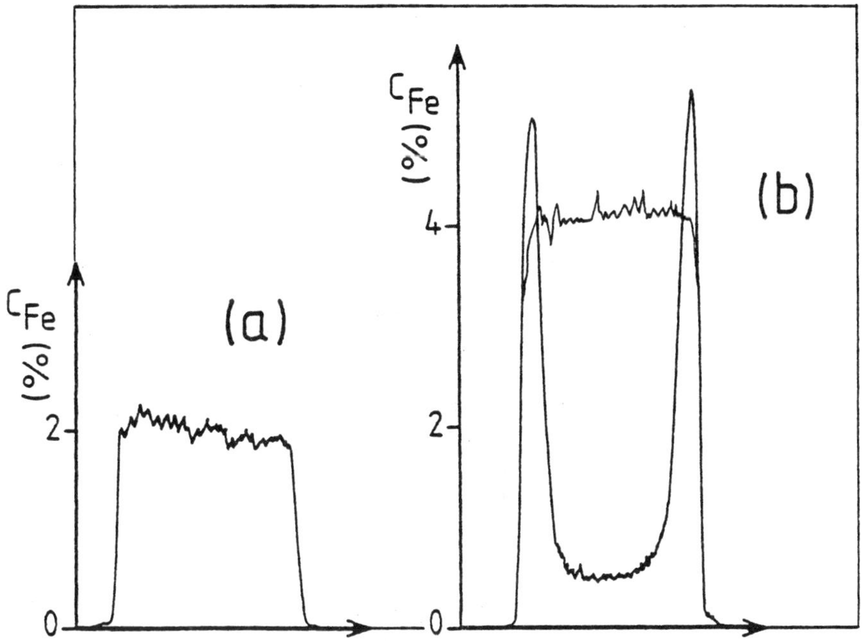

Figure 3 - Concentration profiles across the Nafion membrane thickness

(a) Iron profile for a Fe^{3+} Nafion sample.

(b) Iron and potassium (narrower trace, arbitrary scale) profiles for a Fe^{3+} Nafion sample reexchanged with K^{+}.

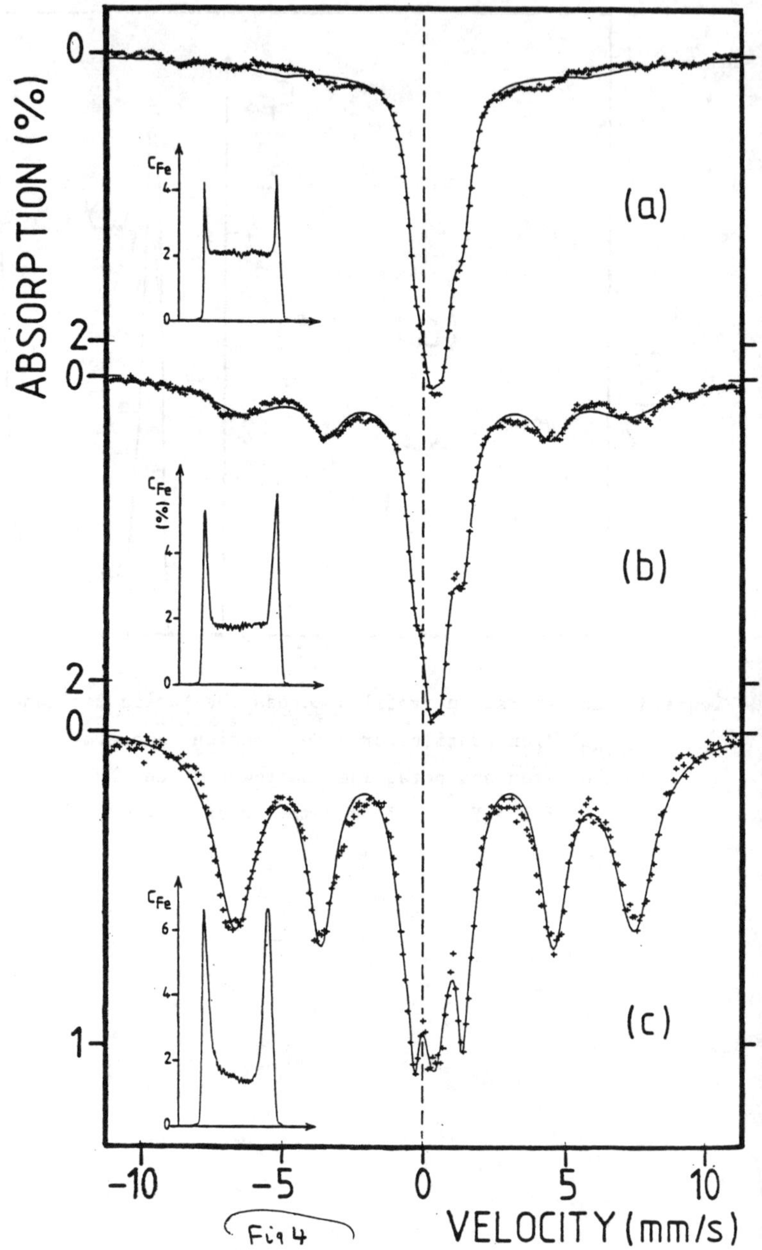

Figure 4 - Mössbauer spectra at 4.2 K and iron concentration profiles for a Fe^{3+} Nafion sample soaked for different times in 0.2M KOH. (a) 15 minutes ; (b) 45 minutes ; (c) 120 minutes.

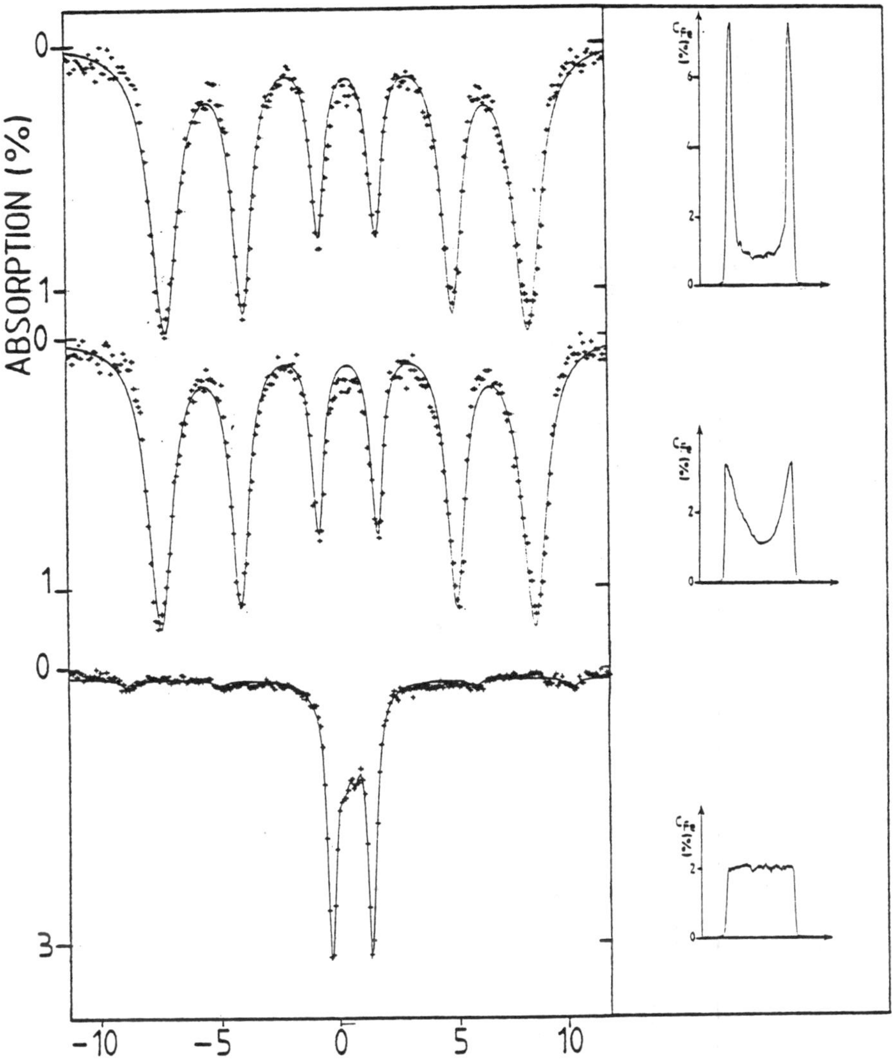

Figure 5 - Mössbauer spectra at 4.2 K and iron concentration profiles for a Fe^{3+} Nafion sample soaked for 40 hours in solutions of NaOH differing concentrations (a) 0.1M ; (b) 6.0M ; (c) 12.0M

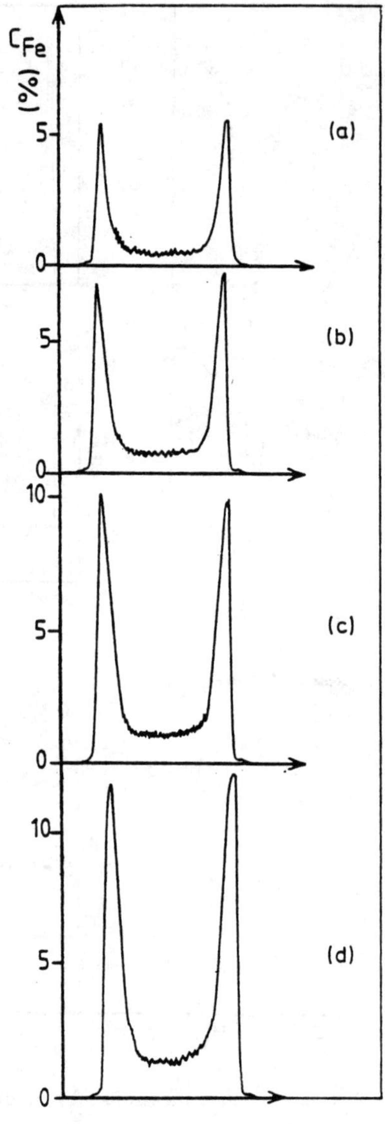

Figure 6 - Iron concentration profiles for a sample after successive precipitati
in 0.2M KOH

(a) 1 precipitation

(b) 2 precipitations

(c) 3 precipitations

(d) 4 precipitations

ION EXCHANGE KINETICS IN ZEOLITE A

D. Drummond, A. De Jonge[†] and L.V.C. Rees[*]

Physical Chemistry Laboratories,

Imperial College of Science and Technology,

London SW7 2AY.

ABSTRACT

The kinetics of Na → Ca exchange in zeolite A has been measured at 25°C. The rates of complete replacement of Na^+ by Ca^{2+} have been obtained and, also, the rates of small incremental replacements over the range of zeolite composition from pure sodium to 90% calcium have been measured. The kinetics of the reverse exchange has also been studied to a limited extent. All of these kinetics have been analysed using the Paterson solutions of the diffusion equation.

In a separate study the degree of hydrolysis of zeolite A as a function of pH has been ascertained.

INTRODUCTION

Zeolite A, in its sodium form, has recently been introduced in extensive quantities as a builder in detergents. In a recent paper[(1)] the high selectivities of Na-A for Ca^{2+} and Mg^{2+} have been demonstrated. In this equilibrium study the thermodynamic quantities

[*]To whom all correspondence should be addressed.

[†]Present Address:- Department of Analytical Chemistry, Veszprem, Schönherz Z. u 12, Hungary.

for these binary exchanges were obtained and polynomial expressions derived from which the binary isotherms at 25 and 65°C could be calculated for any normality of solution phase. The more difficult ternary exchange was also studied for solutions which initially contained Ca^{2+} : Mg^{2+} ratios of 2 : 1, 1 : 1, and 1 : 2. An attempt was made to predict the ternary isotherms from the binary isotherms.

Although a high selectivity for both of these divalent ions is an essential feature in the use of zeolite A as a builder in detergents it is also necessary for the removal of the divalent ions to be extremely rapid to prevent interactions with the surfactants present. Rapid removal of Ca^{2+} ions is readily accomplished if the Na-A cubes have side lengths of ∿ 1 μm. The Ca^{2+} ion concentration is reduced to acceptable levels in ∿ 20s. The removal of Mg^{2+} ions is not so rapid but the higher concentrations of Mg^{2+} which still exist after ∿ 1 min. is not a serious problem.

In this paper the kinetics of $Na^{+} \rightleftharpoons Ca^{2+}$ exchange in zeolite A will be presented and the effect of the presence of Mg^{2+} ions on the exchange will be demonstrated. Also, zeolite A has one other important characteristic which enhances its suitability as a builder. It is readily hydrolysed and so causes no environmental waste problems. The hydrolysis of zeolite A at various pH values will also be reported.

EXPERIMENTAL

In order to follow exchange times much greater than those which apply under detergency conditions larger crystals of zeolite A were required. These large crystals were synthesised using the Charnell method[2]. The crystals obtained were well-formed cubes of side lengths in the range 15 - 18 μm.

The kinetics of exchange were usually followed by means of sodium or calcium ion selective electrodes but some exchange rates were also measured, or compared, radiochemically with ^{22}Na as the radiotracer in the zeolite phase.

RESULTS

In the first series of experiments $Na^{+} \rightarrow Ca^{2+}$ exchange rates were measured at three different solution normalities. In these experiments the ratio of equivalents of cation (in this case Ca^{2+}) in the solution phase, M_s, to equivalents of cation (in this case Na^{+}) in the zeolite phase, M_z, was also varied from 100 : 1 down to 1 : 0.8. The three rates curves obtained are shown in Figure 1. These rate curves are very similar indicating that differences in the parameters which were varied had little effect on the rate of exchange.

The rate curves in Figure 1 were analysed using the Paterson[3] solutions for limited volume diffusion in a similar manner to that described by Danes and Wolf[4]. In these solutions the parameter W is required where W is defined as

$$W = \frac{M_z}{M_s} \times \frac{dCa_z}{dCa_s} \qquad (1)$$

Ca_z and Ca_s are the equivalent cation fractions of Ca^{2+} in the zeolite and solution phases respectively. The required values of W are obtained from the slope of the isotherm at specific values of Ca_s. The values of W used in the analysis of the rate curves were calculated from the polynomial expression for the Na - Ca exchange at 25^oC given by Barri and Rees[(1)] and are plotted in Figure 2. The isotherms for the three normalities used in these rate measurements are shown in Figure 3.

In the rate measurement for an initial solution phase of 3.4×10^{-3} equivalent dm^{-3} of Ca^{2+} and $M_z/M_s = 0.8$ it can be deduced from the isotherm in Figure 3 that Ca_z at the ingoing surface of the zeolite must always be > 0.90 during the time of the exchange. The values of W required for this region of Ca_z are shown in Figure 2 to be approximately constant around a value of 0.15. In Figure 4 $\tilde{D}$ is plotted against F, the fractional attainment of equilibrium for this system where $\tilde{D}$ is the average diffusion coefficient operating from F = 0 to the value of F against which $\tilde{D}$ is plotted.

The corresponding $\tilde{D}$ against F plots for solutions of 0.1 and 0.2 equivalent dm^{-3} are also shown in Figure 4. For these experimental conditions M_z/M_s was 1 : 40 and 1 : 100 respectively and Ca_z at the ingoing surface was now always > 0.975 for the first case and > 0.99 for the second. The values of W required in the analysis of these rate curves can be seen from Figure 2 to be around 0.04 and 0.015 respectively.

The rate of exchange for the 0.2 equivalent dm^{-3} solution was measured using a sodium selective electrode and also radiochemically.

The two $\tilde{D}$ against Ca_z (equal to F in these experiments) curves obtained from these rate measurements are shown in Figure 5. Although small differences can be seen at lower Ca_z values these two curves indicate that the two experimental techniques give very similar rates of exchange.

When the curves of $\tilde{D}$ against F at the different solution normalities are compared (Fig. 4) only minor differences are found when $F > 0.5$. At lower F values differences can be seen in these plots but not too much significance should be placed on these differences. The large, sudden decrease in $\tilde{D}$ when $F < 0.2$ for the 3.4×10^{-3} equivalent dm^{-3} solution is undoubtedly due to the response time of the ion selective electrode at times of exchange < 60s. The extrapolation shown in Figure 4 for this system is probably a better description of the variation of $\tilde{D}$ with F in this region.

No experimental values were obtained at $F < 0.35$ with the 0.1 equivalent dm^{-3} solution system. The extrapolated $\tilde{D}$ values shown in Figure 4 below $F = 0.35$ were calculated from an extrapolation of the $\sqrt{t}$ rate curve from $F = 0.35$ to zero.

When F is very close to 1 only very small changes in the Na^+ concentration occur and the $\tilde{D}$ values calculated from the rate curves in this region are liable to large errors.

The decrease in $\tilde{D}$ as F increases indicates that the differential exchange diffusion coefficient, $\overline{D}$, must decrease with increasing calcium loading of the exchanger.

The reverse exchange kinetics i.e. $Ca^{2+} \longrightarrow Na^{+}$ have also been measured at a solution normality of 0.2 equivalents dm^{-3}. The rate curve obtained is shown in Figure 6 where it can be compared with the corresponding $Na^{+} \longrightarrow Ca^{2+}$ rate curve. As the self-diffusion of sodium has been found to be much faster than that of calcium in zeolite A these forward and reverse exchange kinetics follow the predictions of the Helfferich & Plesset theory[(5)]. In Figure 5 the variation of $\tilde{D}$ with Ca_z for the reverse kinetics may be compared with the corresponding variation for the forward kinetics. In the reverse kinetics $F = (1 - Ca_z)$ so $\tilde{D}$ now increases as the exchange proceeds i.e. on going from pure Ca-A to pure Na-A zeolite and correspondingly the differential diffusion coefficient $\bar{D}$ for this exchange must also increase on going from Ca-A to Na-A zeolite. Comparison of $\tilde{D}$ against Ca_z indicates, however, that for both forward and reverse kinetics $\tilde{D}$ decreases with increasing calcium content of the zeolite and $\tilde{D}$, at any value of Ca_z, is always greater for the forward kinetics than for the reverse kinetics.

In a second series of experiments the $Na^{+} \longrightarrow Ca^{+}$ exchange rates were measured over small 5% changes in Ca_z from pure Na-A up to 83% Ca-A zeolite. The solution phase concentration was maintained at a constant 3.4×10^{-3} equivalents dm^{-3}. Each of the fifteen rate curves obtained were analysed as before using Paterson's solutions. The $\tilde{D}$ values obtained as a function of F from three rate curves are plotted in Figure 7. This figure includes details of the range of Ca_z covered and the values of M_z/M_s involved in each rate measurement. At low values of Ca_z, because of the very high selectivity of Ca^{2+} in zeolite A, the range of W covered in each rate analysis was very large e.g. for Ca_z going from 0.28 to 6.44%, W increased from 0.809 at small F values to 122 near the end of the run where little Ca^{2+} remained in the solution phase.

In Figure 8 the variations of $\tilde{D}$ over each increment of composition covered are plotted against Ca_z to show the overall trend in $\tilde{D}$ on going from pure Na-A to almost pure Ca-A in small incremental steps. Although the trend of $\tilde{D}$ against Ca_z in this figure is similar in shape to that for the corresponding system (3.4×10^{-3} equivalent dm^{-3}) in Figure 4 where exchange from pure Na-A to almost pure Ca-A is covered in one step it is interesting to note that the $\tilde{D}$ values at corresponding Ca_z values are a little greater for the incremental uptakes.

Finally in Figure 9 some preliminary ternary rates of exchange are presented. In these experiments pure Na-A was mixed with solutions containing 1 : 1 and 2 : 1 Ca^{2+} : Mg^{2+} ratios at zero time. The rate of uptake of total divalent ion was measured as a function of time. Figure 9 shows that the rate of uptake for the 2 : 1 Ca^{2+} : Mg^{2+} case was little different from the rate for the corresponding pure calcium solution. When the solution phase, however, was 1 : 1 Ca^{2+} : Mg^{2+} initially, the rate of uptake of divalent ion was now somewhat slower but still much greater than the rate of uptake for the pure magnesium solution phase. Because of the very high selectivity for Ca^{2+} the composition of the zeolite phase at infinite time in both of these cases tended to be almost pure Ca-A. However, it is known from equilibrium studies[(1)] that Mg^{2+} ions are taken up by the zeolite at low divalent ion loadings to be replaced by Ca^{2+} at high loadings even when the solution phase contains 1 : 1 Ca^{2+} to Mg^{2+}. The slower rate found than for the 1 : 1 Ca^{2+} : Mg^{2+} system suggests that Mg^{2+} ions are taken up by the zeolite and block the 8-ring oxygen sites which control diffusion in the zeolite A channel system. In the 2 : 1 Ca^{2+} : Mg^{2+} case insufficient quantities of Mg^{2+} are taken up to cause any great degree of blocking. In future rate measurements in these ternary systems all three ions will be analysed in the solution phase as a function of time to obtain a better insight into the kinetics involved.

HYDROLYSIS OF ZEOLITE A.

In order to speed up the initial mixing in an ion exchange kinetics measurement the zeolite was slurried with water at pH > 8 and then the solution of counter-ion added. However, abnormally high outputs were obtained from the sodium selective electrode which drifted upwards at quite high rates before any counter-ion was introduced. These high concentrations of sodium in the solution phase were confirmed by atomic absorption spectroscopy. A study of this hydrolysis of Na-A zeolite as a function of pH was, therefore, carried out. In order to vary the pH of the zeolite/water slurry and exclude the possibility of ion exchange NMe_4OH was used as the alkali. NMe_4^+ ions are too large to enter the zeolite A channels.

In Figure 10 the release of Na^+ at various initial pH values is shown as a function of time. In these experiments large crystals, synthesised by the Charnell method, were used but similar rates of release were obtained from much smaller crystals indicating that this release was not particle diffusion controlled. In Figure 14 it can be seen that, after 2 hours, 24, 14 and 10% of the sodium in the zeolite is released at pH's of 6.49, 7.96 and 9.06 respectively. After about one hour the rate of hydrolysis seems to attain a constant value. The small amount of hydrolysis at pH 11.87 discounts the possibility of any entrained sodium salts being present in the zeolite channels from the synthesis process.

In the above experiments the pH was also followed as a function of time. It can be seen from Figure 11 that there was an initial positive pH jump due probably to $Na^+ - H^+$ exchange followed by a slow drop in pH after about one hour. The size of the initial jump decreased as the initial pH increased. In Figure 12 the results shown in the previous two figures are combined to demonstrate the degree of hydrolysis after various lengths of contact time plotted against the initial pH. There seems to be a roughly linear relationship between the degree of hydrolysis and the initial pH in this first hour of contact for initial pH's less than 9.

A series of Na - Ca exchange experiments were next carried out by adding Na-A zeolite to a calcium chloride solution of known initial pH and monitoring both the sodium and calcium concentrations in the solution phase with sodium and calcium specific electrodes. If any hydrolysis occurred during these experiments then the equivalent sodium concentration in the solution phase should be greater than the decrease in equivalent calcium concentration. The results obtained are shown in Figure 13. Any value of the ordinate greater than zero should indicate the degree of hydrolysis. Although the results are not conclusive there does seem to be some hydrolysis at an initial pH of 6.81. The initial negative values in the ordinate at small times are probably due to differences in the response times of the respective ion selective electrodes. The presence of calcium ions in the zeolite does seem to have a strong stabilising effect.

A tentative mechanism is shown in Figure 14 to explain these hydrolysis results. The actual form which the aluminium and silicon oxides/hydroxides take is not known. Since the initial Na^+ - H^+ exchange can only account for a small fraction of the observed Na^+ ion release it seems probable that the hydrolysis process is autocatalytic with the release of H^+ continuing the reaction. The rate of hydrolysis eventually decreases due to the stability of H-A and the diffusion of H^+ to the outer solution gives rise to the observed pH drop after one hour.

ACKNOWLEDGEMENT

We should like to thank the Procter and Gamble Company, for the provision of a grant-in-aid and the donation of apparatus which allowed the research reported in this paper to be carried out.

REFERENCES

1. BARRI, S.A.I. and REES, L.V.C., J. Chromatography 201 (1980) 21.

2. CHARNELL, J.F., J. Cryst. Growth 8 (1971) 291.

3. PATERSON, S., Proc. Physic. Soc. 59 (1947) 50.

4. Danes, F. and Wolf, F., Z. Phys. Chem., Leipzig 252 (1973) 15.

5. Helfferich, F. and Plesset, M.S., J. Chem. Phys. 28 (1958) 418.

Figure 1. Rates of $Na^{+} \longrightarrow Ca^{2+}$ exchange in zeolite A at three different solution phase total normalities (TN) and three different ratios of M_z/M_s. F is fractional attainment of equilibrium.

○ TN = 3.4×10^{-3} N ; M_z/M_s = 0.81

● TN = 0.1 N ; M_z/M_s = 2.7×10^{-2}

□ TN = 0.2 N ; M_z/M_s = 1×10^{-2}

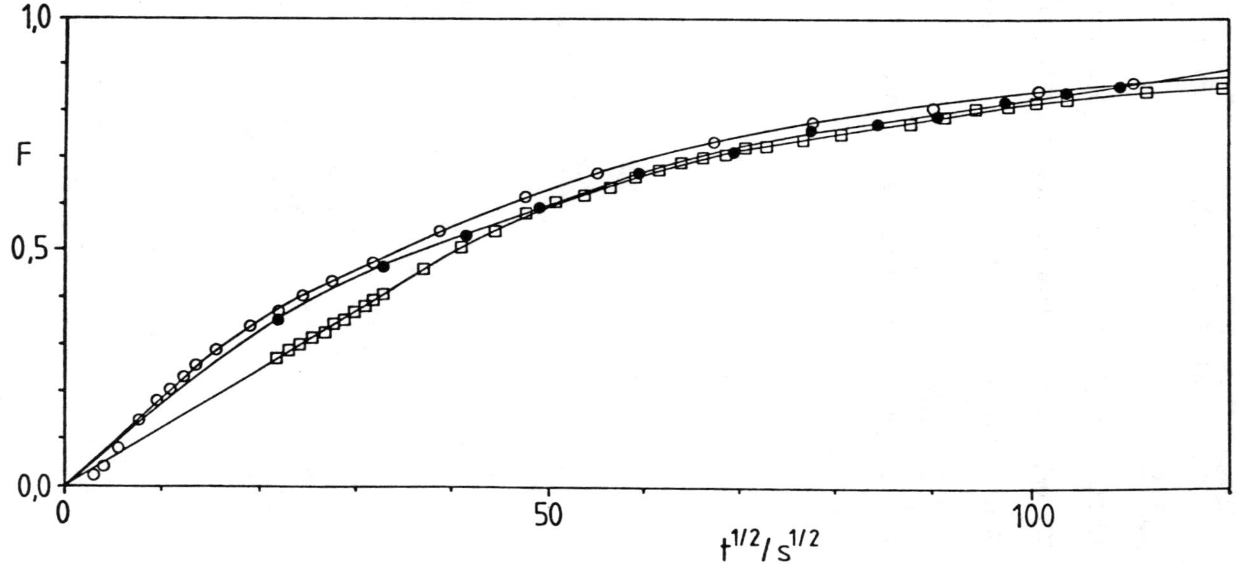

Figure 2. Dimensionless border concentration factor, W plotted as a function of Ca_z for three different total solution phase normalities.

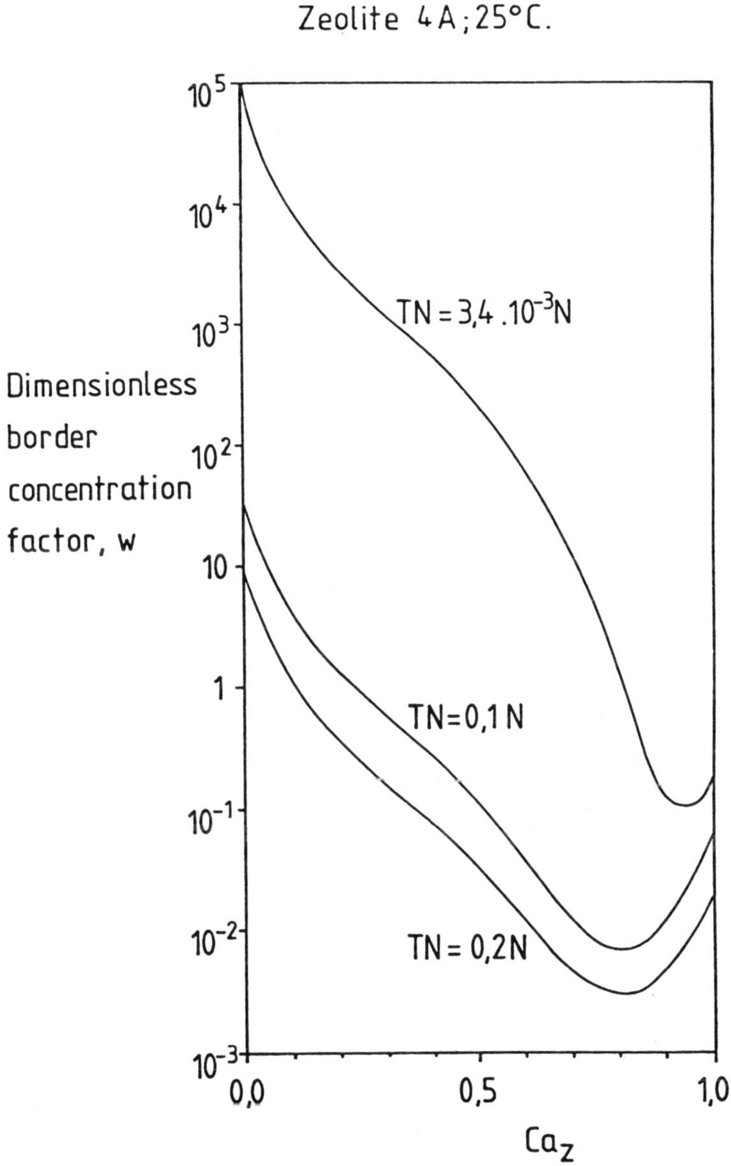

Figure 3. Na - Ca exchange isotherms at 25°C for zeolite A at three different total solution phase normalities. Ca_s and Ca_z are the equivalent cation fractions of calcium in solution and zeolite phase respectively.

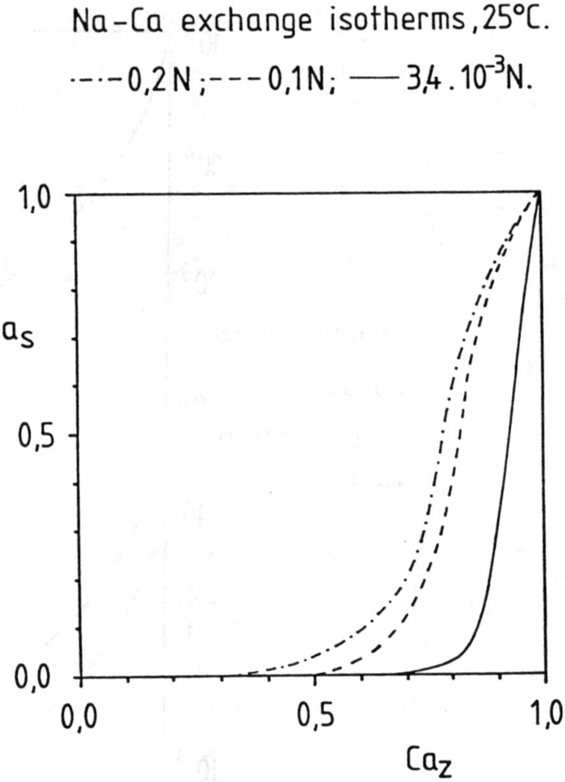

Figure 4. Plots of $\tilde{D}$ against fractional attainment of equilibrium for $Na^+ \rightarrow Ca^{2+}$ exchange in zeolite A at 25°C and three different total solution phase normalities.

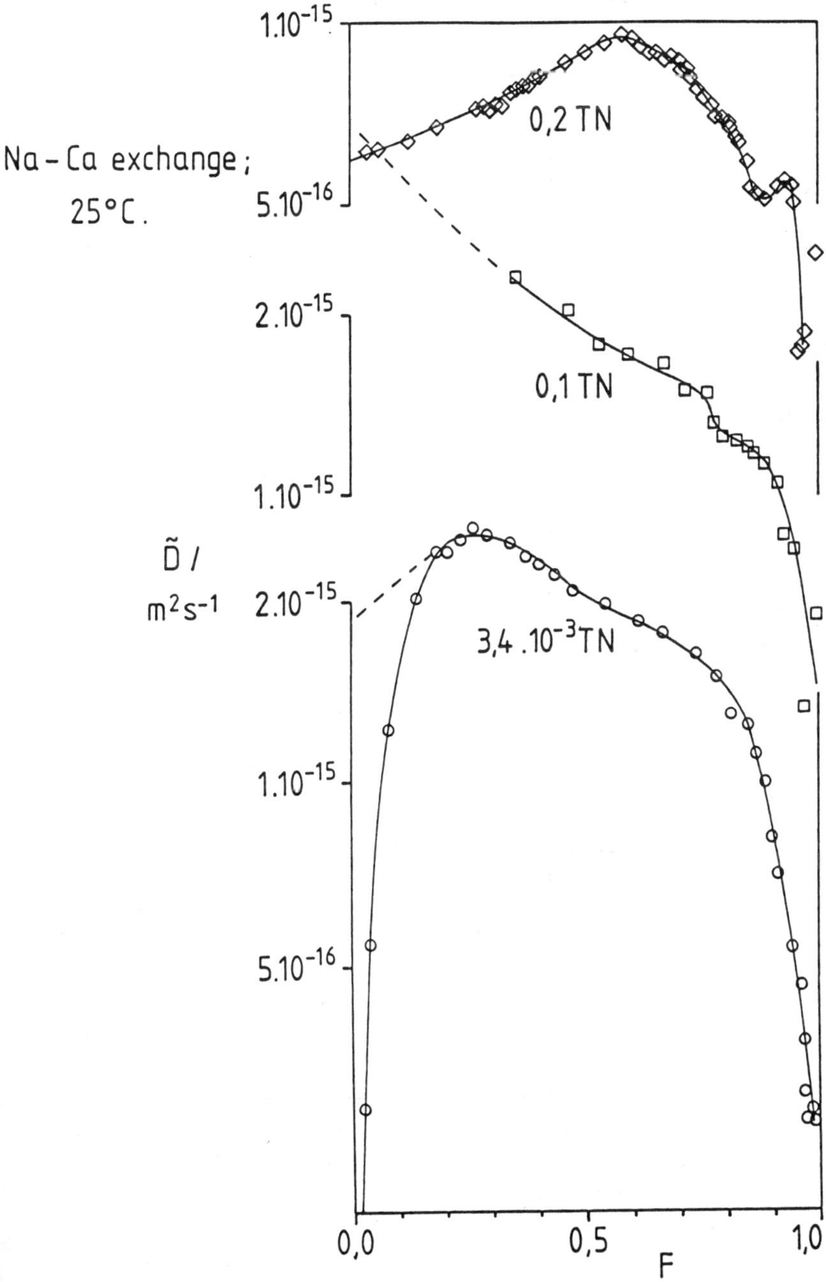

Figure 5. Plots of $\tilde{D}$ against Ca_z for zeolite A at 25°C and total solution phase normality of 0.2 N.

$Na^+ \rightarrow Ca^{2+}$ □ Radiotracer ◇ Electrodes

$Ca^{2+} \rightarrow Na^+$ ○ Electrodes

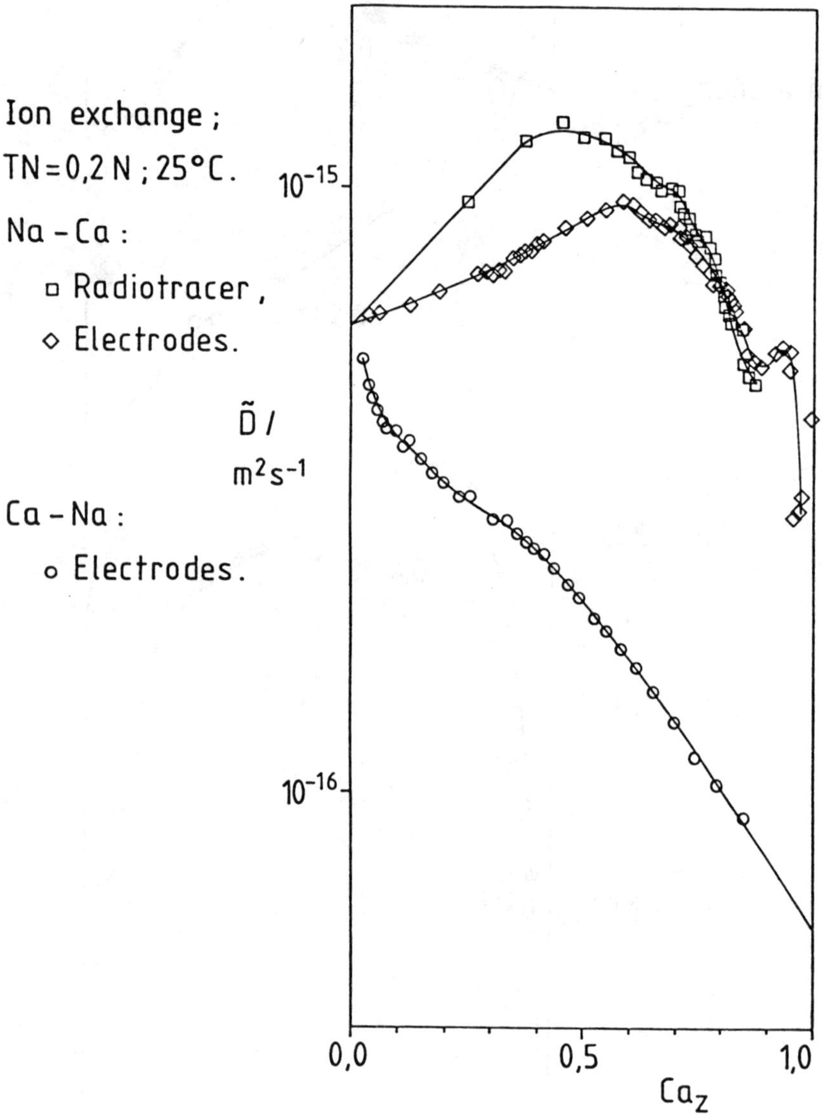

Figure 6. Rates of exchange in zeolite A at 25°C for total solution phase normality of 0.2 N and $M_s/M_z = 0.01$
○ $Na^+ \rightarrow Ca^{2+}$ exchange ● $Ca^{2+} \rightarrow Na^+$ exchange.

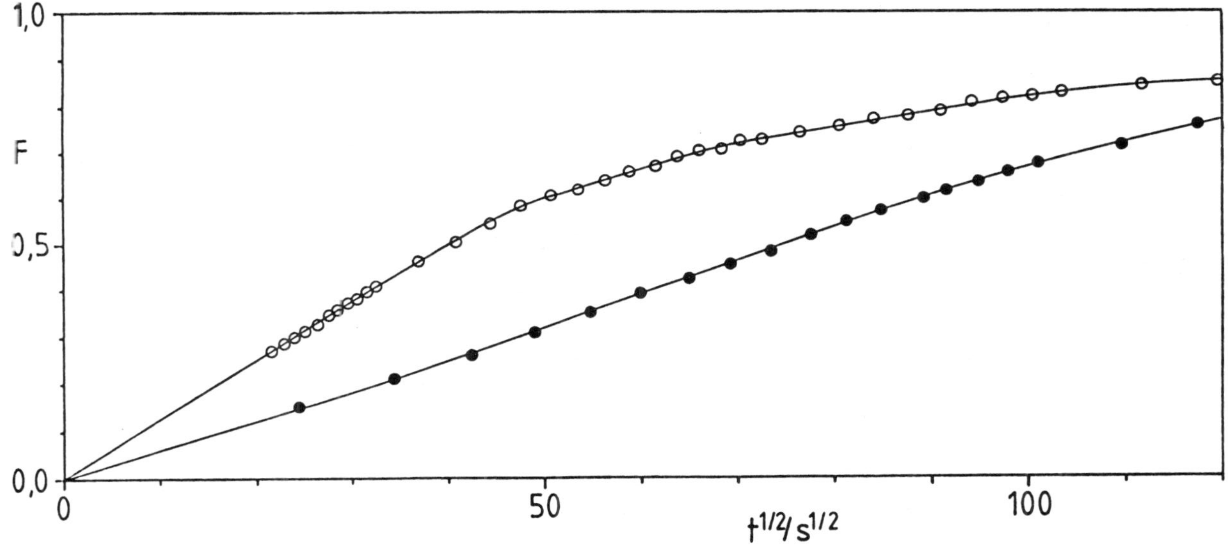

Figure 7. Plots of $\tilde{D}$ against fractional attainment of equilibrium for small incremental uptakes of calcium in zeolite A at 25°C. Total solution phase normality is 3.4×10^{-3} N. The range of Ca_z and value of M_z/M_s covered in each step are indicated.

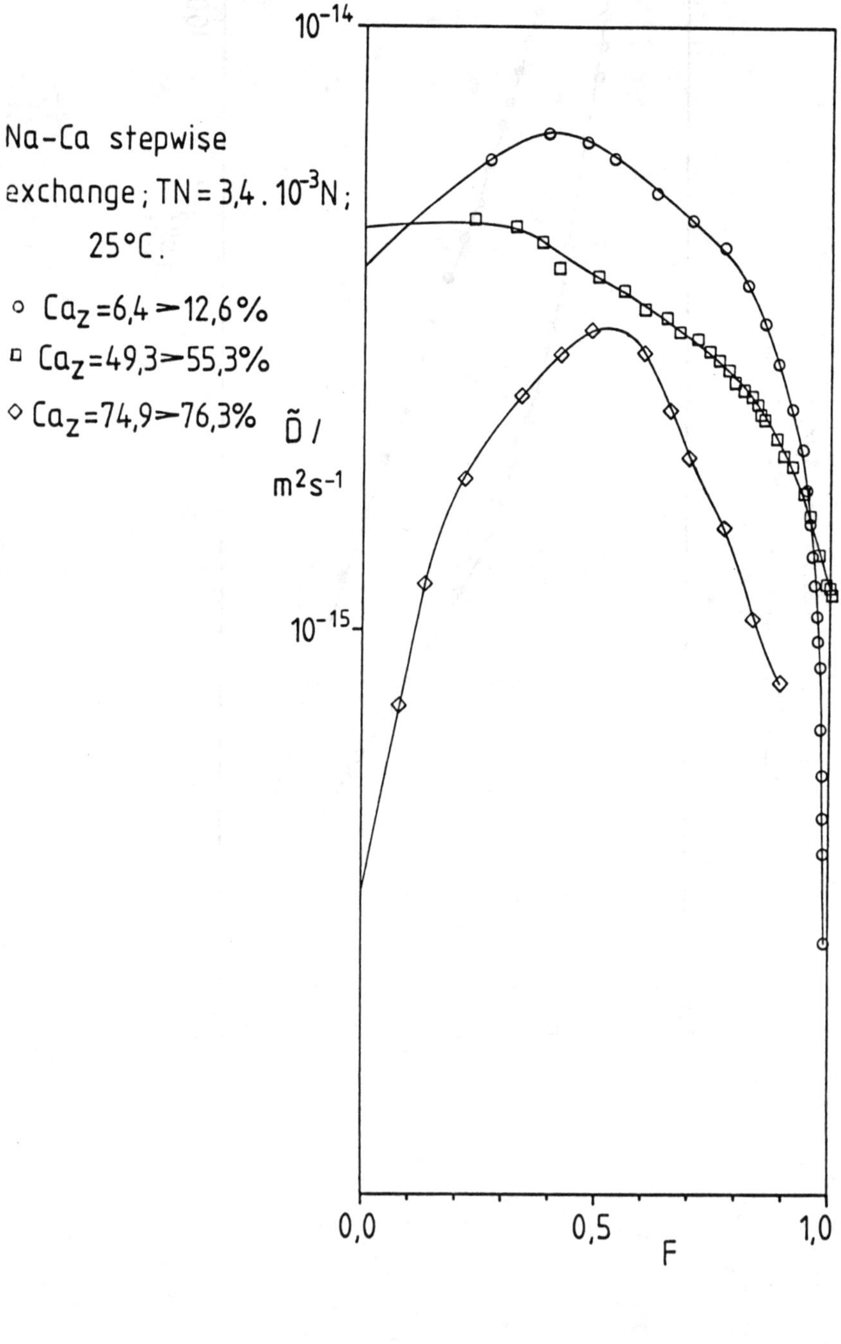

Figure 8. Variation of $\tilde{D}$ over each incremental composition plotted against Ca_z for stepwise $Na^+ \rightarrow Ca^{2+}$ exchange in zeolite A at 25°C. Total solution phase normality is 3.4×10^{-3} N.

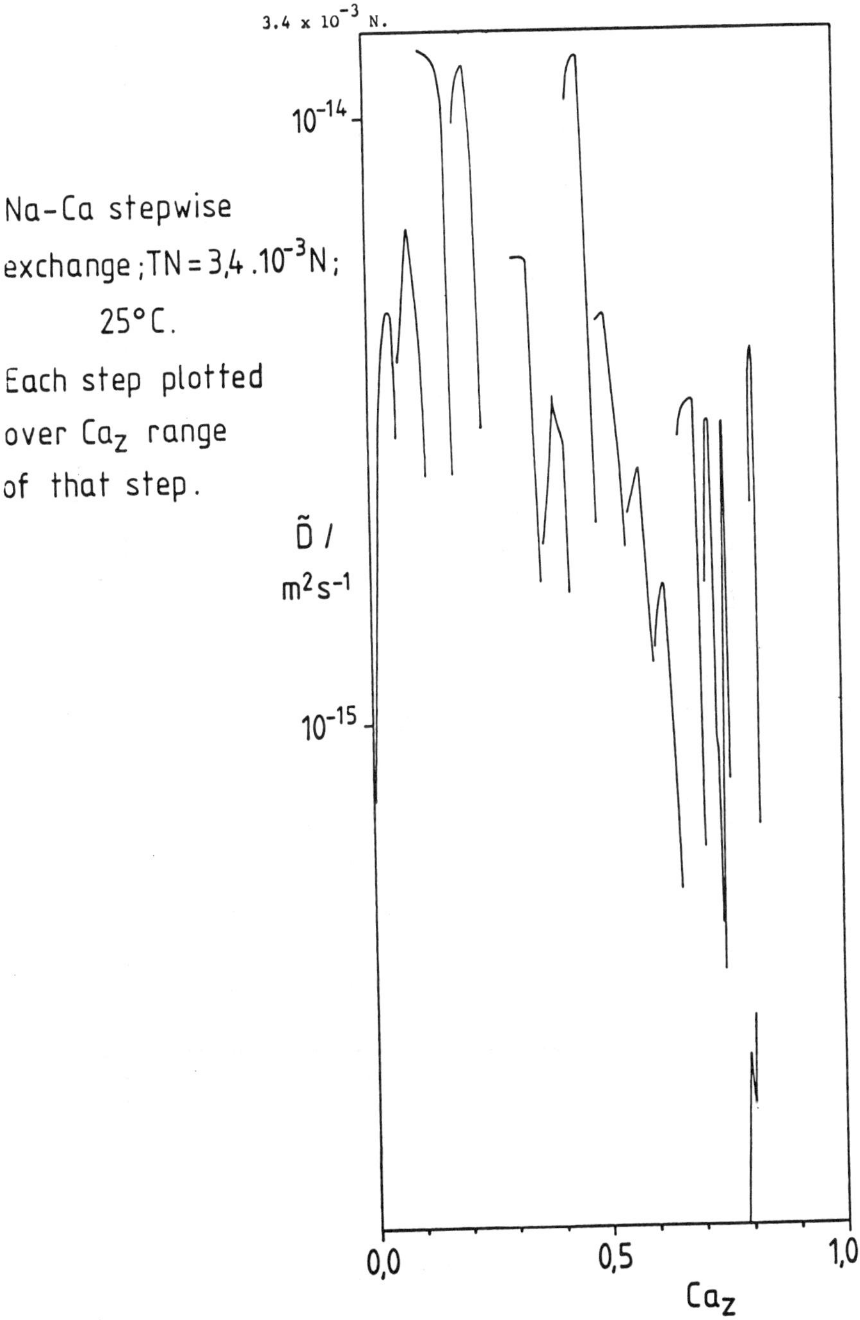

Figure 9. Ternary exchange kinetics in zeolite A at 25°C

$Na^+ \rightarrow Ca^{2+}/Mg^{2+}$

○ Initial solution phase was $(0.5\ Ca + 0.5\ Mg)_s$ and total solution phase normality was 0.25 N

□ Initial solution phase was $(0.67\ Ca + 0.33\ Mg)_s$ and total solution phase normality was 0.2 N.

The binary exchange kinetics are included for comparison.

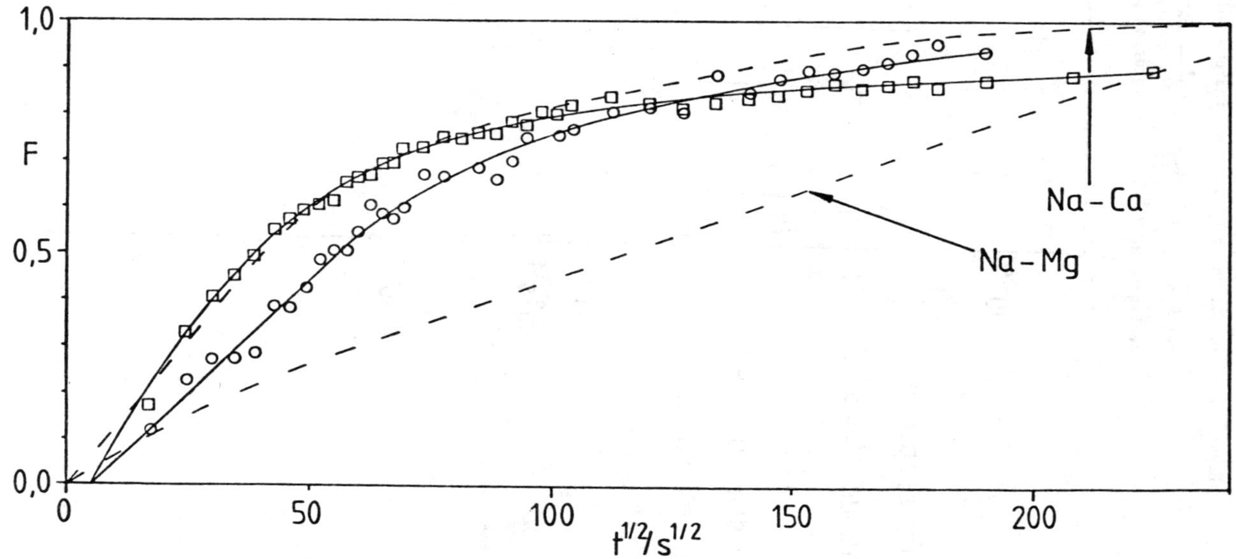

Figure 10. Rate of release of Na^+ ions at 25°C from Na-A zeolite at various initial pH values.

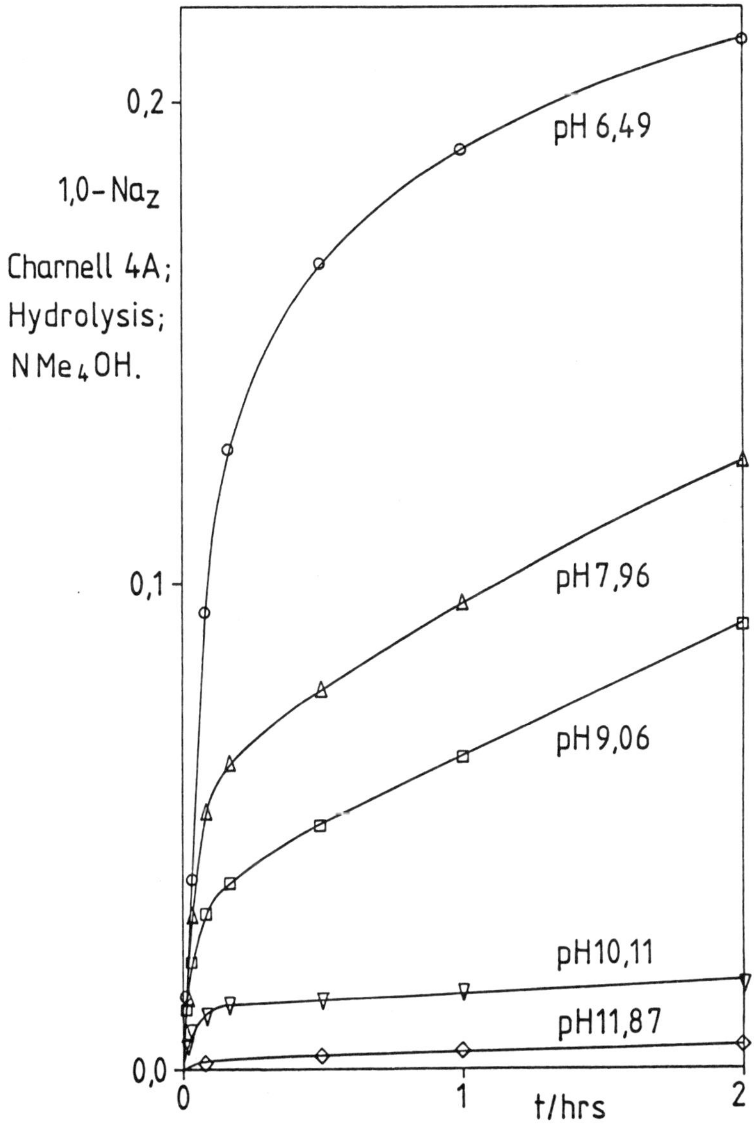

Figure 11. Variation of pH with time when Na-A zeolite is introduced to water at various initial pH values.

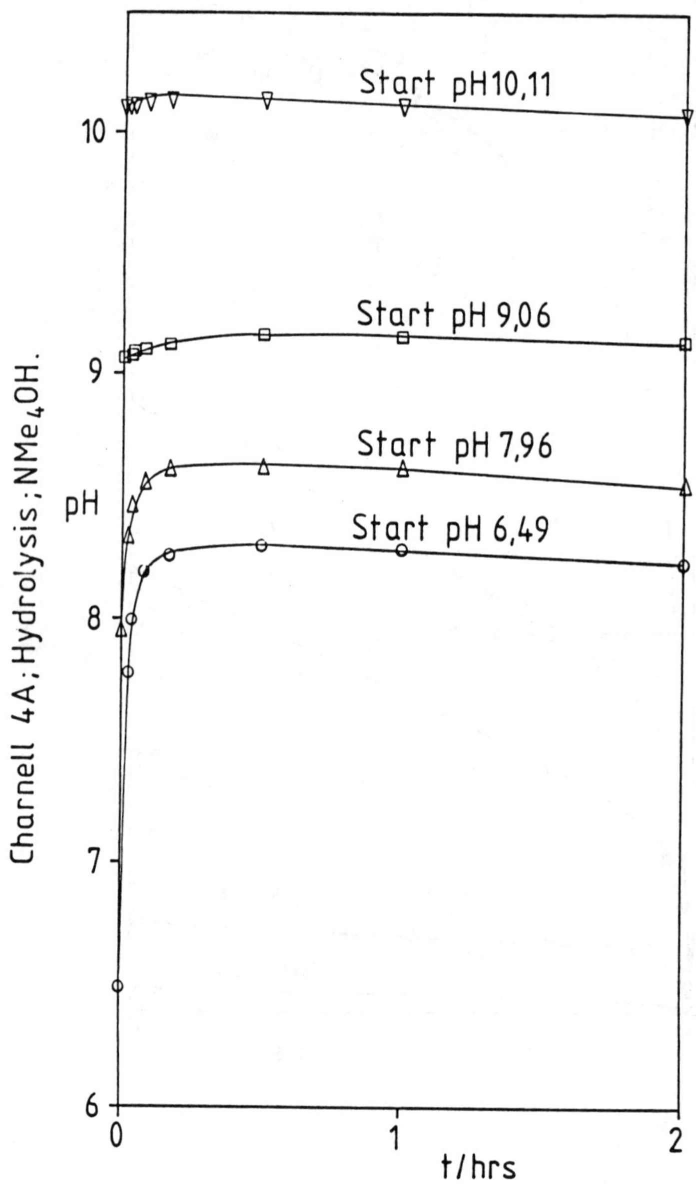

Figure 12. Degree of hydrolysis of Na-A zeolite at 25°C after various contact times plotted against the initial pH.

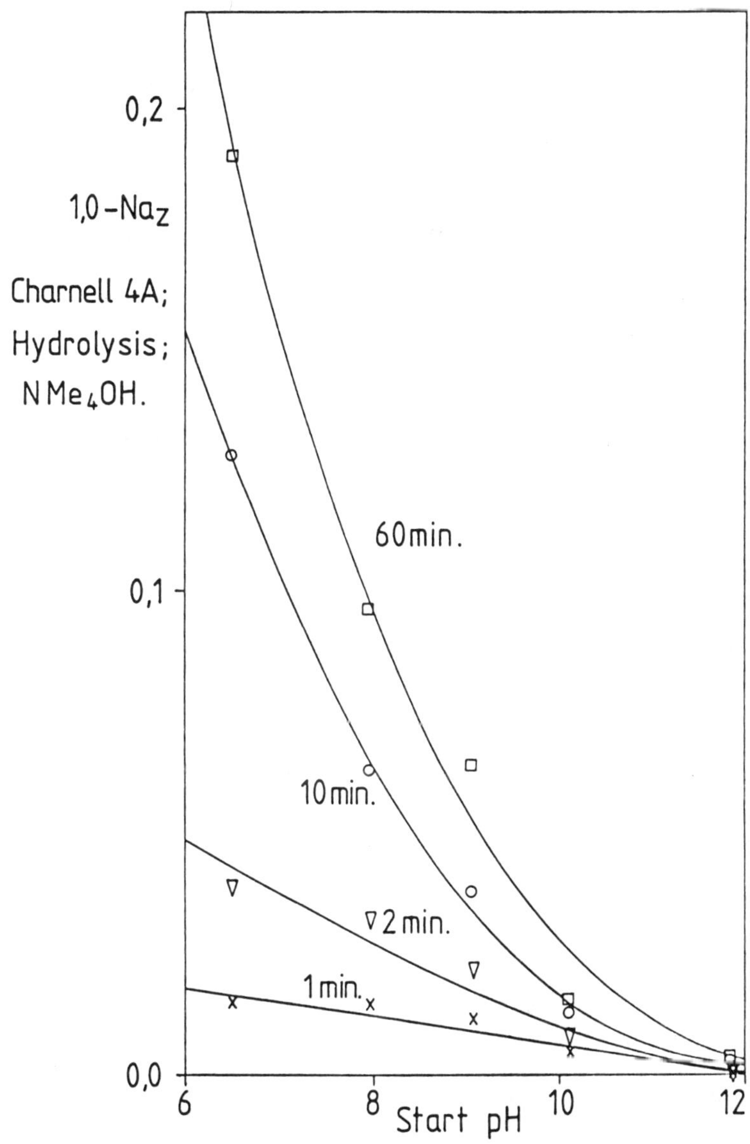

Figure 13. Rate of hydrolysis during $Na^{+} \rightarrow Ca^{2+}$ exchange in zeolite A at $25^{o}C$ and various initial pH values.

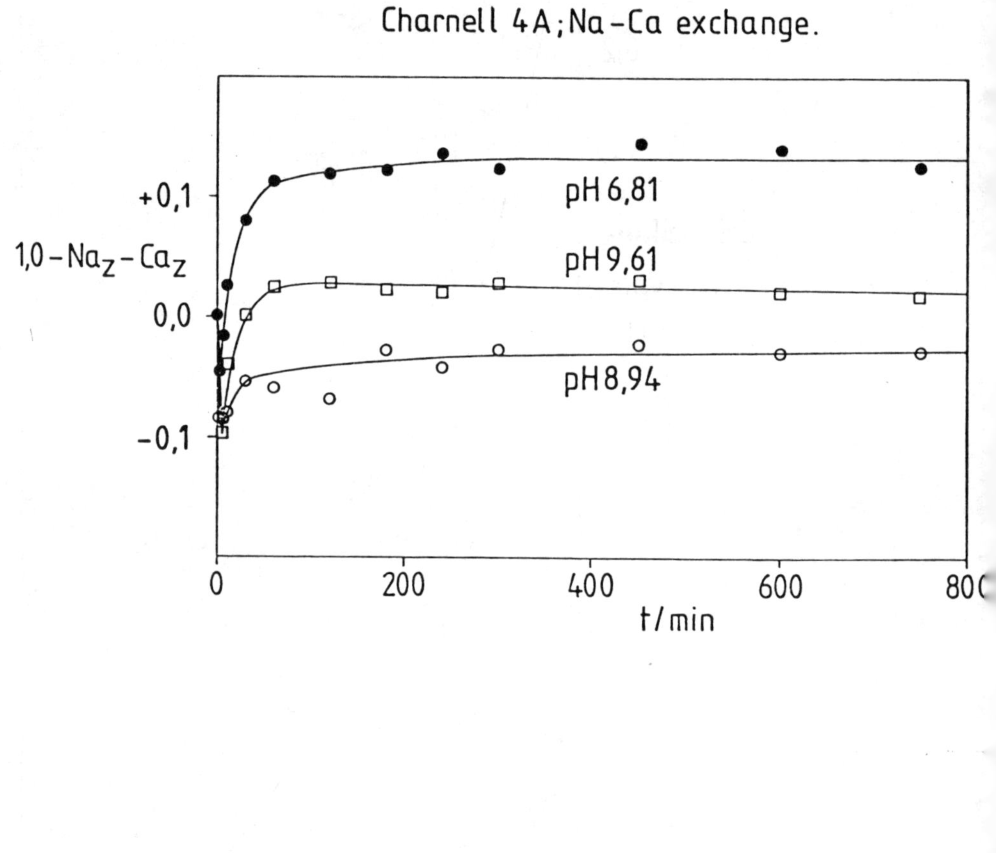

Figure 14. Tentative hydrolysis mechanism of Na-A zeolite.

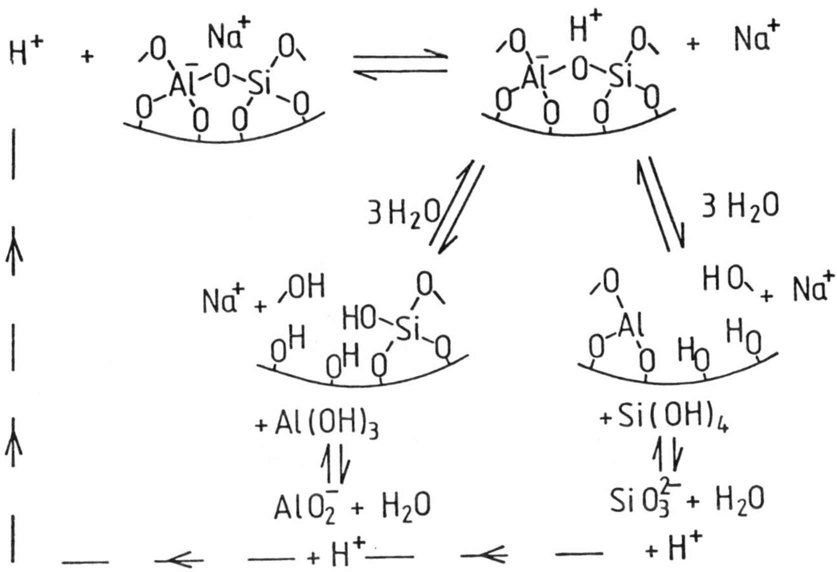

KINETICS OF BULK AND INTERFACIAL IONIC MOTION:

The Microscopic Basis and Limits for the Nernst-Planck-Poisson System

by

Richard P. Buck

The William R. Kenan Laboratories of Chemistry
The University of North Carolina
Chapel Hill, N.C. 27514 USA

ABSTRACT:

Criticisms, of, limitations on, and alternatives to the Nernst-Planck equation for ionic transport are outlined in the context of motion into and through liquids, gells, ion exchanger membranes, ionic crystals and polycrystalline materials. The origin of the equation from the point of view of Langevin, and contributions of Onsager and Schlögl provide a basis for discussion of limitations on applications to non-homogeneous materials. The surprising usefulness and wide applicability are demonstrated using correlations for diverse cases including coupled transport, association (ion pairing and clustering), adsorption and concentrated electrolytes. The microscopic models for the friction coefficient, in terms of atomic properties, are discussed. New material on kinetic boundary conditions, slow ion exchange and ionic overpotential are presented.

OUTLINE

Topic

I. Introduction
   - A. Criticisms of the Nernst-Planck Equation and Its General Applicability
   - B. Alternatives for the Nernst-Planck Equation
   - C. Some Limitations of the Nernst-Planck Equation and Its Applications.

II. Quantitative Macroscopic Transport Theory
   - A. Force-Velocity Equations for Ionic Systems
   - B. Langevin's Analysis
   - C. Onsager's Contribution
   - D. Schlögl's Contribution
   - E. The One-Dimensional Dilute-Solution Nernst-Planck Equation
   - F. The One-Dimensional Nernst-Planck-Poisson System

I. INTRODUCTION

A. Criticisms of the Nernst-Planck equation and its General Applicability

The Nernst-Planck flux equation has a long history in electrolyte transport theory for solids, gels and liquids.[1,2] For transport to and from surfaces, and in homogeneous bulk, the two-term, diffusion-migration equation accommodates ion and neutral species transport over wide concentration ranges. Equations for conductivity and time-dependent currents and voltages can follow easily. At blocked interfaces the Poisson-Boltzmann equation and the Gouy-Chapman theory are natural consequences.[3] In modern electrode kinetics, the Nernst-Planck equation permits calculation of the mass transport-controlled limit for currents. In supporting electrolytes, the migration term can be deleted and the well-known Fick's First Law is found.

The principal arguments against validity of the Nernst-Planck flux equation for bulk transport arise from the macroscopic, smooth nature of the model used in its derivation. Averaging over particles and barriers assumes many spacially-random and energetically-distributed barriers for species jumping. No exact microscopic description of the unitary events occurs, but rather friction coefficients relating non-random 'slow' forces and velocities are found. The medium of transport cannot be chemically-inhomogeneous or coarse-grained in a non-random way. In fact, because the mobile ions are treated as point charges, homogeneity usually means random distribution of materials that are homogeneous on the atomic or molecular scale. The membrane is presumed to be a medium of uniform dielectric constant. The equation in one dimension ignores edge effects (fringing fields) and assumes the membrane is a parallelpiped. Then microscopic theory permits estimation of the transport parameter magnitudes.

Criticism of the Nernst-Planck equation has again arisen recently in connection with Nafion membranes. These materials are grossly heterogeneous, and contain hydrophobic, glassy regions and more hydrophilic, ion-penetrating regions. Now, as previously, it is thought that Nernst-Planck equations should not, or would not, apply to ion-exchange resin beads and membranes, to mosaic membranes, filled membranes, supported membranes, pressed solids such as Ag_2S-AgCl, suspended solids such as resins and inorganic solids in PVC or in silicone rubbers, and liquid ion exchangers in inert supports. Surprisingly, deductions from Nernst-Planck equation theory concerning permeabilities, conductivities and membrane potentials, are very good first approximations.

Evidently "Homogeneity" may also include randomly-distributed coarse-grained matter, in regions large compared with the moving particles. In that case, some averaged transport parameters probably apply. Thus even though use of the Nernst-Planck equation may not be strictly correct, criticisms need not limit the application of the equations. Provided regions of a membrane containing mobile ions are large compared with the ion size, and provided the packing is randomly defective so that jump sites are available, then the Nernst-Planck equation can be used in one dimension for membranes, or in three dimensional systems as required. The use of the activity coefficient automatically includes some effects of size, 'grainyness' and non-uniformity of dielectric constant, although not always in explicit ways.

In addition to a presumption of chemical uniformity of the transport medium (at least in the region of ion motion) physical homogeneity is also required. The phase boundary is considered to be either sharp and laterally smooth, or uniformly diffuse and smooth. It must be free from crevasses and grain boundaries. Although equations can be written for low resistance pathways,[4] the ordinary Nernst-Planck equation applies in the homogeneous regions.

Recent studies of ion transport and ionic states in Nafion membranes suggest possible limitations of the Nernst-Planck equations in dealing with coupled ionic motion and ionic cluster motion. It is perhaps not so well known that the Nernst-Planck equation is not limited to free ion motion, but can be used to describe pairing and clustering.

For multiple-ion transport the Nernst-Planck equations for each ion can be coupled using so-called quasi-electroneutrality, that asserts in any microscopic volume large compared with the interatomic distance there will be obeyed an equation of the type:

$$\sum_M z_M(c_M) + \sum_X z_X(c_X) = 0 \tag{1}$$

z_M and z_X are cation and anion charges; c's are concentrations. This sum applies to counter-ion, co-ions, mobile and fixed, ionized sites. When chemical interactions are present, another kind of coupling or grouping of Nernst-Planck equations for charged and neutral species is used. Consequently ion pairing with mobile and immobile species can be explicitly accounted for. This includes dimerization and clustering of ions.

As we will point out in detail later, the Nernst-Planck equation accounts for standard free energies of infinitely dilute ions in each phase, even in non-uniform regions; it accounts for spacial dependences of standard state energies; and it accounts for electrostatic interactions and size effects of ions via activity coeffieients. The latter can be related back to aqueous standard states using medium effect coefficients;[5] it considers mobilities separately from diffusion coefficients and can account for concentration and space-dependence of mobilities and diffusion coefficients.

B. Alternatives for the Nernst-Planck Equation

There are several other equations and models for ionic and neutral species (including solvent) transport. Irreversible thermodynamic equations, while useful in principle, are quite general and do not yield easily an intuitive interpretation of term magnitudes.[6,7] Similarly, the Stefan-Maxwell equation method[8,9] is "inside-out" relative to Nernst-Planck, and gives little new insight into microscopic processes.

The simple jump-barrier approach applies very well in cases of neutral species transport.[10] It is easy to apply and is intuitively clear. However, introduction of the effect of local electric field on barrier crossings requires arbitrary partitioning of energy contributions to forward and backward processes. Eyring's method for this effect reduces in the linear region to the Nernst-Planck equation. This derivation is given later.

Of these remaining approaches, Scatchard's quasi-thermodynamic equation[11] and the Nernst-Planck equations of motion are closely related. Although both are macroscopic, there is term-wise similarity and the statistical mechanical formalism supports both.[12] Insight into the microscopic processes underlying each term is available with modest additional interpretation.

C. Some Limitations of the Nernst-Planck Equation and Its Applications

1) The point-charge Nernst-Planck equation has intrinsic limitations in dealing with bulk phase, finite-ion size transport. However, these may be of minor consequence. Although the force-velocity equation reflects motion of numbers of particles per volume, Onsager pointed out that energies (electrochemical potentials) and their gradients relate to the "slow" force. Then, activities replace concentrations and effects of particle size and interionic attractions and repulsions can be implicitly accounted for.

2) The Nernst-Planck equation does not deal explicitly with discrete energetic barriers. This is a primary limitation in the interpretation of surface ion exchange kinetics and effects of inhomogeneous barriers in the

bulk. Discrete short-range barriers cause an automatic separation of the transport problem into low-barrier, often equilibrated regions, and the essential rate-determining barrier step (local singularity). Frequently, the principal barrier is a slow surface rate due to intrinsically slow chemical steps, hydrated layers and inhomogeneous regions parallel, and just inside, the membrane surface. The latter is particularly important in determining net flux and current flow through ion exchange membranes and through site-free membranes. For slow, potential-dependent ion transfers, an "activation" resistance comparable with bulk resistance can occur.

(3) The Nernst-Planck equation applies at times greater than 10^{-12} to 10^{-13} sec. Prior to that time, the inertial random fluctuations are not yet averaged. After ensemble and time averaging, the Langevin equation shows the resulting proportionality between "slow velocity" and "slow force" becomes valid. This result gives a generalized derivation of the Nernst-Planck equation.

(4) The proportionality coefficients or "friction" coefficients: mobility and diffusion, are not microscopically defined by the Nernst-Planck equation. They are not properties of a particular ion, but a property of the entire membrane system. Current inquiry on the microscopic character of mobility of ions in polymer structures cannot be satisfied by Nernst-Planck analysis alone. Certainly random walk models are consistent with the ideas of the Nernst-Planck equation.[13] Thermodynamic inferences can be made concerning the microscopic processes that influence measureable entropy and enthalpies of transport. The latter are generally deduced from properties, such as resistance or conductivity, that are derived in terms of the coefficients in the Nernst-Planck equation (See Appendix A).

5) The Nernst-Planck equations do not account explicitly for the magnitudes of mobilities and diffusion coefficiencts in terms of solvent content or cross-linking. Each membrane containing different water (or mixed solvent) contents, for example, has specific friction coefficients. These variations are not predictable by macroscopic theory alone, but the results could be, but are not commonly, included in the transport parameters used.

6) The Nernst-Planck equation does not apply to transport of ions through regions of space comparable in size with the moving species. Thus channel motion of ions through membranes is best described along other related lines. Materials including membranes that are hetergeneous on the molecular scale and that restrict motion of ions to particular regions, e.g., mosaic membranes may be better treated by other means.

II. QUANTITATIVE MACROSCOPIC TRANSPORT THEORY

A. Force-Velocity Equations for Ionic Systems.

For crystals, liquid aqueous electrolytes and membranes of all kinds, an individual ion oscillates about its equilibrium position within a matrix cage. The ion frequently collides with the cage wall molecules: solvent or solid phase. The frequency is about 10^{12} - 10^{13} Hz. Occasionally, ions move to vacancy positions by slipping through a relatively "tight" region. Depending upon the barrier, many, or few, ions exchange positions by moving approximately 1 - 2 Å. In all media the resulting linear velocity of an ion, on average, moving under a concentration gradient or an electric field, is much less than the instantaneous velocity during an oscillation in its own cavity. The reason is that only relatively rarely, in time, is the oscillation-collision process effective in moving an ion out of its region to another. If one considers the complete differential equation of motion in one dimension for an ion in a constant local field along the x-direction, the slow jump process provides a large and effective damping term. The motion, on average, turns out to obey a different equation: one in which the applied force is proportional to average particle velocity, rather than acceleration.

B. Langevin's Analysis[14]

When an individual particle is immersed in a condensed phase at temperature T, it is subjected to random forces. Consider the particle whose center-of-mass coordinate at time t is x(t) with velocity v = dx/dt. Random interactions with many neighboring particles create some fluctuating force F(t) and resulting instantaneous fluctuating velocity v. The particle may be acted upon by other forces such as a local electrostatic field denoted by $\mathfrak{F}(t)$. In the one dimensional case, Newton's second law can be written

$$m\frac{dv}{dt} = \mathfrak{F}(t) + F(t) \tag{2}$$

Using the mean gas particle velocity and the mean intermolecular distance, the random force probably changes direction about once in 10^{-13} sec. If so, the random force probably averages out to zero in 10 to 100 collisions. In addition there must also be a slowly varying part of F(t) that correlates with the slow drift velocity. If so, we can write

$$F(t) = \overline{F}(t) + F'(t) \tag{3}$$

in which F'(t) does average to zero, but $\overline{F}$ does not. Similarly, the total velocity can be written

$$v = \overline{v} + v' \tag{4}$$

where v' averages to zero but $\overline{v}$ does not. The slowly varying part $\overline{F}$ must be some function of $\overline{v}$. If $\overline{F}(\overline{v})$ can be expanded in a power series, then

$$\overline{F} = -\alpha\overline{v} + \text{higher power terms}$$

The slowly varying part of eqn. (2) becomes

$$m\frac{d\bar{v}}{dt} = \mathfrak{F} + \bar{F} = \mathfrak{F} - \alpha\bar{v} \qquad (5)$$

According to Langevin, eqn. (2) can be written

$$m\frac{dv}{dt} = \mathfrak{F} - \alpha v + F'(t) \qquad (6)$$

because the fluctuating terms will average out in a very short time.

In an electric field, a particle with charge ze would obey

$$m\frac{dv}{dt} = zeE - \alpha v + F'(t) \qquad (7)$$

or, on a molar basis

$$M\frac{dv}{dt} = zFE - N^o v + F'(t) \qquad (8)$$

where M = molecular weight; N^o = Avogadro's number and F is the Faraday = $N^o e$.

Taking the mean values of both sides and considering the steady-state situation where $d\bar{v}/dt = 0$, this yields

$$zFE - N^o\alpha\bar{v} = 0 \qquad (9)$$

This is a general result:

$$\text{force} - N^o\alpha v = 0 \qquad (10)$$

and the superbar is omitted. Writing the friction coefficient

$$f_i = 1/N^o\alpha \qquad (11)$$

the usual primitive form of the Nernst-Planck equation is:

$$f_i \times \text{force} = \text{velocity} \qquad (12)$$

The "slow" velocity is associated with drift in an electric field or with diffusion.

C. Onsager's Contribution[15]

For each species of ion and solvent in a mixture, the flux, J, (mol/cm^2s) is related to the colinear force or force component at constant temperature.

The flux in a laboratory coordinate system is slightly different from that measured with respect to a center of mass. Flux with respect to center of mass J' is given by

$$J' = c_i v_i = c_i f_i \times \text{force} \tag{13}$$

If one focuses only on the transport of ions, and presumes that the solvent or crystal environment is motionless, then

$$J = c_i f_i \frac{\partial \tilde{\mu}_i}{\partial x} \tag{14}$$

where c_i is the local ion concentration and $\tilde{\mu}_i$ is the electrochemical potential (an energy) of the ion. If c_i is mol/cm^3, $\tilde{\mu}_i$ is joules, then f_i is u_i and $u_i RT = D_i$ cm^2/s. The electrochemical potential is the average energy of a particle, and it depends on particle density (activity), local potential, and pressure:

$$\tilde{\mu}_i = \mu_i^o + RT \ln a_i + z_i F \phi + p\bar{v} \tag{15}$$

where μ_i^o is a constant (at fixed T); a_i is activity $c_i \gamma_i$; γ_i is the activity coefficient; ϕ is the local potential; p is the local pressure and $\bar{v}$ is the molar volume.

D. Schlögl's Contribution[16]

The general form for flux J_i of one ionic species in the presence of moving solvent is:

$$J_i = c_i v_{solvent}\,(1-\sigma) - D_i \left[\frac{\partial c_i}{\partial x} + c_i \frac{\partial \ln \gamma_i}{\partial x} + z_i \frac{F}{RT} c_i \frac{\partial \phi}{\partial x} + \frac{c_i}{RT}\left(\bar{v}_i - \frac{M_i}{M_s}\bar{v}_s\right)\frac{\partial p}{\partial x}\right]. \tag{16}$$

$\bar{v}_i$ and $\bar{v}_s$ apply to species i and solvent s; M_i and M_s are species and solvent molecular weights. This equation accounts for motion of the center of mass and motion relative to the center of mass. $\bar{v}_i$ and $\bar{v}_s$ are partial molar volumes, and (1-σ) is the Stavermen rejection coefficient term.

E. The One-Dimensional Dilute-Solution Nernst-Planck Equation

For most systems involving bulk rather than membrane or thin film transport, the solvent can be considered stationary. Then,

$$J_i = -D_i\left[\frac{\partial c_i}{\partial x} + c_i\frac{\partial \ln \gamma_i}{\partial x} + \frac{z_iF}{RT}c_i\frac{\partial \phi}{\partial x}\right] \tag{17a}$$

$$= -D_i\left[\frac{\partial c_i}{\partial x} + c_i\frac{\partial \ln \gamma_i}{\partial x} - \frac{z_iF}{RT}c_iE\right]. \tag{17b}$$

For ideal systems which are generally dilute, one can assume γ_i is not a function of distance. Then the dilute solution form, called the Nernst-Planck equation, is found to be

$$J_i = -D_i\left[\frac{\partial c_i}{\partial x} - \frac{z_iF}{RT}c_iE\right]. \tag{18}$$

This form is the usual starting point for calculations of concentration, field and potential profiles in bulk in the absence of solvent penetration.[17] The first term is the gradient producing diffusion. The second term describes motion in an electric field and is called migration. These are illustrated in Fig. 1. When one is considering transport of a single ion in the presence of a high concentration of an inert, static electrolyte, the migration term may be deleted. The reason is that high solution conductivity decreases E to a value that can be ignored. Most of the field changes rapidly through the space charge regions at the electrode surfaces and not through the electrolyte bulk. This situation occurs in many plating processes during which one ion, say Cd^{2+} is plated out while a high concentration of neutral electrolyte, say KNO_3, remains essentially immobile. This is a very practical case because the electrode potential can be controlled such that Cd^{2+} is reduced, but K^+ and NO_3^- are not. The still-more-simplified equation without the migration term is called Fick's first law of diffusion.

F. The One-Dimensional Nernst-Planck-Poisson System

Subsidiary equations are:

for current

$$I = F \sum_i z_i J_i - \epsilon \frac{\partial^2 \phi}{\partial x \partial t} \tag{19}$$

which includes a form of Poisson's equation,

$$\frac{\partial E}{\partial x} = \frac{\rho}{\epsilon} = zF(c_+ - c_-)/\epsilon \tag{20}$$

for a single salt with charge ±z. Time dependences of concentrations and fluxes are found by solving the continuity equation:

$$\frac{\partial c_i}{\partial t} = -\frac{\partial J_i}{\partial x} \tag{21}$$

for times greater than the Debye time. This equation follows from consideration of the change of concentration with time in a small volume as deduced from the flux in one side minus the flux out of the other. This equation is Fick's second law. Both eqns. (14) and (21) can be generalized by introducing the gradient and the divergence, for treatment of 3-dimensional transport problems.

In the bulk of all electrolytes, quasi-electroneutrality prevails. Any hypothetical sphere, large compared with the volume of an ion, will have no net charge inside. Over short distances, electroneutrality can be violated. At surfaces, fluxes are continuous, while concentrations are not. If an ion does not pass through an interface, its flux is zero at that point.

III. Applications to Dissociated (Dilute and Concentrated), Associated and Adsorbing Electrolyte Systems

A. Origin of the Ideal Relations Among Conductivities and Transport Parameters

For a single, dilute, ideal electrolyte containing a single salt $M_{|z_a|}X_{z_c}$ at concentration c_s (mol/cm^3), application of a field causes ions to move. When r.f. fields are used with blocked electrodes, or when reversible electrodes are used, concentration gradients are inevitably formed near the electrode surfaces. In the bulk, away from the electrodes, the concentration can be assumed to remain uniform. In that case, eqn. (18) reduces to the migration term only for transport in the bulk regions:

$$J_i = -D_i \frac{z_i F}{RT} c_i \frac{\partial \phi}{\partial x} \tag{22}$$

where

$$c_i = c_+ = |z_a| c_S \text{ mol/cm}^3. \tag{23}$$

or

$$c_i = c_- = z_c c_S \text{ mol/cm}^3. \tag{24}$$

Then

$$\begin{aligned} I &= J_+ - J_- \\ &= \frac{F^2}{RT} [D_+ z_c^2 c_+ + D_- z_a^2 c_-] \frac{\partial \phi}{\partial x} = \sigma \frac{\partial \phi}{\partial x}. \end{aligned} \tag{25}$$

Various forms can be derived using other parameters with the results accumulated in Table A-1.

B. Coupling of Fluxes in Electroneutral Diffusion of Salts

In uniform electrolytes, conduction of ions produces fluxes of positive and negative ions moving in opposite directions. However, diffusion of charged species from high concentration to low produces fluxes in the same direction and requires that the current be zero. In other words, electroneutrality of diffusion is a result of local fields that build up between moving ions to insure that ions of opposite charge move at the same velocity so that no space charge will develop. If one ion has an intrinsically higher mobility than the other, it tends to move ahead at short times. The resulting space charge then creates a field that pulls on the slower ion, and retards the faster one until they move together, at the same velocity. This coupling of fluxes leads to a salt diffusion coefficient, D, which is a weighted mean. It applies to each moving ion in a single salt solution. For a 1:1 electrolyte, the well-known formula is

$$D = \frac{2D_+D_-}{D_+ + D_-}. \tag{26}$$

More generally, for a salt $M_{|z_a|}X_{z_c}$

$$D = \frac{(|z_a| + z_c)D_cD_a}{z_cD_c + |z_a|D_a} \tag{27}$$

$$D = \frac{RT}{F^2}\frac{(z_c + |z_a|)}{z_c|z_a|}\frac{\lambda_+^0\lambda_-^0}{\lambda_+^0 + \lambda_-^0}. \tag{28a}$$

For two ions, with same sign of charge, moving through a fixed site membrane, electrostatic interactions again determine a mean diffusion coefficient:

$$D = \frac{D_1D_2(z_1^2c_1 + z_2^2c_2)}{z_1^2c_1D_1 + z_2^2c_2D_2} \tag{28b}$$

If each side is maintained at pure ion #1 and pure ion #2, then the simplification $z_1c_1 = z_2c_2$ applies and

$$D = \frac{D_1D_2(z_1 + z_2)}{z_1D_1 + z_2D_2} \tag{28c}$$

C. Transport Equations for Concentrated Single Electrolytes

The force-velocity equation for a single ion, eqn. (17), can be rewritten as

$$J_i = -u_i c_i\left[RT\frac{\partial \ln c_i}{\partial x} + RT\frac{\partial \ln \gamma_i}{\partial c_i}\frac{\partial c_i}{\partial x} - z_i FE\right]. \tag{29}$$

Since, in dilute solution $u_i = D_i/RT$

$$J_i = -D_i\left[\left(1 + \frac{\partial \ln \gamma_i}{\partial \ln c_i}\right)\frac{\partial c_i}{\partial x} - \frac{z_i F}{RT}c_i E\right]. \tag{30}$$

Now it is clear that the diffusion coefficient and the mobility do not bear the same relationship in concentrated electrolytes as in the ideal, dilute case. The apparent diffusion coefficient decreases with total salt concentration because the activity coefficient is a decreasing function of concentration. The correction term $d\ln\gamma_i/d\ln c_i$ is a negative quantity and,

$$D(\text{conc.}) = D(\text{dilute})\left(1 + \frac{\partial \ln \gamma}{\partial \ln c}\right). \tag{31}$$

D. Associated Electrolytes

Coupling of homogeneous diffusion and chemical reactions is a process frequently encountered in electrodiffusion systems in solvents of low dielectric constant where ion pairing is significant.[18] These problems are handled by grouping the flux equations according to the coupling and using the chemical equilibrium constants for ion pairing as the remaining necessary conditions on the system. Thus, for cation i^+ and an anion x^- which form a neutral ion pair with formation constant

$$K_{ix} = \frac{(ix)}{(i)(x)} \tag{32}$$

individual fluxes are grouped

$$J_i^* = J_i + J_{ix} \tag{33a}$$

$$J_x^* = J_x + J_{ix} \tag{33b}$$

The time dependences follow from continuity equations which now include local sources and sinks. For example:

$$\frac{\partial c_i}{\partial t} = -\frac{\partial J_i}{\partial x} + k_{ix}^b c_{ix} - k_{ix}^f c_i c_x \tag{34}$$

$$\frac{\partial c_{ix}}{\partial t} = -\frac{\partial J_{ix}}{\partial x} + k_{ix}^f c_i c_x - k_{ix}^b c_{ix}. \tag{35}$$

One has a similar equation for the rate of change of c_x. Note that the divergence of the grouped fluxes leads to cancellation of the chemical rate terms. The k's are chemical rate constants. As a further example, in dealing with liquid ion exchanger membranes containing trapped anion sites x^-, and permeable cations i^+ and j^+, both of which form neutral ion pairs ix and jx, grouping of fluxes gives

$$J_i^* = J_i + J_{ix} \tag{36a}$$

$$J_j^* = J_j + J_{jx} \tag{36b}$$

$$J_x^* = J_x + J_{ix} + J_{jx}. \tag{36c}$$

At any time, the current is carried internally by i^+ j^+ and x^-, so that

$$I = F(J_i + J_j - J_x) = F(J_i^* + J_j^* - J_x^*). \tag{37}$$

However, at the surfaces, J_x^* must be zero since species containing x do not pass the interface, although none of the flux components of J^*_x needs to be zero. This result means that externally measured current is determined by fluxes of both cation and neutral species through the surfaces. This statement is not true inside the membrane until the steady state is reached where J_x^* is zero everywhere and

$$I = F(J_i^* + J_j^*). \tag{38}$$

Equations 32-38 may also be important for describing processes in fixed site ion exchangers. In those cases, the species containing fixed sites have no mobility. For conversion of resins in one weak acid and ion paired form to another, there is a conflict between the homogeneous, Fick's Law model (eqns. 28a-c) and the reaction-controlled (shell-progressive) models using the kinetic and equilibrium expressions eqns. 32-38.*

E. Transport with Adsorption[19]

Adsorption is general accumulation of material at an interface. Desorption is general removal at an interface. Adsorption can be visualized in three parts: 1) mass transport accumulation because material is carried by diffusion or migration faster than it can pass through the surface; 2) accumulation as diffuse space charge (in the case of ions) because of energy differences of ions in two phases and the resulting fields; and 3) contact-adsorbed or specifically-adsorbed material at a surface (usually an electrode surface) by virtue of chemisorption and physisorption processes.

In thermodynamic terms, accumulation by whatever method is measured relative to the uniform concentration in the bulk of the phase. The excess or deficit carries a + or a - sign from the definition:

$$\Gamma_{salt} = \int_{bulk}^{surf.} (c_{salt} - c^{b}_{salt})dx \qquad (39)$$

When salt is accumulated in the surface vicinity, Γ_{salt} is positive. Only relative excesses are measurable, and the usual reference "excess" is that of the solvent. Relative excesses of ions also can be defined and found, and the definitions are the same as eqn. (39), when ion concentrations are substituted.

During current flow through an electrolyte/membrane interface, the concentration profiles of salts and single ions develop. Eqn. (39) avoids counting decreases and increases of concentration as "adsorption" because both

*Examples will be described in the Proceedings volume NATO Advanced Study Institute: Mass Transfer and Kinetics of Ion Exchange, Maratea (Italy) May 31-June 11, 1982.

concentration terms are understood to be time-dependent. Except in the immediate vicinity of the surface, the two concentrations remain equal, although both change with time, within the Nernst boundary layer. To account for total adsorption, the flux balance equation at the interface is amended. Thus instead of

$$J_i \text{ (at } x = 0) = \bar{J}_i \text{ } (x = 0) \tag{40}$$

one writes

$$\frac{d\Gamma}{dt} = J_i - \bar{J}_i \text{ (at } x = 0) \tag{41}$$

If no salt passes through the interface, then the flux produced by an applied voltage or current, is accounted for in terms of continuously varying $\Gamma(t)$, whose value would be, at any time, given by the integral in eqn. (39). Application of this equation for constant current ion transport at immiscible electrolyte interfaces has been made by Joos and van den Bogaert.[20]

IV. Microscopic Models for Diffusion, Migration and Conductivity

A. Random Walks, Barriers and Absolute Rate Theory[21]

We have already related diffusion coefficients, mobilities, friction coefficients, conductivities and transference numbers on the macroscopic level. The model was "smooth", and particulate properties, particle jumps, and local vibrations of charge carriers, were omitted. A truly detailed, atomistic analysis of transport from first principles is not feasible. Instead, it is possible that some of the factors can be accounted for generally in terms of probabilities of significant events. There have been developed a number of instructive micro-models which introduce these statistical considerations and the relations of improbable events to the general theory of absolute reaction rates. The latter provides an intuitive grasp of transport process kinetics from which activation entropies and activation heats (enthalpies) can be interpreted.

The starting point is the barrier-free, one-dimensional jump of uncharged, point (dimensionless) species. Following this is the introduction of an energetic barrier and barrier-limited jump processes. In Fig. 2, imagine a section of short length, d, in a dilute concentration gradient of concentrations c_l and c_r (l = left, r = right) related by

$$c_l = c_r - d\frac{\partial c}{\partial x} \tag{42}$$

The length d is typically a jump distance of molecular dimension, $\sim 10^{-8}$ cm. Each species is thought to vibrate at the Debye frequency ($\nu = kT/h$), about 10^{12} - 10^{13} Hz, and there are no barriers because of shape, size, repulsion or lack of vacancy locations. The jump frequency is also ν. Each jump can be forward or backward. Considering only the planes at x = 0 and x = d, the concentrations in each plane are n_l and n_r. The flux from left to right (positive x) is

$$J = 1/2(n_l - n_r)\nu. \tag{43}$$

But all planes from x = 0 to x = d can also product jumping species that cross the plane x = d, so, including these species, the overall flux becomes

$$J = 1/2(c_l - c_r)\nu d = -1/2d^2\nu\frac{\partial c}{\partial x} \tag{44}$$

and from Fick's law, one identifies

$$D = 1/2d^2\nu. \tag{45}$$

In three dimensions, jumps of frequency ν lead to

$$D = 1/3d^2\nu. \tag{46}$$

Often, these equations are written:

$$D = gd^2\nu = gd\bar{v} = g\bar{v}^2\tau \tag{47}$$

where g is a 'geometric' factor and τ is the jump time. Introduction of the derivative into eqn. (45) is justified because the spacing is narrow, the concentrations are nearly equal, and the concentration profile is essentially linear.

One kind of statistical barrier is provided by space filling from the finite size of ions that were assumed to be only point charges in the foregoing. As a result there is a limited number of vacancy spots for jumping which decreases the diffusion and mobility coefficients. In addition, orientation of non-spherical particles for optimal jumping and the pushing apart of nearby molecules to form partial vacancies can be considered. A way of expressing all of these 'entropic' effects is to include the mole fraction of vacancies, x_v and the probability, w_v, of a jump to a particular vacancy. Probability values depend on local structure, and numbers of nearest neighbors have to be estimated or determined. The frequency of an adjacent location jump is diminished from ν to $\nu w_v x_v$, so that D is modified according to

$$D = gd^2 \nu w_v x_v = gd^2 \nu \exp(\Delta S^*/R) \tag{48}$$

where ΔS^* is the activation entropy.

Each jump in a liquid is unrelated to previous jumps. However, for solid lattices with few mobile species and few vacancies, the apparent diffusion coefficient can differ for (1) tracer diffusion, (2) diffusion in a concentration gradient, and (3) diffusion-migration in an applied field. The argument is made that tracer diffusion jumps in crystals are correlated, in the sense that forward-backward jumps are more likely than forward-forward jumps. In a field, mobile ions and vacancies are forced to move in opposite directions, while in tracer diffusion a random process brings ions and vacancies together. In conductivity experiments it is more probable that an ion will find a vacancy and continue its forward progress.

$$D_T = fgd^2 \nu \exp(-\Delta G^*/RT) \tag{49}$$

where

$$f \sim (1 - 2/n); \quad n = \text{number of nearest neighbors} \tag{50}$$

which depends on crystal structure. Then D_T (measured) will be less than D (calculated from conductivity). Two examples are given by Shewmon.[17]

A real barrier involving lattice strain through forced expansion or compression by jumping species leads to an energy of activation ΔH^*. In absolute rate theory,[22] the jump frequency is considered to be the Debye value reduced by the activation free energy term:

$$\nu_J = \nu \exp(-\Delta G^*/RT) \sim kT/h \exp(-\Delta G^*/RT) \tag{51}$$

$$= kT/h(Fu'/Fu) \exp(-\epsilon_o/RT) \tag{52}$$

where Fu'/Fu is the partition function ratio for the moving species at the top of the barrier divided by the function for the initial state. ε_0 is the 0°K barrier height. Then

$$D = D_0' \exp(-\Delta G^*/RT) = D_0 \exp(-\Delta H^*/RT) \tag{53}$$

with

$$D_0 = gd^2\nu \exp(\Delta S^*/R). \tag{54}$$

In this relation from absolute rate theory, the transmission coefficient is assumed to be unity and tunneling is ignored.

B. Eyring's Extension and an Alternate Derivation of the Nernst-Planck Equation

The situation is altered when barrier-free and barrier-limited one-dimensional jumps of charged species are considered. Fig. 3 shows a possible jump barrier with and without the addition of a small local field

$$E = (\phi_l - \phi_r)/d \tag{55}$$

resulting from an applied potential. Local potentials are designated ϕ_l and ϕ_r. The assumption is made that ΔG^* is modified according to fractions of the particles energy change, W, because of the local potentials.

$$G^*_{\text{left}} = G - \alpha W \tag{56a}$$
$$= G - \alpha zF(\phi_l - \phi_r)/RT \tag{56b}$$
$$= G + \alpha zFdE/RT. \tag{56c}$$
$$G^*_{\text{rt}} = G + (1-\alpha)W \tag{57a}$$
$$= G + (1-\alpha)zF(\phi_l - \phi_r)/RT \tag{57b}$$
$$= G - (1-\alpha)zFdE/RT. \tag{57c}$$

The partitioning factor, α, on the work done on the particles zFEd, occurs because part of the work presumably raises the initial state, and part stabilizes the final state. The particle charge is z (with sign).

Using the same procedure for the derivation of eqn. (44), the left to right flux is

$$J = 1/2\nu_J d\{c_l \exp[\alpha zFEd/RT] - c_r \exp[-(1-\alpha)zFEd/RT]\} \tag{58}$$

and since the two concentrations are related by eqn. (42), then with expansion of the exponentials, one finds that the flux is approximately

$$J \sim -1/2d^2\nu_J \frac{\partial c}{\partial x} + 1/2d^2\nu_J czFE/RT. \tag{59}$$

This is the two-term Nernst-Planck equation involving the diffusion and migration parts. As before, we identify

$$D = 1/2d^2\nu_J \tag{60}$$

and from eqn. (13), the mobility must be

$$u = 1/2d^2\nu_J/RT \tag{61}$$

and the relationship u = D/RT follows directly.

V. Interfacial Ion Transfer Kinetics and Kinetic Boundary Conditions on Nernst-Planck Equation Transport

A. Resistance Sources for Membranes in Addition to Those Deduced from the Nernst-Planck Equation

In non-associated electrolytes, the Nernst-Planck equations, applied to multiple ion transport, give the definition of normal AC and DC resistances, conductances and conductivities (see Table A-1). For homogeneous electrolytes and homogeneous site-free membranes with rapid, reversible ion exchange at the bathing electrolyte/membrane interfaces, the AC and DC resistances are the same theoretical expression. For ideal fixed-site ion exchangers, the AC and DC resistances are also the same when a single counter ion is in the membrane and in the bathing solutions. However, for biionic and mixed electrolyte bathing solutions, there can be two different DC resistances depending on DC current direction. Rectification can also occur.

Ideal liquid ion exchangers with a single counter ion also show different AC and DC resistances since the trapped exchanger contributes to the AC conductivity, but not to the DC value. There are further complications when generation and recombination of charge carriers (associated electrolytes) are involved. But all of these transport cases can be treated by the Nernst-Planck equation method, to yield expressions for the gross electrical membrane properties. These results are possible because the bulk processes described by the Nernst-Planck equations are the only processes involved in determining the equivalent circuit elements.

Another but less well known process contributes to AC and DC resistances. This is the surface resistance from surface film (inhomogeneous surface regions) and from slow ion exchange kinetics. Surface films that contribute significantly to resistances are inhomogeneous regions of higher than normal bulk resistance. "Activation" resistance by virtue of slow ion transfers is believed to occur because of slow solvent sheath reorganization, including solvent stripping as ions pass from one phase to the other.[23,24]

B. Qualitative Visualization of Overpotential at Immiscible Electrolyte/ Ion Exchanging Interfaces

The 'construct' used in electrochemical rate theory for metal electrodes is shown in Figures 4 and 5 juxtaposed with a likely construct for immiscible ionic electrolyte interfaces. The latter ionic interfaces can include ordinary salt extraction examples with multiplicity of ions crossing the interface, interfaces with ions constrained to one phase such as a liquid ion exchanger systems, and conventional fixed-site ion exchanger electrolyte interfaces. Because of the close resemblance, the electrolyte/solid ionic conductor also belongs in this latter category. At the equilibrium electrolyte/metal interface, the well-known subdivision of space includes the constant potential, bulk electroneutral region, the diffuse layer characterized by space charge and potential curvature, the compact layer with or without contract-adsorbed ions and with an approximately linear potential profile, and finally, the metal. Space charge in the metal is confined to the surface, and so potential curvature is great near the interface on the metal side. In the simplest models, potentials are taken to be constant in places parallel to the surface.

For ion-transfer interfaces, the same subdivisions of space normal to the surface are expected. Instead of one space charge region, we expect two.[25] Potential profiles are expected to be curved, but modified by fixed sites in ion exchangers in the same way that fixed dopant charges modify the space charge potential profiles in semiconductors. For mobile site ion exchangers and extraction interfaces, the space charge regions are probably Gouy-Chapman-like and therefore similar to those in the bathing electrolyte phase. Although capacitance data at liquid/liquid and liquid/solid ionic conductor interface systems are rare, it is reasonable that some alignment of solvent dipoles at an interface, and possibly some similar alignment in the hydrophobic phase are appropriate. One has also the possibility of specific or contact adsorption when ions arise from electrolytes in each phase, and the resulting space charge distributes itself away from the adsorbed ions in a typical diffuse region within the same phase. But, in addition, there is also the possibility of surfactant adsorption which produces sheets of adsorbed charge and/or diffuse charge in each phase, even though the surfactant is added to only one of the phases initially. The more hydrophobic ion remains in the lower dielectric constant phase while the more hydrophilic ion equilibrates mainly in the aqueous phase, for example. In all of these cases, regions of linear potential profiles can be expected to connect with the curved profiles of the space charge in either or both phases.

In Fig. 4 (upper) the simplest arrangement of electrolyte, space charge, compact layer and metal, reversible to cations, is shown. Contact adsorption is not shown. The Outer Helmholtz and Inner Helmholtz Planes (OHP and IHP), and the interface separating the regions are marked. For illustration, the potential ϕ_M is relatively negative to ϕ(electrolyte), taken as zero. ϕ_2 and ϕ_1 correspond to the positions OHP and IHP. The single ion chemical potentials μ_+ and μ_- are $\mu^o + RT\ln a_i$, respectively. In bulk electrolyte, they are not necessarily equal since μ_+^o and μ_-^o are not equal. Curvature occurs because of the space charge region in which a_+ exceeds the bulk value, while a_- is repelled and is less than bulk value. Since the electrode is reversible to cations, addition of $F\phi(x)$ to $\mu_+(x)$ produces the required constancy of the electrochemical potential $\tilde{\mu}_+ = \tilde{\tilde{\mu}}_+$ in both phases. Little can be said about the compact layer where cations do not exist at equilibrium in this example.

In Fig. 4 (lower), both monovalent cations and anions are extracted. In this example anions are preferred to cations in the organic phase so that the bulk potential $\bar{\phi}$ is relatively negative to ϕ(aqueous), which is taken as zero. Again, potentials at the indicated places are shown. Chemical potentials of the single ions are illustrated and these functions bend in response to potential. Of course, electrochemical potentials are uniform in each phase, and throughout the space charge regions.

At equilibrium, regardless of the kinetics of ion transfer, those ions free to exchange in real time, obey a condition of constant energy throughout the regions of space where the ions exist.[26] In Fig. 4 this total energy is the uniform ionic electrochemical potential. Without inclusion of the local concentrations, the ion's chemical energies are curved in the space charge regions, and are similar in shape to the energy band pictures of electrons in semiconductors. Thus, in Fig. 5 in uniform electroneutral bulk phases without current flow, the chemical and electrochemical potentials are constant, but different values. In the space charge regions, the energies $\mu^o + zF\phi$ bend in the opposite direction from the chemical potential $\mu^o + RT\ln a_i$. This is exactly the ionic analog of the electron energy band drawings, i.e., as the energies bend down at the surface, electrons accumulate. In electrodic theory, the curvature in the space charge regions is accounted for in a corrected surface region concentration of reactive species (the static psi effect).[24]

In the compact layers, there is presumably an energy barrier to ion exchange which determines the rate, but not the equilibrium condition. These hypothetical barriers determine the forward and backward ion fluxes at equilibrium (the ion exchange current density). In Fig. 5 the barriers are shown approximately parabolic within the compact regions. These begin at the surface so that the effect of space charge is already accounted for in a continuous way. At zero current (I=0), the barrier determines the exchange current density. When the potential of the metal or membrane is made more negative, the barrier height is diminished for the forward reaction (left to right) for cations, but is increased for negative ions. For a net applied voltage $(\bar{\phi}-\phi)-(\bar{\phi}-\phi)_{I=0}$, which can be written $\Delta\phi-\Delta\phi^{eq.}$, the barrier is lowered to $\alpha(\Delta\phi-\Delta\phi^{eq.})$ as in the theory of Butler-Volmer-Erdey Gruz,[24] which is related to the earlier treatment of bulk transport by Laidler and Eyring.[22] Although the barriers are shown as curved (parabolic), triangular barriers suffice to describe the major effects, such as constancy of the transfer coefficient α.

Quantitative Description of Ion Transfer Kinetics

Kinetic factors can play a significant role at the interfaces. For a neutral species, interfacial crossings are frequently first order

$$J_i = k_f c(t) - k_b \bar{c}(t). \tag{62}$$

However, for ion crossings or ion attachments, the same overall equation applies, but the individual rate constants are potential dependent[27]

$$k_f = k_f^{eq} \exp\left[-\alpha \frac{zF}{RT}(\Delta\phi - \Delta\phi^{eq})\right] \tag{63a}$$

$$k_b = k_b^{eq} \exp\left[(1-\alpha) \frac{zF}{RT}(\Delta\phi - \Delta\phi^{eq})\right] \tag{63b}$$

For a single ion crossing, the equilibrium state is defined by the equality of the electrochemical potentials on both sides of the interface.[26] When rearranged, one finds

$$K_{\pm} = \left(\frac{k_f}{k_b}\right)^{eq} = \frac{\bar{a}}{a} \exp\left[\frac{zF}{RT}\Delta\phi^{eq}\right] = \exp\left[(\mu_{\pm}^{0} - \bar{\mu}_{\pm}^{0})/RT\right] \tag{64}$$

from which follows the well-known result:

$$\Delta\phi^{eq} = \left(\frac{RT}{zF}\right) \ln\left[\frac{K_{\pm} a}{\bar{a}}\right] \tag{65}$$

Equation (64) defines the single-ion-extraction coefficient (single-ion-partition coefficient) in terms of rate constants and in terms of the intrinsic thermodynamic stabilities of ions in each phase in their standard states ($\mu_{\pm}^{0}$). If two ions of equal magnitude, but opposite charge, are partitioned at equilibrium

$$\Delta\phi^{eq} = \left(\frac{RT}{zF}\right) \ln\left[\frac{K_{+} a_{+} + \bar{a}_{-}}{K_{-} a_{-} + \bar{a}_{+}}\right] \tag{66}$$

$$K_{+} K_{-} = \frac{\bar{a}_{+}\bar{a}_{-}}{a_{+} a_{-}} = K^{2}(\text{salt}) \tag{67}$$

then

$$\Delta\phi^{eq} = \frac{RT}{2zF} \ln \left[\frac{K_+ \gamma_+ \bar{\gamma}_-}{K_- \gamma_- \bar{\gamma}_+} \right]. \tag{68}$$

The single-ion interfacial potential difference eqn. (65) has the usual logarithmic Nernst form.[27] However, when an interfering ion of the same charge is also present, the equilibrium interfacial potential difference shows a form related to (65), viz.,

$$\Delta\phi^{eq} = \frac{RT}{zF} \ln (a_1 + K_{i\text{exch}}\, a_2) + \text{constant} \tag{69}$$

where $K_{i\text{exch}}$ is the ion exchange constant for the reaction

$$\text{ion 1} + \overline{\text{ion}}\ 2 \rightleftarrows \text{ion 2} + \overline{\text{ion}}\ 1. \tag{70}$$

The relation exists

$$K_{i\text{exch}} = \frac{K_\pm (\text{ion 1})}{K_\pm (\text{ion 2})} \tag{71}$$

so that ion exchange is merely a manifestation of the relative extraction coefficients that reflects the intrinsic relative stabilities of ions in the two phases. Depending on relative magnitudes $K_{i\text{exch}}\, a_2$ and a_1, large changes in $\Delta\phi^{eq}$ can occur.

When ions of opposite sign cross an interface, the equilibrium partitioning leads to very small changes in the interfacial potential difference. If activity coefficients and single-ion-partition coefficients were independent of interionic attraction effects, the change in interfacial potential difference by (68) would be zero. If two sets of salts are simultaneously partitioned, the equilibrium potential change can be larger. The exact expression for the interfacial potential difference for two salts MX and NY is

$$\Delta\phi^{eq} = \frac{RT}{2zF} \ln \frac{K_M a_M / \bar{\gamma}_M + K_N a_N / \bar{\gamma}_N}{K_X a_X / \bar{\gamma}_X + K_Y a_Y / \bar{\gamma}_Y} \tag{72}$$

Changing the activity of even one salt by a factor of 10 can cause significant sub-Nernstian changes in interfacial potential difference.

The rate of single-ion crossing an interface with potential-dependent barriers is found by combining (62) for the first-order reaction with (63). For well-stirred bathing solutions

$$J_{\pm} = \frac{1}{zF} i^0 \{\exp[\pm\alpha F(\Delta\phi - \Delta\phi^{eq})/RT] - \exp[\pm(1-\alpha)F(\Delta\phi - \Delta\phi^{eq})/RT]\} \tag{73}$$

in which the equilibrium exchange current density is given by [28]

$$i^0 = zF(k_f^{eq})^{1-\alpha}(k_b^{eq})^{\alpha}(C)^{1-\alpha}(\overline{C})^{\alpha}. \tag{74}$$

However, for imperfectly stirred solution, the first exponential term in (73) must be multiplied by $c(0,t)/c$ and the second by $\bar{c}(0,t)/\bar{c}$. The first term gives the flux of + or - ions moving left to right from aqueous to film phase. The second term gives the return direction flux.

The current-voltage curves from eqn. (73) for Na^+ and Cl^- are illustrated in Fig. 6. The equations predict exponential current increases away from the equilibrium potential. This behavior is different from a reversible case that would show a nearly vertical I-$\Delta\phi$ curve crossing the axis at the equilibrium potential. The effective resistance at the equilibrium or steady-state potential, for slow surface rate control is

$$R_R = R_0 = \frac{2RT}{zFI^0} = \frac{2RT}{z^2F^2k_0(\bar{a}_+a_+)^{1/2}} \tag{75}$$

for single-cation transport, and

$$R_R = R_\partial = \frac{2RT}{F^2[z_+^2k_0^+(\bar{a}_+a_+)^{1/2} + z_-^2k_0^-(\bar{a}_-a_-)^{1/2}]} \tag{76}$$

when ions of both signs transfer slowly with rate constants k_0^+ and k_0^- at unit activities.

D. Digital Simulation of Nernst-Planck-Poisson Equations

The implicit simulation procedure which we have developed[29] can determine concentration, field and potential profiles throughout a membrane in the transient and steady states or equilibrium states. We include profiles in the space charge regions inside and outside of a bathed membrane. The impedance-frequency response of virtually any steady state system can also be computed. The examples will illustrate some of the capabilities of the technique.

The efficient numerical solution of systems of coupled non-linear partial differential equations has been made possible only in the past 10 to 15 years with the advent of large and fast digital computers. In 1965, Cohen and Cooley[30] developed a simulation procedure for the complete diffusion-migration problem. Although their procedure was explicit, they introduced the system of reduced units still employed in most electrochemical simulations, and they included space charge via the displacement current equation (as we do in this treatment). Feldberg's 1969 chapter in the Bard series[31] provides an excellent background for electrochemical simulations.

In 1975 Sandifer and Buck[32] introduced a mixed explicit-implicit method similar to that of Cohen and Cooley. The procedure is implicit in the electric field variable and allows the reduced dielectric constant to assume any value (including zero). The explicit concentration calculations, however, place an upper limit of 1/2 on the reduced mobilities, making the procedure time-consuming and costly, especially for thick systems. In 1974, Joslin and Pletcher[33] demonstrated how distance grid expansion could be used to economize on the total number of grid points and calculations in both explicit and implicit simulations. We have employed this concept in our treatment by tailoring the distance grid system to each specific problem.

Examples: Fixed and mobile site ion exchange membranes under biionic conditions

We will now present some examples relevant to the ion exchange membrane field and consider three ionic species M^+, N^+, and X^- contained in a membrane with biionic boundary conditions (i.e., all M^+ in the left bathing solution and all N^+ in the right). The ion X^- is a site constrained to the membrane which may be fixed or mobile. There is no applied DC current in any case, but because of the unsymmetrical biionic boundary conditions, M^+ and N^+ have equal and opposite fluxes in the steady state. The biionic conditions have found wide experimental use in determining selectivity properties of ion exchange membranes.[34] The simulation procedure allows one to generate the steady-state potentials and impedance frequency responses for a wide variety of such cases.

Fig. 7a shows the steady-state concentration profiles for a mobile site membrane with infinite extraction rates (equilibrium condition) for M^+ and N^+. The biionic boundary conditions give rise to the linear profile of the site X^- and the expected curved profiles for M^+ and N^+, which have a diffusion coefficient ratio of $(D_{M^+}/D_{N^+}) = 10$. This results in a total potential of $\phi = 2.3$ in agreement with the theoretical expression for such systems.[34] Note that most of the interfacial space charge and potential is at the right interface. The impedance plane diagram in Fig. 7b shows a slightly broadened bulk arc (arc 1) because of the non-uniform concentration distribution. The finite Warburg (arc 3) is of normal shape, which is expected whenever the membrane is permeable only to ions of one sign (permselective). The interfacial rate arc (arc 2) is absent in this case due to the infinite extraction rates.

Fig. 8a shows the same mobile site membrane, except that the interfacial rates are sufficiently slow that the fluxes of M^+ and N^+ are under partial kinetic control instead of pure diffusion control as above. The interfacial space charge remains unchanged, but the bulk profiles are less steeply polarized than in the equilibrium extraction case above. The diffusional potential is thus decreased, and the total potential drops from $\phi = 2.3$ to $\phi = 1.66$. The impedance plot in Fig. 8b shows a bulk arc which is less broadened than the bulk arc in Fig. 7b, since the bulk concentrations are less polarized. The interfacial rate arc (arc 2) now appears because of the slower rates. In analyzing this arc we note that the total surface resistance is the same at both interfaces, but the double layer capacitance is greater on the right (since the total salt concentration of M^+X^- and N^+X^- is higher). This slight asymmetry is not sufficient to distort the shape of the rate arc.

Fig. 9a shows a fixed site membrane with biionic boundary conditions and fast (equilibrium) interfacial rates for both M^+ and N^+. The concentration profiles and steady state potential are in close agreement with theory.[34] The space charge is equal and of the same sign at both interfaces. The impedance diagram shows only one broadened arc. The interfacial and finite Warburg arcs are absent because of the fast interfacial rates and the fact that the diffusion coefficient of X^- is equal to zero.

Fig. 10a shows the effect on this system of lowering the interfacial rates for both M^+ and N^+. The effects are quite similar to those in the mobile site case of Fig. 8a: the space charge remains constant and the fluxes become partially rate controlled, thus lowering the concentration polarization and the diffusion potential. The impedance plot in Fig. 10b shows all three arcs - a broadened bulk arc, the interfacial rate arc, and a severely depressed Warburg-like diffusional arc. Depressed diffusional arcs are characteristic of systems in which both current carrying ions are permeable, and at least one ion is rate-limited at the interfaces. The diffusion coefficients and rate parameters determine the size and shape of such arcs.

E. Prevalence of Slow Interfacial Ion Exchange Processes

The presence of slow, potential-dependent electron transfer processes at interfaces is expected as a general and common occurrence in electrochemistry. Yet, slow potential-dependent ion transfers at ion exchanging interfaces have not been recognized or identified until the late 1960's.[35,36] In part, identification of slow ion transfers at interfaces is difficult when the bulk resistance is equal to or larger than the corresponding activation resistance. Secondly, the first systems studied (Fig. 11), proton exchange surface rates, resulted from slow transport through inhomogeneous surface layers. This problem is particularly prevalent when studying so-called 'grain boundary' transport rates for ions. Nearly impenetrable surface layers possess an equivalent overpotential process which appears in impedance plots, in transient I-t or ϕ-t plots and in steady state I-ϕ (Tafel region) plots, as surface rates. These slow surface ion exchange rates due to mechanical barriers, appear as activation barriers. Etching and sputtering experiments can be used to distinguish between slow rates resulting from inhomogeneous surface layers from those that are 'true' activation barriers.[36] Of course, this distinction is artificial in the sense that slow surface electron transfer rates are attributed to the compact layer region which also is an inhomogeneous region.

A third reason for failure to find slow ion transfers for liquid-like systems of electrolyte/bathed synthetic fixed-site ion exchangers and liquid ion exchangers, is the high water content of these ion exchangers. Systems with moderate or no solvent sheath reorganization during ion crossings are presumed to give rapid, reversible ion transfer kinetics.

The detection of interfacial rates using two-electrode impedance experiments requires that the equivalent surface rate resistance be comparable to bulk resistance of the speciment. For example, if the surface activation resistance is 5% or more of the bulk value, a 'kinetic' semicircle appears in the impedance plane plot (Fig. 12) and the steady state Tafel plot will be distinguishable from the background resistance. The theoretical basis and examples of kinetic semicircles and Tafel plots are given in the literature.[37-41] For a few systems, a reference electrode can be inserted directly into the solid membrane. In those cases, the effect of bulk resistance is reduced.

The aqueous proton rates at Li-Ba-Al-Silicate glass membrane interfaces show two semicircles in the impedance plane.[36] The low frequency 'kinetic' arc can be decreased by surface etching. Baucke[42] subsequently showed the nature of the high resistance region[43] formed as outward diffusion of Li^+ and inward diffusion of H^+ consume vacancies. This process leads to an ion mixture region of high resistance. Similarly DC studies of the Na^+/β-alumina/Na^+ (non-aqueous) interface at elevated temperatures give little or no indication of surface resistance.[44] However, addition of moisture with accumulation of H^+ in the β-alumina surface produces marked surface overpotential and Tafel behavior (Fig. 13). Transport of other monovalent ions show similar behavior. Breiter and Farrington[45] have also reported interpretation of AC and DC measurements on cells with M/M^+ β-alumina interfaces. Slow interfacial rates appear to be involved at temperatures below 97°C.

Cells involving metal contacts are complicated by the possible presence of mixed conductivity by ions and electrons. Thus, studies of symmetric cells by DC and AC methods using $Ag/Ag_2S/Ag$ for example, give results complicated by surface transport by Ag^+ and e^-. The Wagner-Hebb method[46,47] uses one or both ionically-irreversible (blocked) contact electrodes. Then the electronic transport can be separated in cells such as $C/Ag_2S/C$. By analogy, ionic transport can be emphasized using cells of the type Ag^+ (aq.)/Ag_2S/Ag^+(aq.) or Ag^+(aq.)/ Ag_2S/Ag. Rhodes and Buck[48] have shown from impedance measurements on ionic contact cells that a slow surface rate exists for Ag^+/Ag_2S (Fig. 14). But the effect is obscured by internal diffusion (Warburg character) contributions to the overall impedance. Possible slow exchange rate was previously suggested by Cammann and Rechnitz.[49] Breiter and Farrington[45] have summarized and interpreted the principal work using mixed conductivity cells. Rates of generation of interstitial Ag^+ are invoked and those processes can appear as "Warburgs" or as a surface rate, depending on the transport rate from bulk to surface.

Evidence for surface rates in liquid/liquid interface systems is easier to obtain.[50] Indeed, inhomogeneous surface rate effects can be found, measured and eliminated by subsituting hydrophilic for hydrophobic film supports. One such case is the NO_3 exchange measured from impedance data.[51] Using Aliquat nitrate (tricaprylmethylammonium nitrate) in nitrobenzene solvent, impedances were measured from DC to 5 kHz. Platinum, blocking electrodes at controlled spacing were used to obtain system data on the geometric time constant from the high frequency semicircle. These results are shown in Fig. 15a. Derived quantities, conductivities, dielectric constants, heats of transport and ion pairing constants, are extensive and already published. The impedances of free-standing liquids held between the hydrophobic polymer sheets, provides different plots as shown in Fig. 15b. The polymers are contacted by KNO_3 bathing solutions. The films behave as inhomogenerous surface resistances and provide a "kinetic" semicircle. Water permeable membranes as supports give the beginning of a Warburg impedance. An example of free-standing liquid between dialysis films is shown in Fig. 15c. Again, KNO_3 bathing electrolytes contact the membranes. In the conventional ion selective electrode configuration for liquid ion exchangers, a single microporous membrane is saturated with exchanger liquid. If the membrane has defined pores, the liquid is held mainly in the pores. Then the resistance is proportional to the pore cross-sectional areas, since transport via pores is a low resistance pathway. However, nearly homogeneous membranes such as PVC and PVC-acrylonitrile partially dissolve in the nitrobenzene solvent. Then transport occurs throughout the membrane bulk and the enthalpy of transport is greater than that for Aliquat nitrate in nitrobenzene solvent.

A more interesting factor occurs in the PVC-acrylonitrile support-membrane case. Not only are the geometric and Warburg regions present, (the latter is only partially formed), but a "kinetic" sunken semicircle occurs as well. These results are shown in Fig. 16a. The middle semicircle is surface kinetic, because it is independent of support membrane thickness. As the membrane is soaked in electrolyte for many hours, water penetrates into the membrane along strands of the support material. Water does not affect transport parameters of the Aliquat nitrate itself. This fact was noted by comparing dry vs. wet data. Also, the bulk geometric time constant remains the same in Fig. 16b, even though R_∞ (high frequency resistance) is decreasing. As R_∞ decreases and geometric capacitance, C_g, increases, the geometric time constant remains constant only when the transport mobilities remain constant.

Gavach, Seta and Henry,[52] and later Gavach, D'Epenoux and Henry[53] measured ϕ-t curves for constant current transport of quaternary ammonium ions from water to nitrobenzene. They were careful to use salts with common cations, but whose anions were relatively well-restricted to each phase (tetraphenylborate and bromide, for organic and aqueous phases, respectively). In these cases nearly all of the applied current is carried by the cations. Interpretation of the ϕ-t curves is not difficult, and ions from tetramethyl to tetrabutylammonium are kinetically controlled. Studies of this subject have also been initiated by Boguslavskii, Frumkin and Gugeshashvili,[54] and by Koryta and Vanysek.[55,56]

We have repeated and extended this work in a series of recently submitted papers. A systematic survey of the tetraalkylammonium interfacial ion-crossing rates and potential-dependenced was completed. The kinetics obey laws that are best interpreted by assuming that the overpotential appears across a compact layer believed to exist because of self-alignment of water and nitrobenzene molecules at the liquid/liquid interface. Formation of a resistive surface layer occurs in part by classical transferrence of bound waters of hydration that are shed by ions crossing into the organic phase. Examples of our results are shown in Figs. 17 and 18. This work has not yet been extended to kinetic measurements on electrolyte-bathed liquid ion exchangers of the commercial types.

SUMMARY

1) The Nernst-Planck equation is limited, in principle, to point charge transport in homogeneous and uniformly-grainy phases of constant dielectric permittivity. Physical uniformity, freedom from grain-boundary and other low-resistance pathway transport, is required.

2) The Nernst-Planck equation presumes random barriers of randomly-distributed heights to control particle transport. Singularities (high barriers regions) in surface and bulk regions are not explicitly handled. Nernst-Planck equations do not apply to individual events at very short times. Averaging over many unitary events is required to give the linear force-velocity rather than force-acceleration form.

3) The Nernst-Planck equation does not deal with microscopic origins of friction coefficients (diffusion and migration coefficients), activity coefficients and standard energies of species. The equation is parametric and other methods must be used to calculate or correlate these parameters with system properties and composition.

4) The Nernst-Planck equationdoes not deal easily with systems involving molecular-size channels and pores and low vacancy-content materials.

Despite these limitations, the Nernst-Planck equation is remarkably useful. It applies to transport in solids, liquids and gels. It can account for space, time and concentration-dependent parameters: mobilities, diffusion coefficients and standard chemical potentials. It can handle internal chemical kinetics and equilibria such as ion pairing and clustering. It can account for adsorption, space-charge generation, and concentrated and diluted systems. The activity coefficient, treated as an adjustible parameter, permits great flexibility in correcting transport parameters to account for particle size and non-uniformity effects.

APPENDIX

A. Parameters Defined by the Nernst-Planck Equations and Applicable to Many Ionic Conductors

Ionic conductors, unlike electronic conductors, will not obey Ohm's Law over a wide frequency range of applied voltage and field, in experiments with arbitrarily chosen electrodes. However, conditions of frequency and electrode composition can be found from which a resistance, R, for a uniform electrolyte can be computed. The resistance will be independent of frequency for a reasonable range depending on the system. For electrolyte solutions, it is 100 to 10^4 or 10^5 Hz, typically for well-designed experiments. From the cell dimension, length, l, area A, or from a standard cell constant,

$$\sigma = \frac{l}{RA} = \frac{\text{const.}}{R} \qquad \text{(A-1)}$$

The dominant macroscopic parameters determining σ are the concentrations of charge carriers, and their individual mobilities. Without doing the derivations outlined earlier, the numerous basic relationships that follow from the force-velocity equations are summarized in Table A-1. These results are idealized because of the assumption of validity of the Einstein relations for dilute electrolytes between mobility, u, and diffusion coefficient, D,

$$u(\text{cm}^2\text{mol/Js}) = D(\text{cm}^2/\text{s})/RT. \qquad \text{(A-2)}$$

In this Table, a single, non-associated electrolyte is considered at concentration c_s(mol/cm^3). Conventional concentrations in mol/liter or mol/dm^3 are 1000 c_s. Actual ion concentrations are c_+ and c_-, while equivalent concentrations (proportional to charge/cm^3) are c'_+ and c'_-. For intrinsic ionic conductor solids, two mobile ions may have comparable activities and equal concentrations, and thereby resemble liquid electrolyte solutions. However, this situation is not likely to occur at room temperature, and most solid ionic conductors have majority and minority carriers. The equations in this Table still apply when appropriate species concentrations are used.

Mobilities defined later in terms of fluxes and forces have numerous equivalant forms. The chemical mobility u is fundamental from the definition of forces as gradients of electrochemical potentials. Thus, for ideal, e.g., dilute systems, u and D are easily related. However, the other common mobilities are velocities (cm/s) per unit force (dynes), or per unit field (volt/cm). These mobilities are designated u' and u*.

TABLE A-1

Transport parameters for uniform electrolyte solutions of a single salt and related solids

Salt $M_{|z_a|}X_{z_c}$ at concentration c_S (mol/cm^3)

Ionic concentrations

$C_+ = C_{\text{cation}} = |z_a| c_S$ mol/cm^3

$C_- = C_{\text{cation}} = z_c c_S$ mol/cm^3

Equivalent concentrations

$C'_+ = z_c |z_a| c_S$ equiv./cm^3

$C'_- = |z_a| z_c c_S$ equiv./cm^3

Ionic mobilities

$u(\text{cm}^2\ \text{mol/s joule}) = D(\text{cm}^2/\text{s})/RT$

$u'(\text{cm particle/s dyne}) = uN^\circ/10^7 = D/kT$
$= \text{velocity/force of 1 dyne}$

$u^*(\text{cm/s per volt cm}^{-1}) = u'ez \times 10^7$
$= u'Fz \times 10^7/N^\circ$
$= uFz$

(N° is Avogadro's number).

Steady currents and specific conductivity σ S/cm

$$I = \frac{F^2}{RT}[D_+ z_c^2 C_+ + D_- z_a^2 C_-]\frac{d\phi}{dx} = \sigma \frac{d\phi}{dx}$$

$$I = F^2[u_+ z_c^2 C_+ + u_- z_a^2 C_-]\frac{d\phi}{dx}$$

$$I = F[u_+^* z_c C_+ + u_-^* |z_a| C_-]\frac{d\phi}{dx}$$

I is a current density

Conductance: $\sigma A/l$

Conductance for 1 eq./liter ($C'_+ = C'_- = |z_a| z_c c_S = 10^{-3}$ mol/cm^3

TABLE A-1 (continued)

$$\sigma' = \frac{\sigma}{10^3 C'_+} = \frac{\sigma}{10^3 C'_-} = \frac{F^2}{10^3 RT}[D_+ z_c + D_- |z_a|] \quad \text{(ohm cm eq. liter)}^{-1}.$$

Equivalent conductance Λ for a cell spacing 1 cm, area 100 cm²

$$\Lambda = 1000\,\sigma' = \lambda_+ + \lambda_- = \frac{F^2}{RT}[D_+ z_c + D_- |z_a|] \quad \text{(ohm eq./cm}^3)^{-1}$$

$$= \frac{\sigma}{C'_+} = \frac{\sigma}{C'_-}$$

$$\lambda_+ = \frac{F^2}{RT} D_+ z_c = F^2 u_+ = F u_+^*$$

$$\lambda_- = \frac{F^2}{RT} D_- |z_a| = F^2 u_- = F u_-^*$$

$\sigma = \Lambda C'_+ = \Lambda C'_-$ only approx. $\Lambda^\circ C'_+ \gamma_+ = \Lambda^\circ C'_- \gamma_-$.

Transference numbers

$$t_+ = \frac{D_+ z_c^2 C_+}{D_+ z_c^2 C_+ + D_- z_a^2 C_-} = \frac{D_+ z_c}{D_+ z_c + D_- |z_a|} = \frac{\lambda_+}{\Lambda}$$

$$t_- = \frac{D_- z_a^2 C_-}{D_+ z_c^2 C_+ + D_- z_a^2 C_-} = \frac{D_- z_a}{D_+ z_c + D_- |z_a|} = \frac{\lambda_-}{\Lambda}$$

In most thermodynamic literature, a salt is described as $M_{\nu_m} X_{\nu_x}$ where $\nu_x = z_c$ and $\nu_m = |z_a|$. This extra nomenclature has been deleted here.

Conductivities in Siemens/cm relate current densities (amp/cm^2) and fields (volts/cm). The relationships can apply locally even in non-uniform electrolytes. However, the basic relationships are derived from currents passing through uniform electrolytes or ionic crystals. The so-called "equivalent" conductivity of a salt is introduced in Table A-1. The salt equivalent conductivity is composed of equivalent conductivities of ions that are convenient quantities proportional to D, u, u', and u* for individual ions. The equivalent conductivities of ions λ_+ and λ_- and salt Λ, are experimental conductances corrected linearly to a 1 eq/liter solution between electrodes of area 1000 cm^2 spaced 1 cm apart. Because of non-ideality, primarily because of ion pair and interionic attraction, Λ values depend upon the concentration measured, i.e., the linear correction does not give constant values of Λ except when very dilute solutions are measured.

In the limit of zero concentration, the quantity Λ^0 may be found by extrapolation.

B. Temperature Dependence of Conductivities

For a single salt in a liquid, or in a crystal, the ideal conductivity is

$$\sigma = \frac{F^2}{RT}[D_+ z_c^2 C_+ + D_- z_a^2 C_-]. \tag{B-1}$$

Temperature dependences of conductivities reflect the temperature coefficients of both the concentrations and the mobilities or diffusion coefficients. In the most general single salt case, both concentration and mobility terms depend on temperature. This complication is shown by many systems, liquids and solids: saturated aqueous solutions, saturated non-aqueous solutions, solutions of weak electrolytes and intrinsic ionic conductors. For these, the temperature dependences are written explicitly:

$$D = D_0 \exp(-\Delta H^*/RT) \tag{B-2}$$

and

$$C_+ = \exp(-\Delta G/2RT) \tag{B-3}$$

where ΔG may be ΔG_S or ΔG_F for crystals, ΔG (solubility) or ΔG (dissociation) for solutions. The conductivity expression for these 'intrinsic' cases is explicitly:

$$\sigma = \frac{F^2}{RT}[D_{o+} z_c^2 \exp(-\Delta H_+^*/RT - \Delta G/2RT) + D_{o-} z_a^2 \exp(-\Delta H_-^*/RT - \Delta G/2RT)]. \tag{B-4}$$

This equation is more general than is needed for many cases involving temperature-independent concentration. These are the extrinsic crystal examples and solutions of normal, unsaturated, strong electrolytes. Solutions of constant, temperature-independent electrolyte composition obey a simplified form in which the ΔG terms are essentially zero. Crystals doped with equal concentrations of vacancy and interstitial-producing ions (so-called compensated crystals) also show temperature-independent concentrations of charge carriers over some accessible range of temperature. The ΔG term can again be deleted. Ordinary doping by vacancy formers or by interstitial formers fixes the majority ionic charge carrier. Although the minority carrier is present, its contribution to the conductivity is negligible and may be ignored. Then an entire term in eqn. (B-4) may be deleted for the minority carrier, and the ΔG term is nearly zero for the majority carrier.

Examples of specific systems of interest in the membrane sensor field, and related semiconductors are shown in Table B-1. First consider the various strong and weak electrolytes in water. The concentrations of strong electrolyte charge carriers will be nearly constant with increasing temperatures. Only a small dilution effect occurs on the molarity scale because of volume expansion. For a weak electrolyte, the number of ions increases like an intrinsic semiconductor. For acetic acid, $\Delta G_{diss} \sim 6.5$ kcal/mol, and for a saturated solution of AgCl, $\Delta G_{sol} \sim 13.6$ kcal/mol. Electrolytes can be doped (buffered), and it is possible for mixtures of acetic acid and sodium acetate to show concentrations of charge carriers that are nearly constant at increasing temperatures. The temperature coefficients for the mobilities are generally 4.0 - 4.2 kcal/mol for most aqueous electrolytes, while HCl solutions are abnormally low, about 3.5 kcal/mol. As we will discuss later, the ΔH's of activation for mobility are determined by the formation of solvent 'holes'. For non-aqueous electrolyte solutions, ionic mobilities have activation energies, ΔH^*, related to the heats of vaporization, viz., the heat of vaporization $\Delta H_{vap}/n$; $1/n$ is related to the fraction of a 'hole' required to form in an average ionic jump process. The number n is between 2.5 and 4 for many solvents.

For the solids listed, the selection was made to include a few used in sensor research and sensor devices. The Li-Ba-La silicate glasses are high quality (e.g., wide range, high stability) pH glasses. The charge carriers are Li^+ ions primarily, and their total concentrations are fixed. However, it is not clear that all of the carriers move, or that the carrier concentrations are entirely independent of temperature. The Na-Al-silicates are famous sodium transporting materials in which Na^+ is the sole charge carrier. Once again, it is possible that ion pairing does occur and that the actual number of charge carriers increases slightly with temperature, even though the total Na^+ concentration is fixed. The mobilities are very temperature-dependent. For pH glasses $\Delta H^* \sim 17-23$ kcal/mol while for the sodium-sensing glasses $\Delta H^* \sim 13-17$ kcal/mol.

For the common ionic solids, temperature-dependences of conductivity or diffusion show both the high temperature, intrinsic range, and the low temperature 'structure sensitive', impurity-controlled, extrinsic range. In Fig. B-1, the high temperature range (left side) has both carrier and mobility of substitutionally doped $CdCl_2$-NaCl vacancy Na^+, which is the fixed-concentration, majority carrier. The temperature coefficient in the low temperature range corresponds to the mobility of Na^+ vacancies, and includes any effects of dissociation of Cd^{2+}-vacancy pairs.

If one starts with an interstitial (or negative ion vacancy)-rich crystal, the conductivity depends on interstitial (or vacancy) mobility. Addition of a proper dopant reverses the majority carrier species from interstitial to vacancy (for Frenkel cases), or from negative to positive ion vacancy (for Schottky cases). The mobility of each majority species determines the conductivity in turn. Teltow's famous experiment[57] of $CdBr_2$-doping of AgBr in Fig. B-2, shows the initial conductivity by Ag_i^+ at first decreases with added $CdBr_2$. Then because vacancies are quite mobile, the conductivity increases again.

TABLE B-1

Comparison of activation parameters of liquids and solid ionic conductors

System	Example	Charge carriers	Carrier concentration dependence on temp.	Mobility dependence on temperature	
Strong aqueous non-associated, electrolytes	KCl 1M	K^+, Cl^-	constant†	$\exp(-\Delta H^*/RT)$	$\Delta H^* = 4.0 - 4.2$ kcal/mol for most electrolytes
	HCl 1M	H^+, Cl^-	constant	$\exp(-\Delta H^*/RT)$	$\Delta H^* = 3.5$ kcal/mol (abnormally low)
Weak aqueous, associated, electrolytes	Acetic acid 1M	H^+, OAc^-	increases as $\exp(-\Delta G_{diss}/2RT)$ for HOAc $\Delta G_{diss} \sim 6.5$ kcal/mol	$\exp(-\Delta H^*/RT)$	$\Delta H^* = 4.0 - 4.3$ kcal/mol
Weak non-aqueous associated, electrolytes	Quaternary ammonium salts	R_4N^+, NO_3^-	increases as $\exp(-\Delta G_{diss}/2RT)$	$\exp(-\Delta H^*/RT)$	$\Delta H^* = \Delta H_{vap}/n$ $2.5 < n < 4$
Wet, fixed-site, ion exchangers	Na^+, RSO_3^-	Na^+	constant		
Mobile-site ion exchangers	Aliquat$^+$ NO_3^-	Aliquat$^+$, NO_3^- under a.c. NO_3^- only under d.c.	constant		
Silicate glass	Li, Ba, La silicate	Li^+	constant	$\exp(-\Delta H^*/RT)$	$\Delta H^* \sim 17 - 23$ kcal/mol
	Na, Al silicate	Na^+	constant		$\Delta H^* \sim 13 - 17$ kcal/mol
Frenkel crystals	AgCl	V_{Ag^+} and Ag_i^+	increases as $\exp(-\Delta G_F/2RT)$	$\exp(-\Delta H^*/RT)$	
	($CdCl_2$-doped)	V_{Ag^+} (vacancy)	for intrinsic case, constant, for extrinsic $\Delta G_F/2 = 16.6 - 19.5$ kcal/mol		$\Delta H^* \sim 6.9 \pm 0.7$ kcal/mol (vacancy)
	(Ag_2S-doped)	Ag_i^+ (interstitial)			$\Delta H^* \sim 3.46 \pm 0.2$ kcal/mol (interstitial)
Electronic semiconductors	Si (As-doped)	h^+	increases as $\exp(-\Delta E_g/2RT)$ for intrinsic cases, constant for doped, extrinsic cases	$T^{-3/2}$	
	Si (B-doped)	e^-			

†Excluding the small volume expansion effect.

C. Correlation and Calculation of Transport Parameters

Experimental evidence for the validity of these models for liquid electrolytes requires some additional analysis for viscosity as a rate process. Since the slipping of solvent molecules past each other is a common feature of both diffusion and viscosity, it turns out, not surprisingly, that the coefficient of viscosity and the diffusion coefficient are closely related, and the activation energies are nearly identical. Considering the same barrier model with shear forces rather than electric field forces, it follows that

$$\eta = \frac{hN^\circ}{V} \exp(\Delta G^*/RT) = \left[\frac{hN^\circ}{V} \exp(-\Delta S^*/R)\right] \exp(\Delta H^*/RT) . \qquad \text{(C-1)}$$

N^o is Avogadro's number, V is the molar volume, so that N^o/V is the reciprocal volume of one molecule. Experimentally[22] for over 100 associated and non-associated liquids, it has been found that ΔG^* is a constant fraction of the of the vaporization energy ΔE_{vap}, viz.,

$$\frac{\Delta E_{vap}}{\Delta G^*} = 2.45 \pm 0.5. \tag{C-2}$$

The conclusion drawn is that creation of holes (void volumes) for jumping is necessary, and that the process resembles vaporization. In water and other associated liquids ΔH^* and ΔS^* are both larger than for non-associated liquids ($\Delta H^* \sim 4.1$ kcal/mol). However, increases in both quantities are compensating, so that ΔG^* remains constant for many solvents. For diffusion of water in water, $\Delta H^* \sim 5.3$ kcal/mol, while for ions in water, values lie within $\sim$4.0 - 4.2 kcal/mol.

Assuming that the activation processes are the same for diffusion and viscosity and using eqn. (48), it is possible to derive the relation

$$D = \frac{d^2N^o}{V}\,\frac{kT}{\eta} \tag{C-3}$$

From self-diffusion in water V/d^2N^o is about 1.4×10^{-7} cm, so

$$D\eta = \frac{kT}{1.4 \times 10^{-7}} \tag{C-4}$$

or

$$\Lambda\eta = \frac{1.55 \times 10^{-7}|z|}{1.4 \times 10^{-7}} = 1.1|z|. \tag{C-4}$$

The constancy of the product is remarkable for a given solvent, and decreases appropriately with solvents of larger molar volumes.[22]

The relations of D and η in absolute rate theory have historical hydrodynamic parallels as well. For spherical particles of large size relative to molecules of the solvent medium, the velocity in cm/s is

$$v - \text{force(dynes)}/6\pi\eta r \tag{C-6}$$

for a moving particle with radius, r, in cm in a medium with η, the coefficient of viscosity, in dyne s/cm^2 (or poise). The various relations with viscosity follow:

$$u' = 1/6\pi r\eta \tag{C-7}$$

$$u = 10^7/6\pi r\eta N^o \tag{C-8}$$

$$u^* = F|z|10^7/6\pi r\eta N^o \tag{C-9}$$

$$D = kT/6\pi\eta r. \tag{C-10}$$

This equation differs from eqn. (C-3) by a multiplicative factor that cannot be theoretically established.

To summarize, diffusion is a hopping over barriers by means of random accumulation of thermal energy. Migration is assisted diffusion in which one direction is preferred by application of a field. Although in this paper, mechanisms of motion in crystals are not treated in detail, it is worth noting that barriers for jumps need not be the same in magnitude, although similarities of processes exist. For example, generation of an interstitial ion may be a much more energetic process than the barrier magnitude between successive jumps among interstitial positions.

The primary quantities of transport that can be correlated with thermodynamic data are the enthalpies or heats of transport. These are the ΔH^* values of absolute rate theory that refer to the activation energies required to bring a mobile defect species in a solid through the "activated complex" state. Presumably, this state resembles the "tight" or "squeezed" point through which an ion passes when going from vacancy to vacancy. In liquids, the activated state has been thought to resemble the removal of a solvent molecule to form a vacancy for access by the moving ion. However, these two apparently different interpretations may be readily resolved by noting the similarity of processes. A mobile ion pushing apart lattice point ions to jump, is similar to an ion in solution waiting for solvent molecules to separate. In fact, the processes may be identical. In one case it is viewed as an active process, while the latter is passive.

It will not be surprising to find that both solid and liquid ion transport systems have activation energies which are a fraction of the energy required to create a vacancy. For liquids this quantity ΔH^* is approximately

$$\Delta H^* \sim (0.25 - 0.4)\Delta H_{evap} \tag{C-11}$$

while for solids

$$\Delta H^* \sim 0.30\Delta H'_f \tag{C-12}$$

where $\Delta H'_f$ is the heat of formation of the crystal from the gaseous elements. $\Delta H'_f$ is not the standard heat of formation of a salt from the elements in their standard states ΔH^o_f, but these quantities can be related:

$$M(gas,\ 1\ atm) + X(gas,\ 1\ atm) \rightarrow MX(s,\ 1\ atm)\ H'_f \tag{C-13}$$

$$\Delta H'_f = \Delta H^o_f - \Delta H_D - \Delta H_s \tag{C-14}$$

where ΔH_D and ΔH_s are dissociation and sublimations energies in the cases of salts whose anion element is a gaseous dimer (O_2, Cl_2, etc.), and the elemental cation is a solid.

For crystals with the NaCl structure, Swalin[58] has presented a convincing correlation figure for cation diffusion activation energy and modified heat of formation in Fig. (C-1). There is also a similar correlation for anion diffusion. When the crystal structures are the same, it is expected that entropic effects should be about the same for different compounds. Indeed, Swalin shows that free energies of activation for cation diffusion also correlate well with the modified formation energy.

Calculation of overall heats of transport, from first principles, for defect solids has not given reliable results. The reason is that the energies of moving ions must be known both initially and in the saddle point, the position of highest energy. These quantities are required for application of absolute reaction rate theory. The saddle point, or activated complex state, is the more difficult to determine and to analyze theoretically. Even the energies of ions in lattice positions cannot be calculated accurately; the increased energy of an ion in a "strained" position without the benefit of, symmetrically arranged neighbors is more difficult to calculate reliably.

There are, in addition to the states corresponding to the mobility activation heat, states corresponding to lattice and defect sites for which the heat of generation of defects must be computed. Mott and Littleton[59,60] developed a method for calculating the values of ΔH_F and ΔH_S for subsequent computation of defect species concentrations, but they recognized the additional difficulty in calculating ΔH^*. To illustrate the problem of calculating the energy of removal of an ion from a lattice site to form a vacancy, consider the energy of rearrangement resulting from both small shifts of position and charge of nearby ions, the so-called polarization energy. When an ion is removed to infinity from a Schottky defect crystal, the energy required is not the lattice energy per molecule. It is considerably less because the neighbor ions relax, usually away from the cavity. Second row neighbors also move slightly. There is also some dipolar character induced between concentric spheres of ions. Calculation of the polarization energy gained requires in addition to lattice spacings, values for the ionic polarizabilities. When the calculation is done, the quantities determined are ΔH_S^+ and ΔH_S^-. From the lattice energy, U, the defect generation energy is computed according to

$$\Delta H_2 = \Delta H_S^+ + \Delta H_S^- - U \qquad \text{(C-15)}$$

ΔH_S is the difference between large values (over 100 kcal/mol). Calculation of the entropy term is a further problem so that ΔG_S is an uncertain quantity.

SYMBOL LIST

A	- area
a_i	- activity of an ion or neutral species
a_+, a_-	- activities of cations and anions, mol/ liter or mol/cm^3
c_i , c_k etc.	- concentrations of species, mol/liter or mol/cm^3
c'	- concentration of species, eq/liter or eq/cm^3
$c(x)$	- concentration of species as function of distance
c_+, c_-, c_s	- concentration of cation, anion, salt resp.
d	- membrane thickness
D, D_+(or D_c), D_-(or D_a)	- diffusion coefficients for salts, anions and cations
D_1, D_2	- diffusion coefficients for ions of same sign
e	- absolute electron charge
E	- electric field
eq	- equilibrium value
f_i	- friction coefficient
f	- correlation factor
$F(t)$	- local force
$\mathcal{F}(t)$	- external force
F	- Faraday constant
g	- geometric factor
G	- Gibbs Free Energy
h	- Planck's constant
H	- enthalpy
I	- current density
i^o	- exchange current density
J, J^*	- flux of species, grouped flux
K_+	- extraction (partition) coefficient for single species
K	- thermodynamic salt extraction coefficient
K_{iexc}	- thermodynamic ion exchange constant

k	- Boltzmann constant
k_o, k_o^+, k_o^-	- standard surface rate constants for slow surface processes
k^f, k^b	- rate constants for slow homogeneous processes
k_f, k_b	- rate constants for slow surface processes
M	- molecular weight
m	- species atomic or molecular weight
N^o	- Avogardro's number
p	- local osmotic pressure
R	- gas constant, or measured resistance
R_θ, R_R	- 'activation' or surface resistance
S	- entropy
T	- temperature
t	- time or transference number
u_i	- mobility of species
$\bar{v}_i$, V	- partial molar volume
v	- velocity of a species
w_v	- jump probability
x	- distance variable for one-dimensional transport
x_v	- vacancy concentration
z, z_i, z_a, z_c	- ion charge with sign, a = anion, c = cation
α	- proportionality, transfer coefficient
ε	- dielectric permittivity = $\varepsilon_r \hat{\varepsilon}_o$
ε_o	- " " of free space (rationalized)
ε_r	- conventional dielectric constant
η	- coefficient of viscosity
γ_i	- activity coefficient
Γ	- surface excess
Λ, λ	- equivalent conductance
ϕ	- local inner potential of a phase
ϕ_o	- " " " " " " that is electroneutral

$\Delta\phi$ - potential difference, interfacial or bulk

ρ - space charge density

ω - frequency, rad/sec

μ_i, μ - chemical potential of a species

$\tilde{\mu}_i$, $\tilde{\mu}$ - electrochemical potential of a species

μ_i^o, μ^o - standard chemical potential of a species

σ - conductivity

τ - time constant

ν - frequency of event

ν_J - jump frequency

REFERENCES

1. N. Lakshminarayanaiah, "Transport Phenomena in Membranes", Academic Press, New York, 1969.
2. R.P. Buck, in Critical Reviews of Analytical Chemistry, 5, 323 (1975), CRC Press, W. Palm Beach, Fl.
3. R. Parsons, Equilibrium Properties of Electrified Interfaces in "Modern Aspects of Electrochemistry", Vol. 1, J.O'M. Bockris and B. Conway (eds.), Butterworths Sci. Pub. Co., London, 1954, pgs. 103-179.
4. P.G. Shewmon, High Diffusivity Pathways in "Diffusion in Solids", McGraw-Hill Book Co., N.Y. 1964, pg. 164.
5. O. Popovych, Critical Reviews of Analytical Chemistry, 1, 73 (1970), CRC Press, Cleveland, Ohio.
6. S.R. De Groot, "Thermodynamics of Irreversible Processes", North-Holland Pub. Co., Amsterdam, The Netherlands, 1951.
7. P. Meares, J.F. Thain and D.G. Dawson, Transport Across Ion-Exchange Resin Membranes - Frictional Model of Transport, in "Membranes", Vol. 1, G. Eisenman (ed.), Marcell Dekker, N.Y. 1972 pgs. 55-124.
8. E.N. Lightfoot, R.E. Stafford and D.R. Stone, in Critical Reviews of Bioengineering, 1, 69 (1971, CRC Press, Cleveland, Ohio.
9. E.L. Cussler, J. Am. Inst. Chem. Eng., 17, 1300 (1971).
10. H. Eyring, R. Lumbry and J.W. Woodburg, Record Chem. Prog. (Kresge-Hooker Sci. Lib.) 10, 100 (1949).
11. G. Scatchard, J. Am. Chem. Soc., 75, 2883 (1953).
12. R.J. Bearman and J.G. Kirkwood, J. Chem. Phys., 28, 136 (1958).
13. R.P. Buck, Sensors and Actuators, 1, 137 (1981).
14. F. Reif, "Fundamentals of Statistical and Thermal Physics", McGraw-Hill Book Co., New York, 1965, pgs. 560-567.
15. L. Onsager and R.M. Fuoss, J. Chem. Phys., 36, 2689 (1932)
16. R. Schlögl, "Stofftransport durch Membranen", Steinkopff, Darmstadt, 1964

17. J. Newman, Transport Processes in Electrolytic Solutions, in "Advances in Electrochemistry and Electrochemical Engineering" Vol. 5, P. Delahay and C.W. Tobias (eds.), Wiley, N.Y. 1967 pgs. 87-134.

18. J.P. Sandblom, G. Eisenman and J.L. Walker, Jr., J. Phys. Chem., 71, 3862,3871 (1967).

19. P. Delahay, J. Phys. Chem., 70, 2373 (1966).

20. P. Joos and M. van den Bogaert, J. Colloid Interface. Sci., 56, 206 (1976).

21. P.G. Shewmon, Atomic Theory of Diffusion in "Diffusion in Solids", McGraw-Hill Book. Co., 1963 pgs. 40-81

22. S. Glasstone, K.J. Laidler and H. Eyring, "The Theory of Rate Processes", McGraw-Hill Book Co., New York, 1941.

23. A.J. Bard and L.R. Faulkner, "Electrochemical Methods", Wiley, New York, 1980.

24. J. O'M.Bockris and A.K.N. Reddy, "Modern Electrochemistry", Vol. 2, Plenum Pu New York, 1970.

25. E.J.W. Verwey and K.F. Niessen, Phil. Mag., 28, 435 (1939).

26. R.P. Buck, Theory and Principles of Membrane Electrodes in "Ion-Selective Electrodes in Analytical Chemistry", Vol. 1, Plenum Pub. Co., N.Y. 1978 pgs.

27. R.P. Buck, Application of Electrodics Models to Ion Exchange, Ionic Semicondu Systems in "Ion Exchange: Interfacial and Transport Properties", R.S. Yeo and R.P. Buck (eds.), The Electrochemical Society, Pennington, N.J. 1981 pgs. 16-

28. O.R. Melroy and R.P. Buck, J. Electroanal. Chem., (in press 1982).

29. T.R. Brumleve and R.P. Buck, J. Electroanal. Chem., 90, 1 (1978).

30. H. Cohen and J.W. Cooley, Biophys. J., 5, 145 (1965).

31. S.W. Feldberg, Digital Simulation in "Electroanalytical Chemistry", Vol. 3, A.J. Bard (ed.), Marcel Dekker, N.Y. 1969 pgs. 199-294

32. J.R. Sandifer and R.P. Buck, J. Phys. Chem., 79, 384 (1975)

33. T. Joslin and D. Pletcher, J. Electroanal. Chem., 49, 171 (1974).

34. F. Helfferich, "Ion Exchange", McGraw-Hill Book. Co., New York, 1962 pg. 323-

35. R.P. Buck and I. Krull, J. Electroanal. Chem., 18, 387 (1968).

36. J.R. Sandifer and R.P. Buck, J. Electroanal. Chem., 56, 385 (1974).

37. J. R. Macdonald, J. Chem. Phys., 54, 2026 (1971); 58, 4982 (1973); 61,3977 (1974).

38. J.R. Macdonald, J. Electroanal. Chem., 32, 317 (1971); 53, 1 (1974).

39. J.R. Macdonald, J. Electroanal. Chem., 70, 17 (1976).

40. J.R. Macdonald and P.W.M. Jacobs, J. Phys. Chem. Solids, 37, 1117 (1976).

41. J.R. Macdonald, Interpretation of AC Impedance Measurements in Solids in "Superionic Conductors", G.D. Mahan and W.L. Roth (eds.), Plenum Pub. Co., N.Y. 1976 pgs. 81-

42. F.G.K. Baucke, J. Non-Crystalline Solids , 14, 13 (1974).

43. A. Wikby, J. Electroanal. Chem., 39, 103 (1972).

44. G.C. Farrington, J. Electrochem. Soc, 123, 591 (1976); J. Electroanal. Chem., 76, 165 (1977).

45. M.W. Breiter and G. C. Farrington, Ionic Transport and Electronic Exchange at Solid Electrolyte Interfaces in "Electrocatalysis on Non-Metallic Surfaces", Nat'l. Bureau of Standards Spec. Pub. #455, U.S. Gov't Printing Off., Nov. 1976, pgs. 323-340.

46. C. Wagner, Galvanic Cellswith Solid Electrolytes Involving Ionic and Electronic Conductivity in "Proceedings of the 7th Meeting of CITCE", Butterworths Sci. Pub. Co., London, 1957, pgs. 361-377

47. M.H. Hebb, J. Chem. Phys., 20, 185 (1952); Phys. Rev., 83, 879 (1951).

48. R.K. Rhodes and R.P. Buck, Anal. Chim. Acta., 110, 185 (1979).

49. K. Cammann and G.A. Rechnitz, Anal. Chem., 48, 856 (1976).

50. D.E. Mathis and R.P. Buck, J. Membrane Sci., 4, 379 (1979).

51. D.E. Mathis, F.S. Stover and R.P. Buck, J. Membrane Sci., 4, 395 (1979).

52. C. Gavach, P. Seta and F. Henry, Bioelectrochemistry and Bioenergetics, 1, 329 (1974)

53. C. Gavach, B. D'Epeneoux and F. Henry, J. Electroanal. Chem., 64, 107 (1975).

54. L.I. Boguslavskii, A.N. Frumkin and M.I. Gugeshashvili, Elektrokhimya 12, 856 (1976).

55. J. Koryta, P. Vanysek and M. Brezina, J. Electroanal. Chem., 67, 263 (1976).

56. J. Koryta, Electrochimica Acta, 24, 293 (1979).

57. J. Teltow, Ann. Phys., 5, 63 (1949).

58. R.A. Swalin, Correlation of Thermodynamic and Atomic Transport Properties in "Thermodynamics", Vol. II, Proc. Symp., Vienna, July 22-27, 1965 IAEA and IUPAC, 1966 pgs. 197-211

59. N.F. Mott and M.J. Littleton, Trans. Faraday Soc., 34, 485 (1938).

60. N.F. Mott and R.W. Gurney, "Electronic Processes in Ionic Crystals", Dover Pub's, Inc., New York, 2nd Ed., 1964.

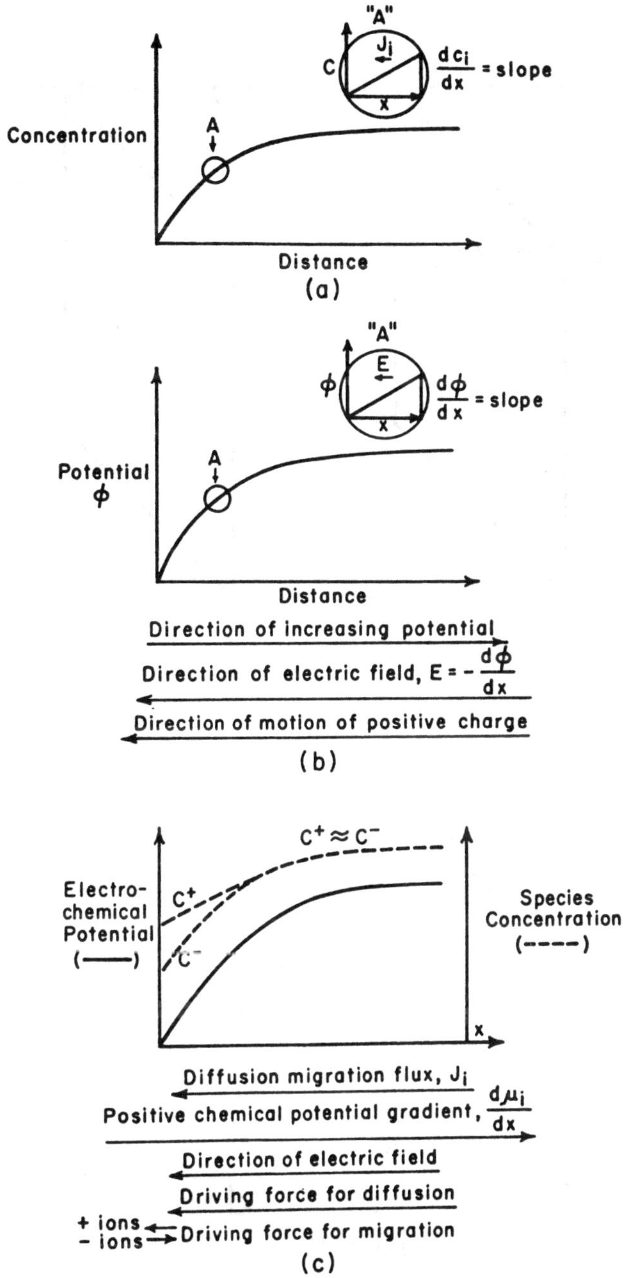

Figure 1. Illustration of diffusion migration coupling for a single-salt, binary electrolyte cell. Positive ions are reduced at the left by application of a voltage (negative at the left, positive at the right). The resulting effects in the salt solution are development of (a) a salt concentration gradient and profile and (b) a potential profile and field. In (c), the species concentration profiles and electrochemical potentials are illustrated, with the directions of derived quantities. R.P. Buck, in H. Freiser (ed.), Ion-Selective Electrodes in Analytical Chemistry, Plenum, New York, 1978.

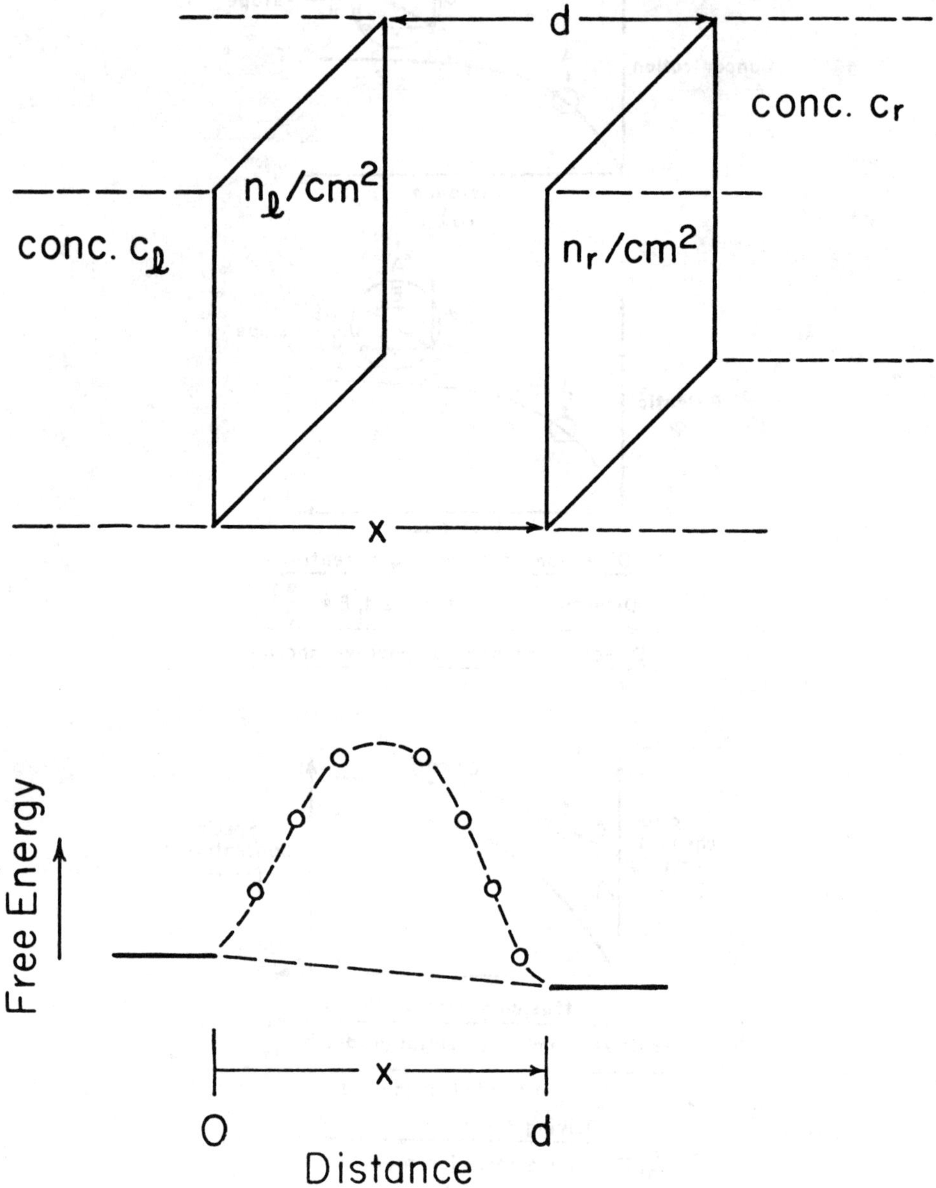

Figure 2. Geometry of the random walk thought experiments. Jumps occur from or to the region of width d. Positive fluxes are computed from left to right. Lower, a barrier-free process follows the decreasing free energy path between initial and final states. A barrier is shown by -O-.

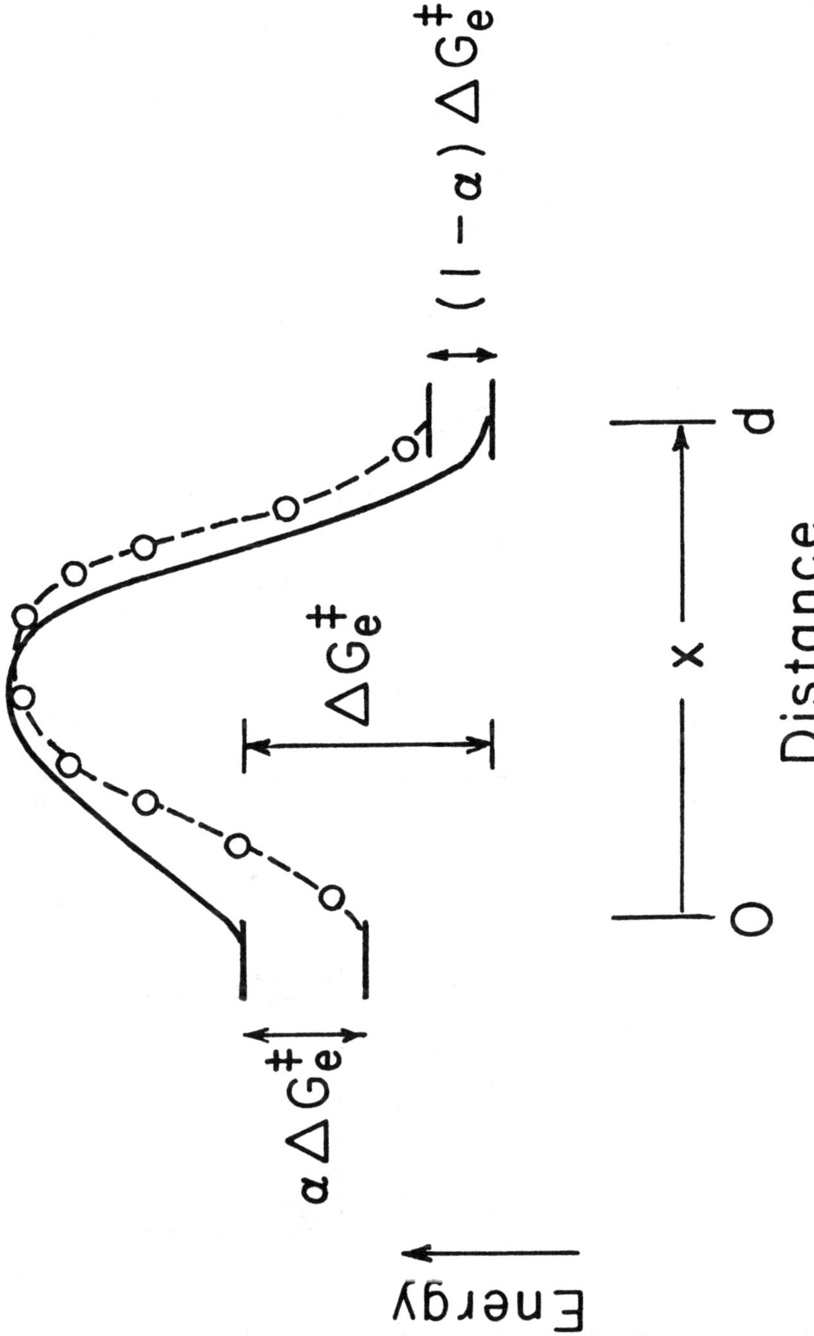

Figure 3. A reaction coordinate schematic showing an initial barrier (solid) between initial and final states for a charged species. A field corresponding to a new barrier lowering of ΔG_e^* is shown partitioned (-0-) by amounts $\alpha\Delta G_e^*$ and $(1-\alpha)\Delta G_e^*$ applied to initial and final states.

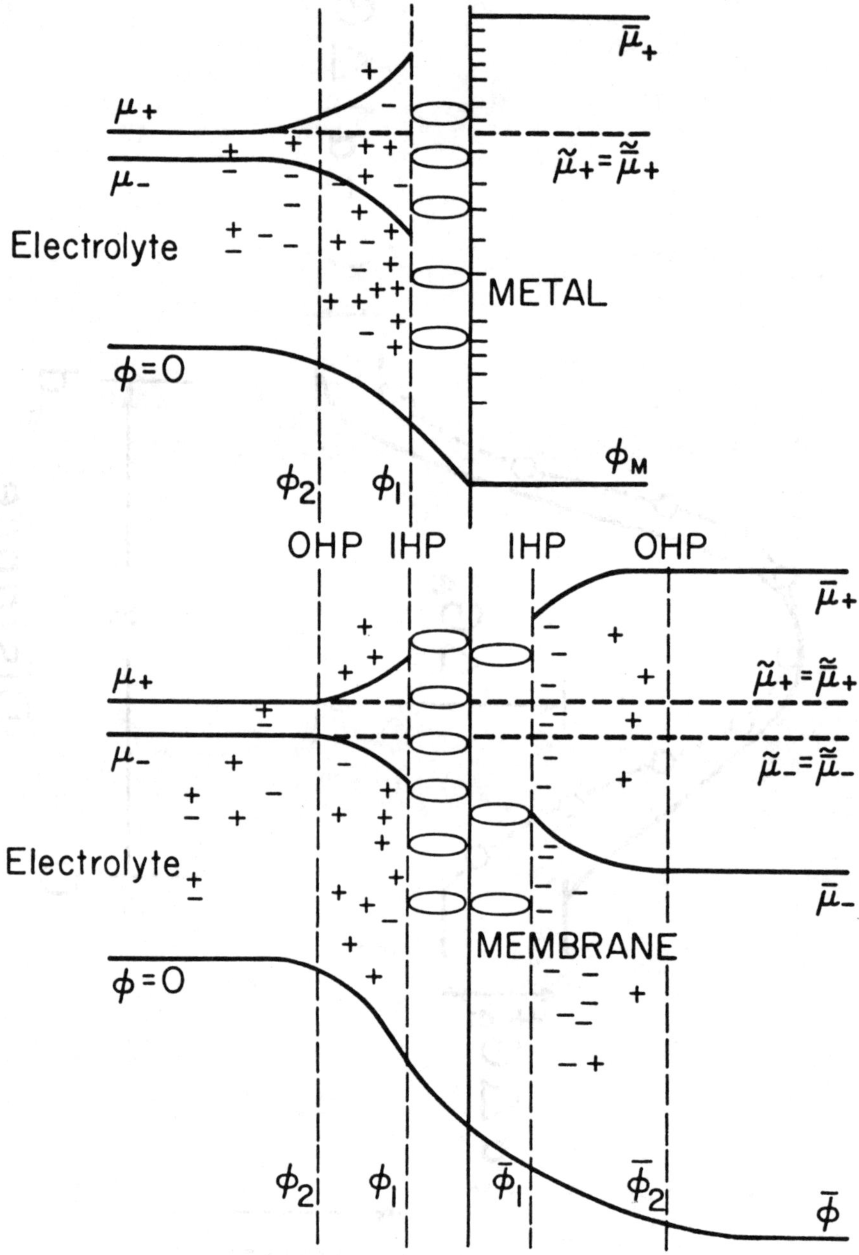

Figure 4. Comparative Interfacial Structure and Ionic Chemical Potentials for Single Salts. Upper part: Electrolyte/metal interface; lower part: electrolyte/immiscible ionic conductor phase, extraction, liquid or or solid ion exchanger. Interface, Inner Helmholtz Plane, Outer Helmholtz Plane and compact layers are shown.

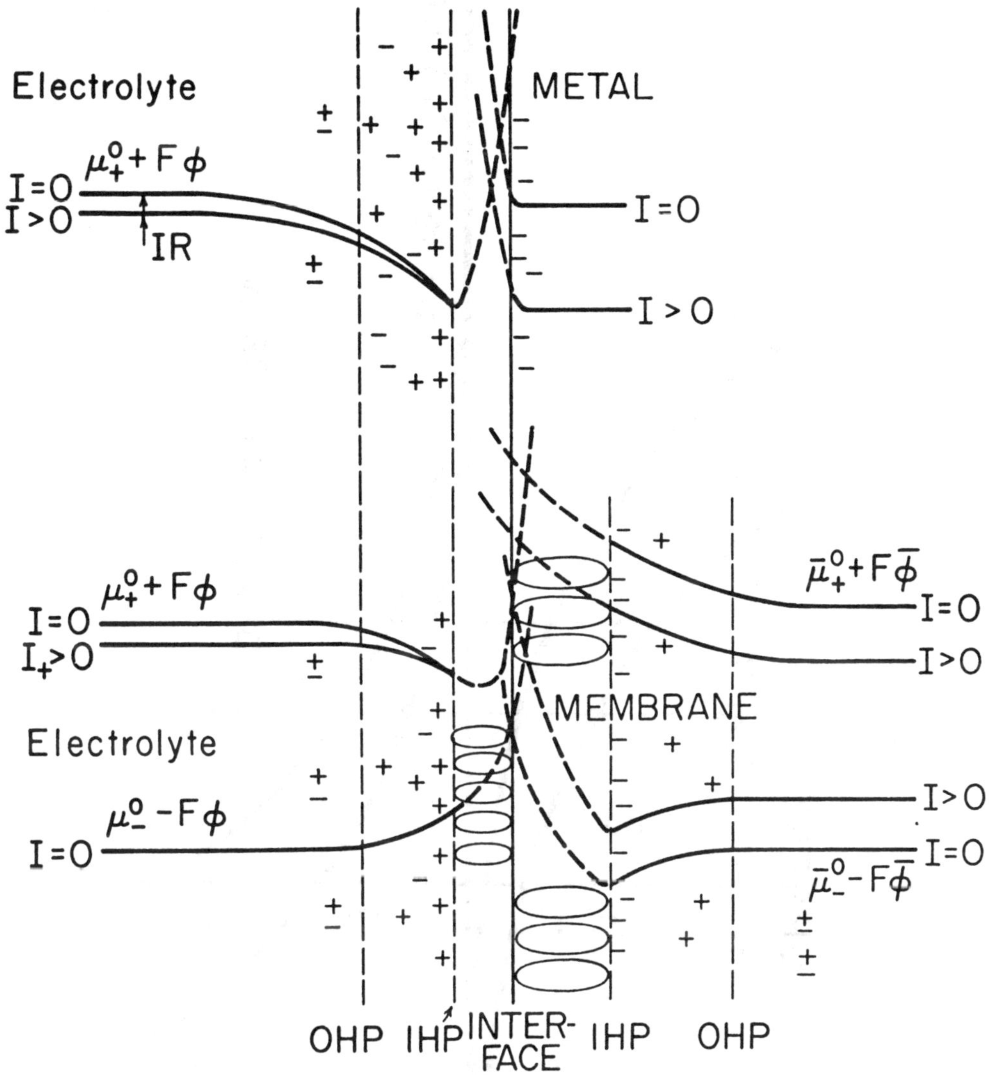

Figure 5. Comparative Interfacial Structure and Ionic Energy (Excluding Activity Term). Upper part: electrolyte/metal interface; lower part: electrolyte/immiscible ionic conductor phase, extraction, liquid or solid ion exchanger. Interface, Inner Helmholtz Plane, Outer Helmholtz Plane and compact layers are shown.

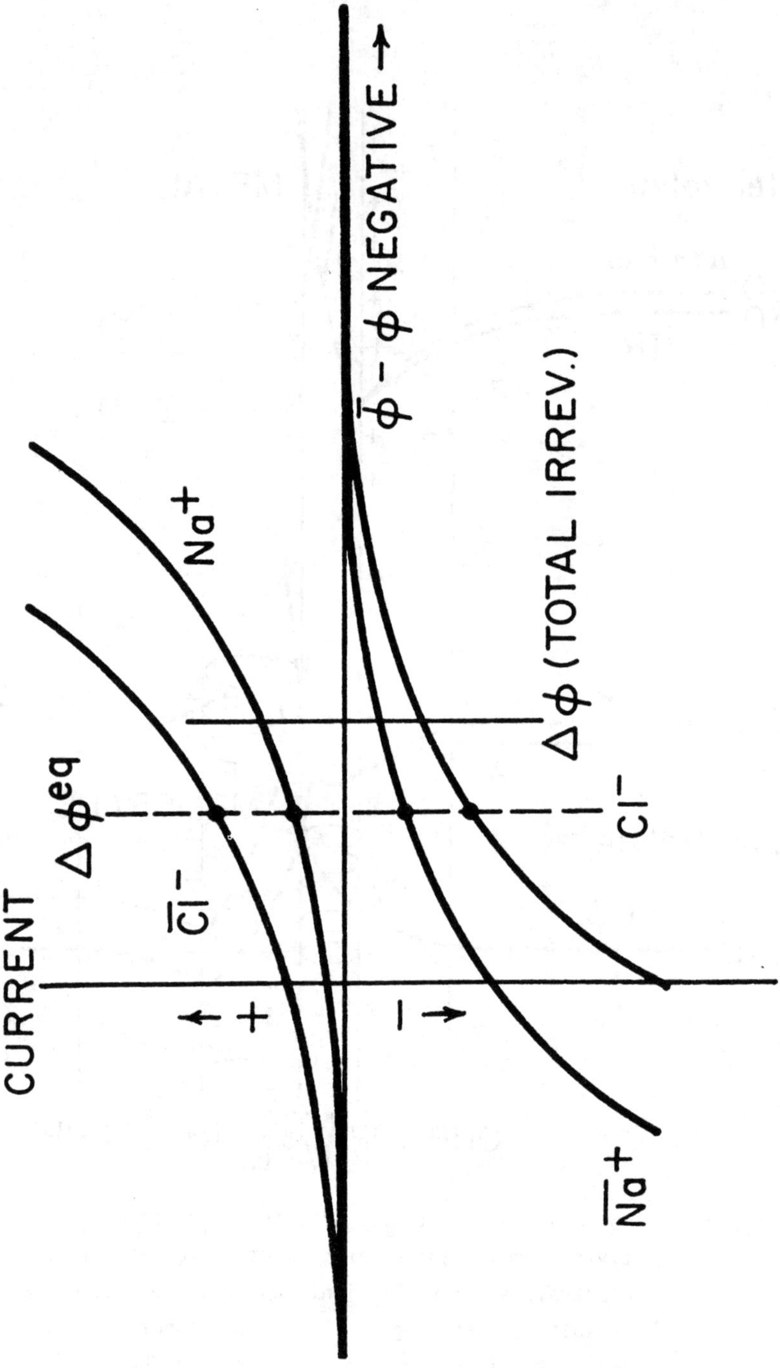

Figure 6. A schematic drawing of single-ion irreversible current-voltage curves for ion crossings. Positive current is plus charge moving from left to right (from aqueous to organic or membrane phase), or negative charge moving from right to left. Each curve is concentration dependent from the exchange current term, and the preexponential surface concentration term in (73) and (74). The equilibrium interfacial potential difference term is $\bar{\phi}-\phi$. The totally irreversible case corresponds to no back reaction of ions from the organic phase.

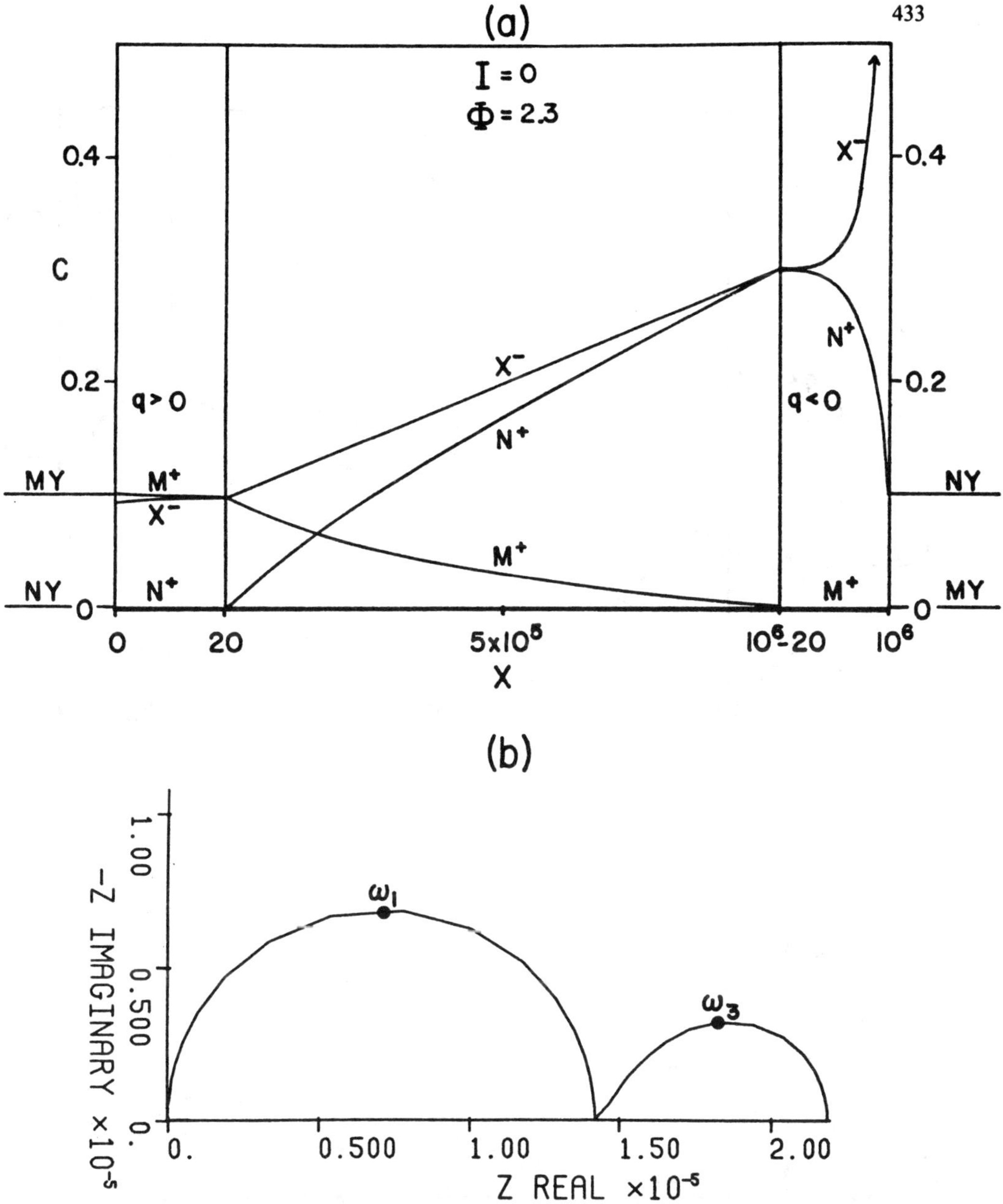

Figure 7. (a) Reduced concentration profiles for mobile site ion exchange membrane with biionic boundary conditions and near infinite extraction rates for M^+ and N^+. $D_{M+} = 1.0 \times 10^{12}$, $D_{N+} = D_{X-} = 1.0 \times 10^{11}$, $d = 1.0 \times 10^6$, $K = 5.0$, $I = 0$, $K_{bL,R} = k_{fL,R} = 1.0 \times 10^{12}$ for M^+ and N^+, $k_{bL,R} = k_{fL,R} = 0$ for X^-. The surface concentration of X^- at the right interface is off scale. (b) Complex impedance plot for system in Fig. 7a. $\omega_1 = 1.4 \times 10^{10}$, $\omega_3 = 2.0$ (Ref. 29).

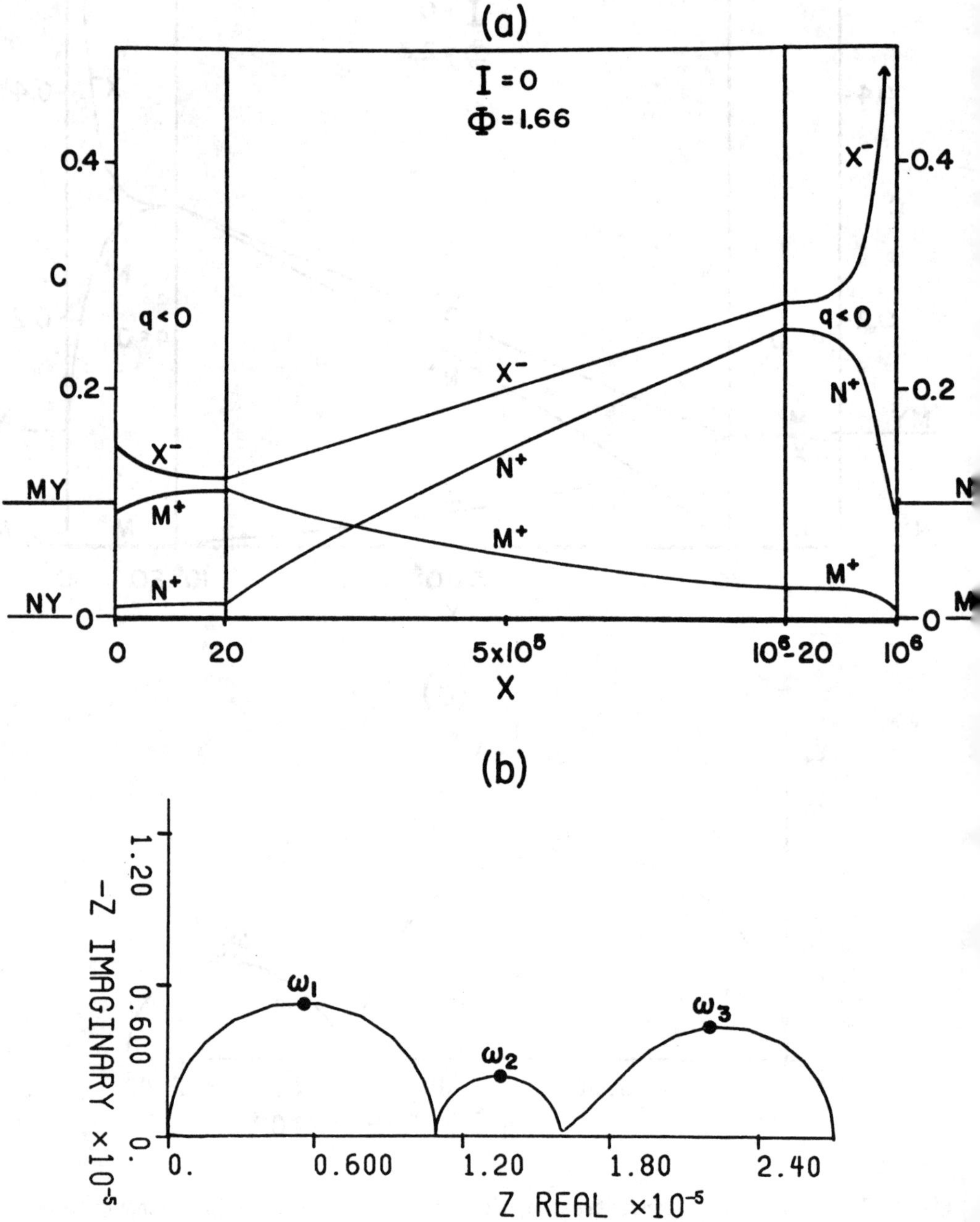

Figure 8. (a) Reduced concentration profiles for mobile site ion exchange membrane with biionic boundary conditions and finite extraction rates for M^+ and N^+. All input parameters as in Fig. 7a except $k_{bL,R}$ = $k_{fL,R}$ = 4.0 x 10^6 for M^+ and N^+. (b) Complex impedance plot for system in Fig. 8a ω_1 = 1.8x10^{10}, ω_2 = 2.6x10^5, ω_3 = 1.3. (Ref. 29).

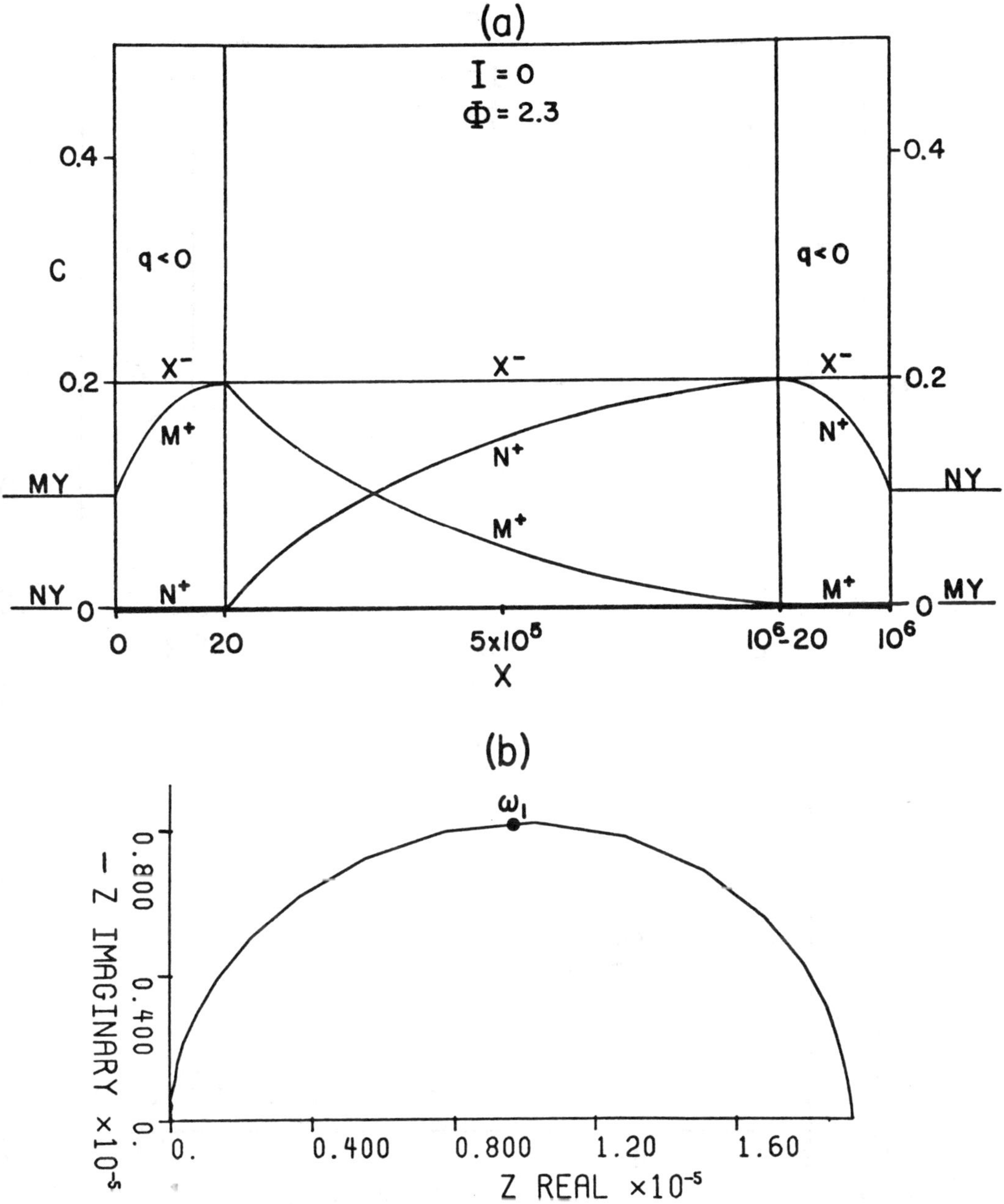

Figure 9 (a) Reduced concentration profiles for fixed site ion exchange membrane with biionic boundary conditions and infinite extraction rates for M^+ and N^+. All input parameters as in Fig. 7a except $D_{X^-} = 0.0$. (b) Complex impedance plot for system in Fig. 9a $\omega_1 = 8.0 \times 10^9$, (Ref. 29).

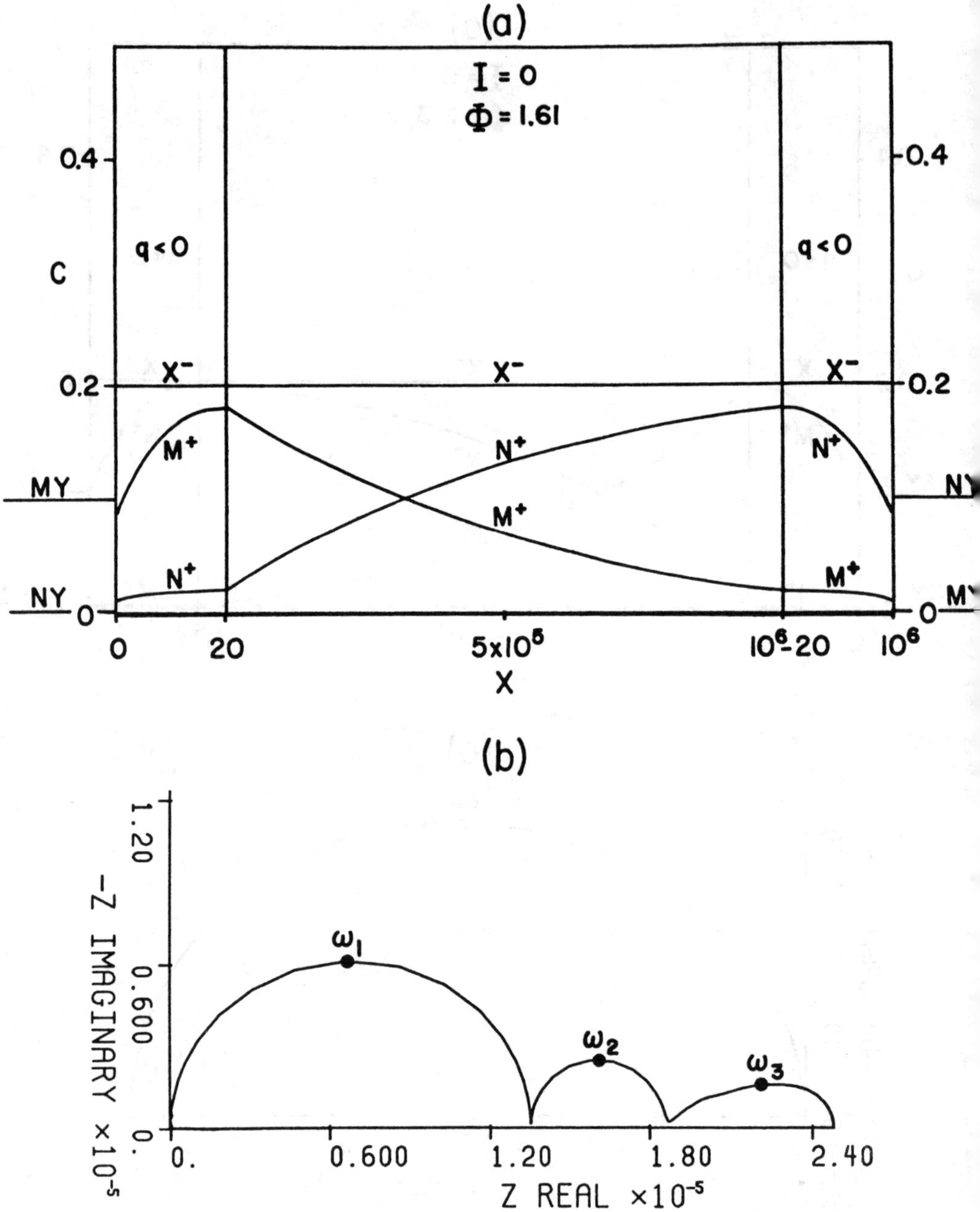

Figure 10 (a) Reduced concentration profiles for fixed site ion exchange membrane with biionic boundary conditions and finite extraction rates for M^+ and N^+. All input parameters as in Fig. 7a except $D_{X^-} = 0.0$. $k_{bL,R} = k_{1L,R} = 4\times10^6$ for M^+ and N^+. (b) Complex impedance plot for system in Fig. 10a $\omega_1 = 1.3\times10^{10}$, $\omega_2 = 5.0\times10^5$, $\omega_3 = 5.0$. (Ref. 29).

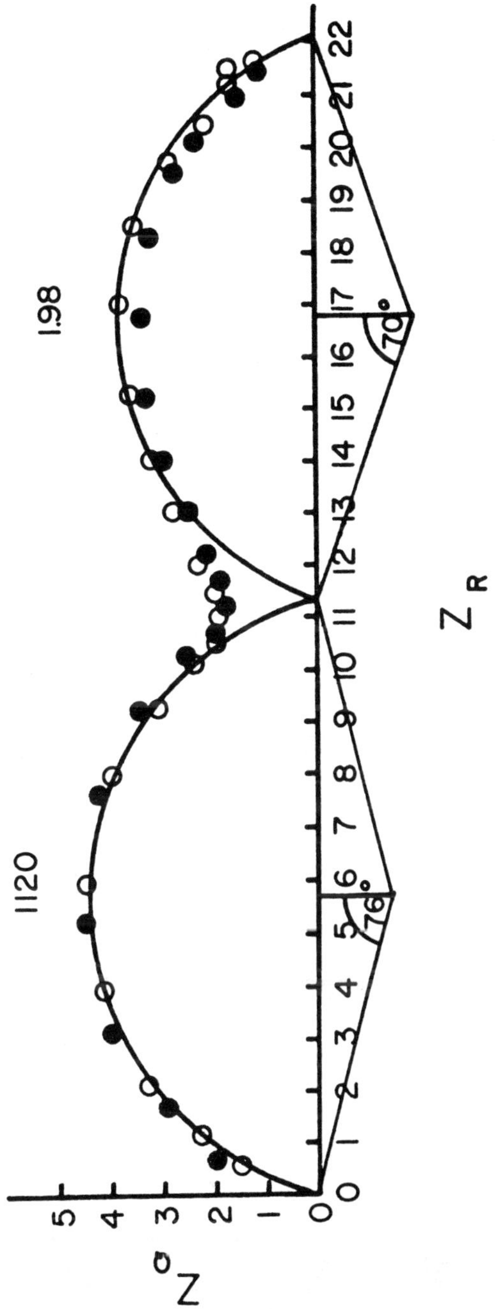

Figure 11a. Complex impedance plane representation of G.P. glass data at 45C. Data are experimental points and circles are theoretical. Z_R and Z_I are given in MΩ. ω at peak Z_I is written above each semicircle. (Ref. 36).

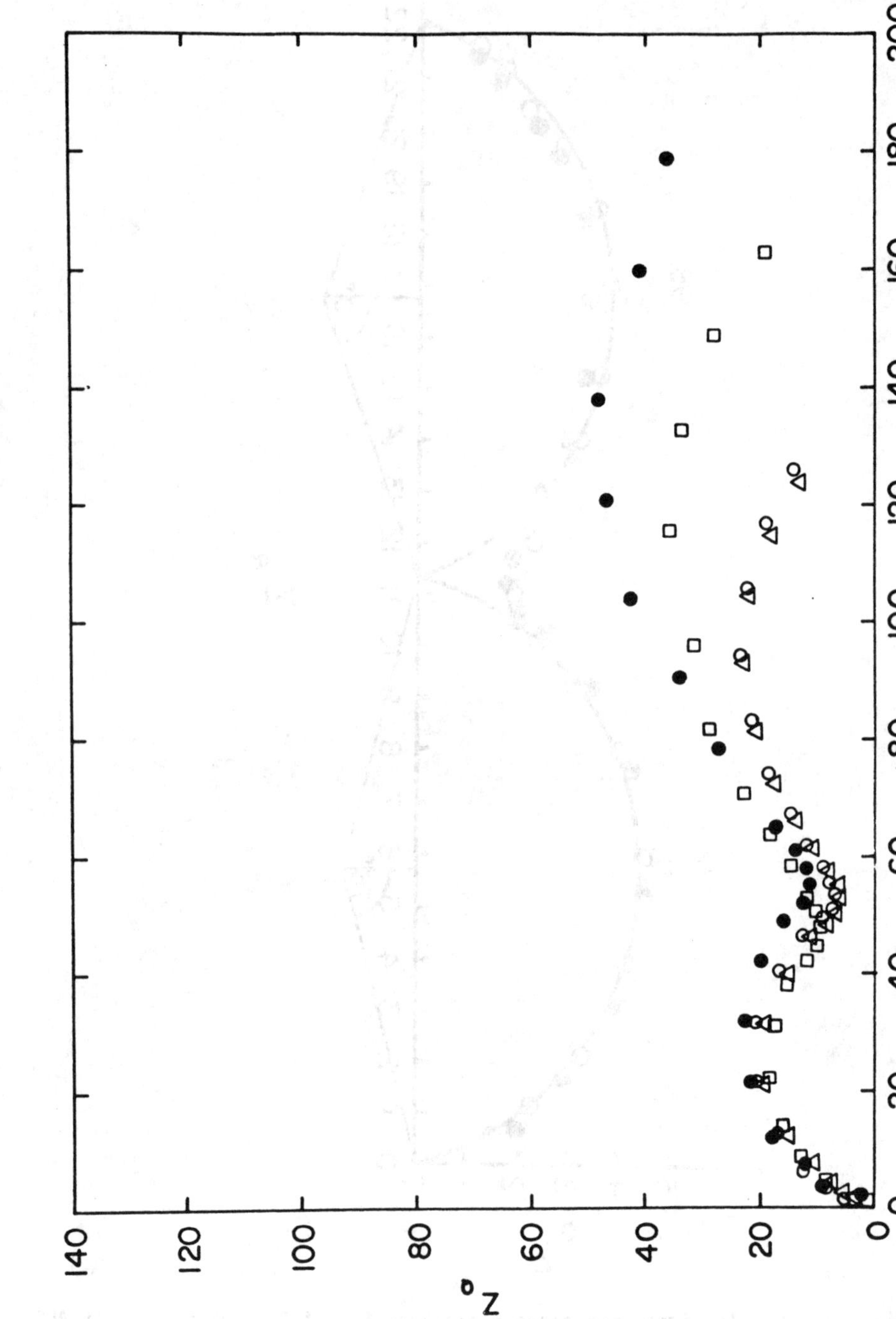

Figure 11b. Complex impedance plane representation of the results of the etching experiments. Z_R' and Z_I are given in MΩ. (●) Before etching (KCl solution contact); (□) before etching (Hg contact); (O) after 10% HF treatment (Hg contact); (Δ) after 10% HF and 18 M H_2SO_4 treatment (Hg contact). All measurements were made at 25C except the measurement denoted by (●) which was made at room temperature. (Ref. 36).

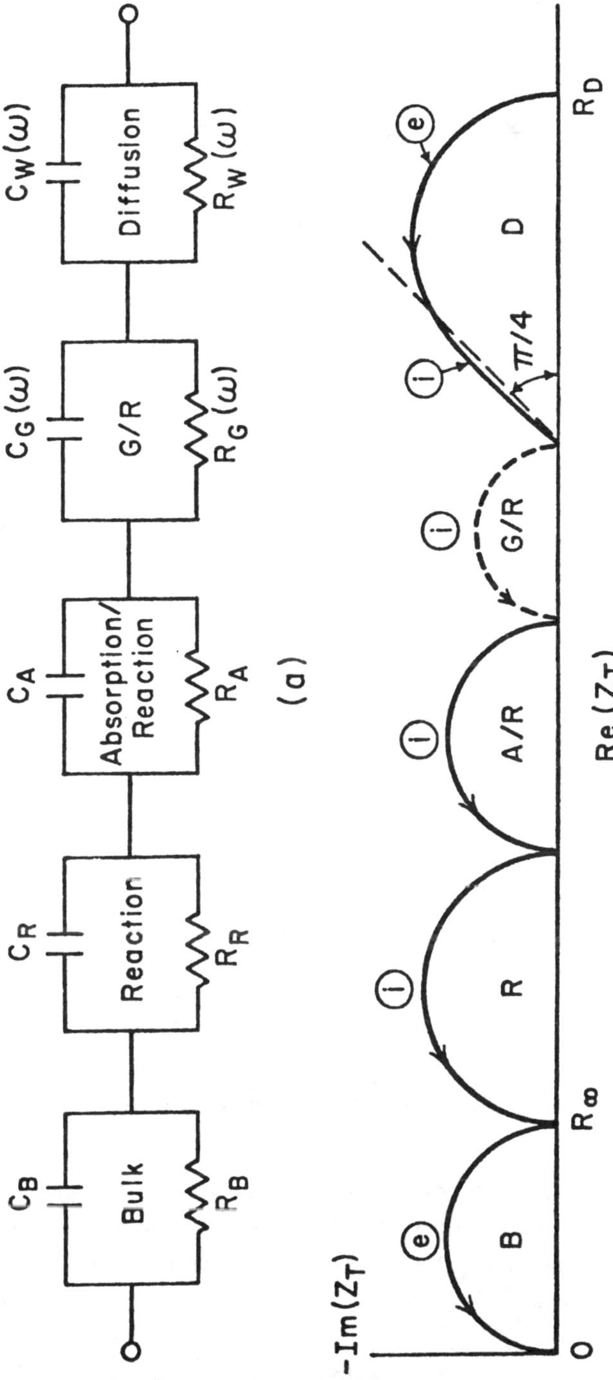

Figure 12. (a) Equivalent circuit following from the recent models (Ref. 41); (b) Possible Z_T (ω) complex plane plot consistent with the circuit in (a).

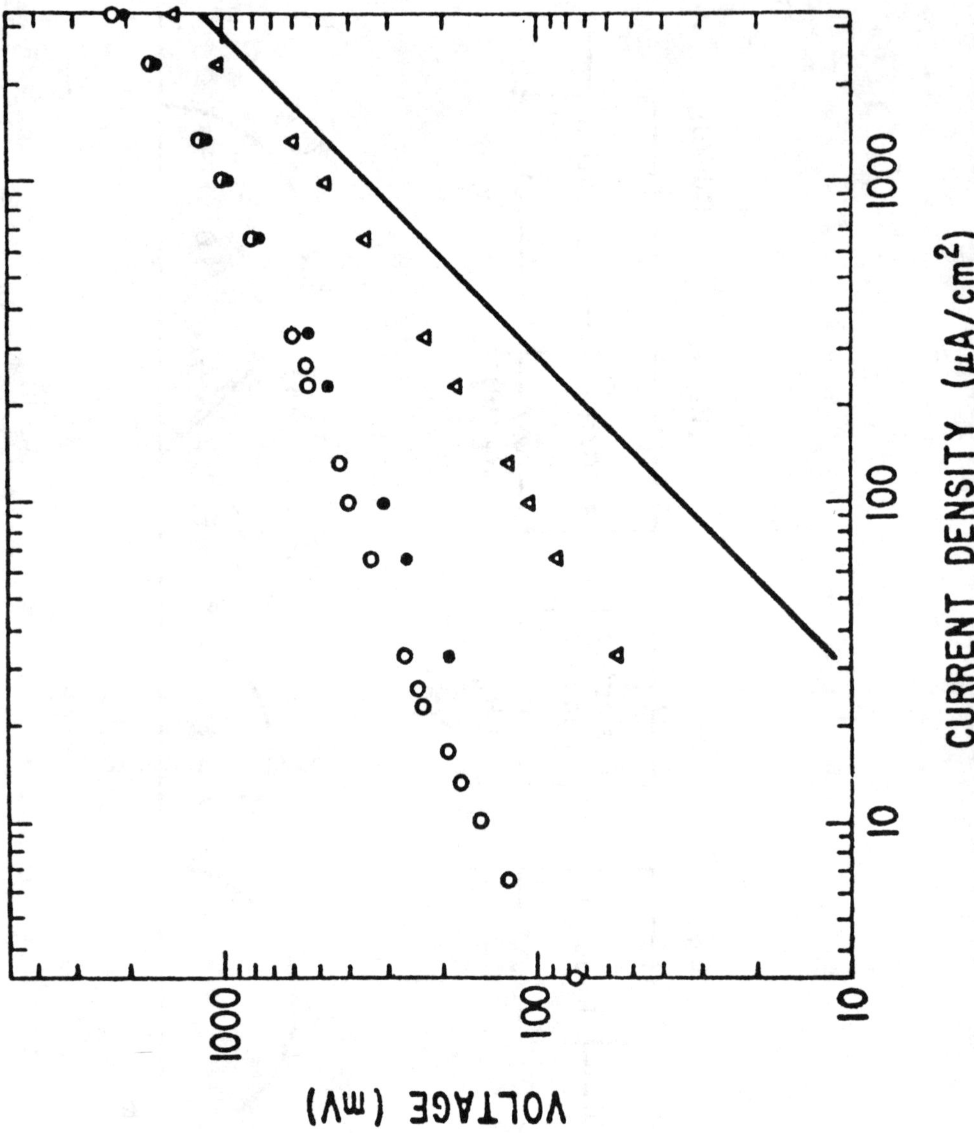

Figure 13. Na^+ exit from β-alumina into 0.010M Na^+; total voltage vs. current density; straight line is background voltage drop; o interface hydrated 20 minutes, • interface hydrated 28 hours, Δ dry interface; 23C. (Ref. 44).

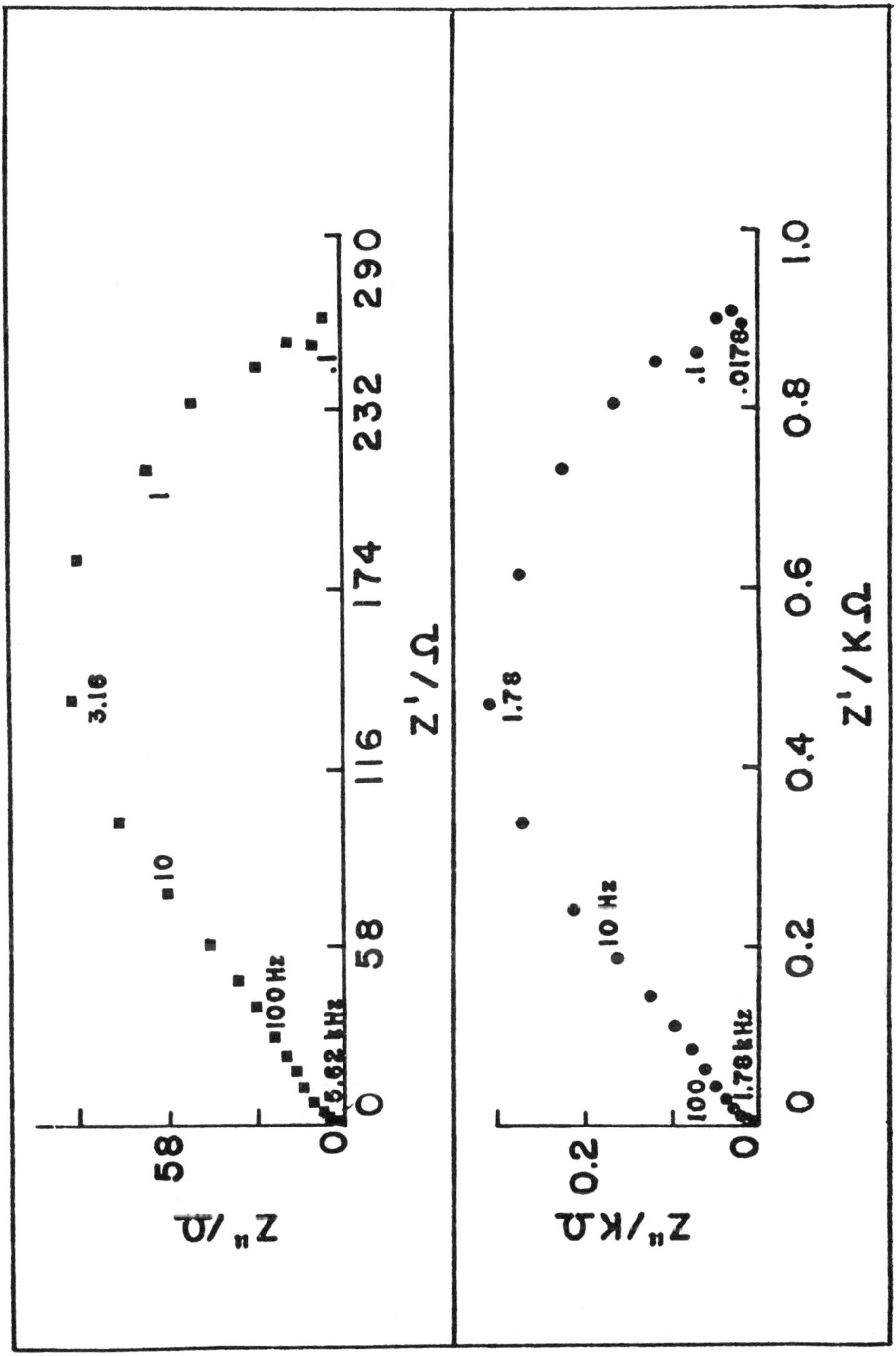

Figure 14. Impedance plane plots at a rotation rate of 36 rps in (■) 10^{-4}M $AgNO_3$ for (A) cold-pressed Ag_2S membrane (excess S^{2-}) and (B) hot-pressed Ag_2S membrane (stoichiometric). (Ref. 48).

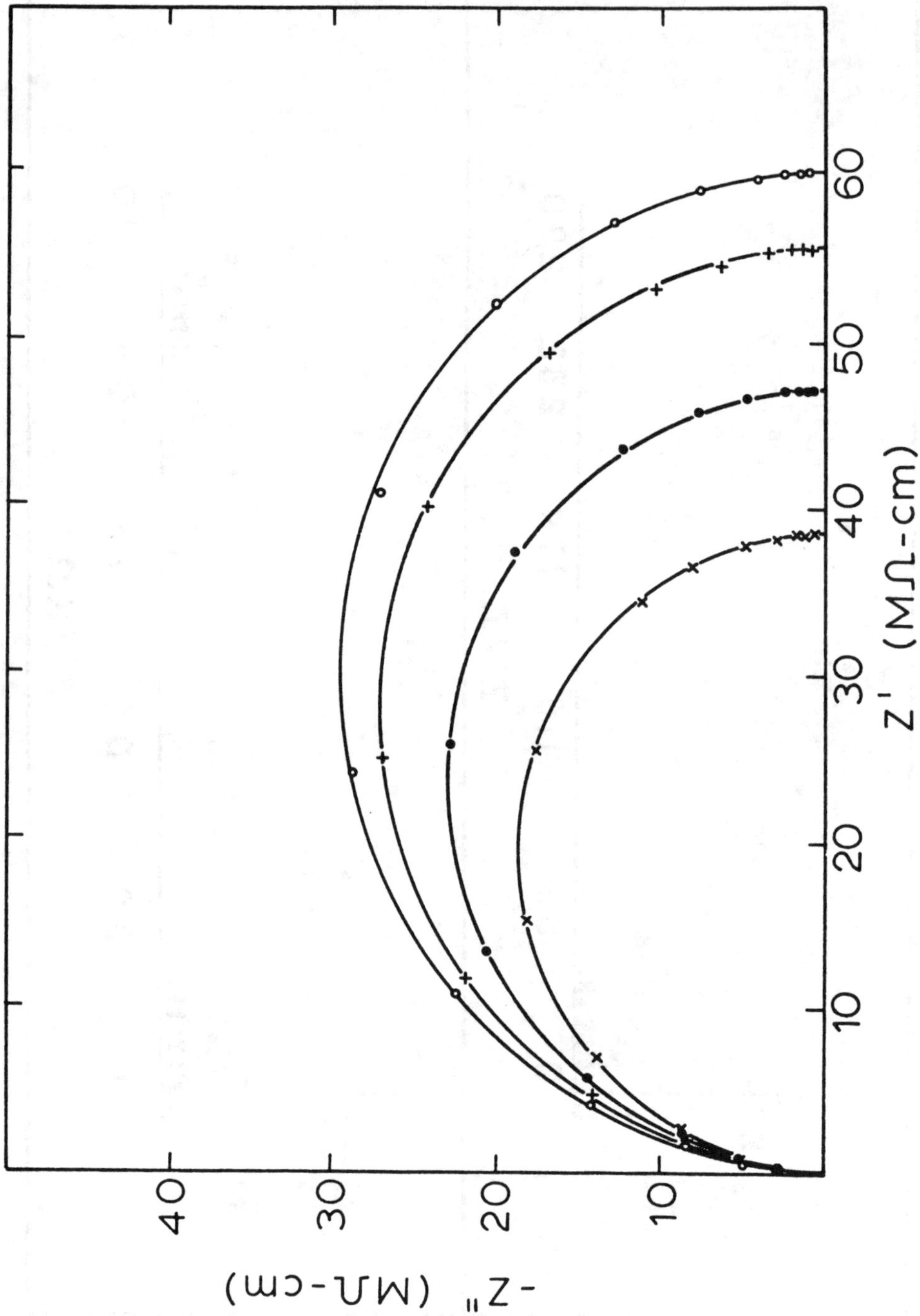

Figure 15a. Impedance plane plot of Z_I vs. Z_R for purified nitrobenzene at four temperatures (o) 20C; (+) 25C; (•) 30C; (x) 35C. (Ref. 50).

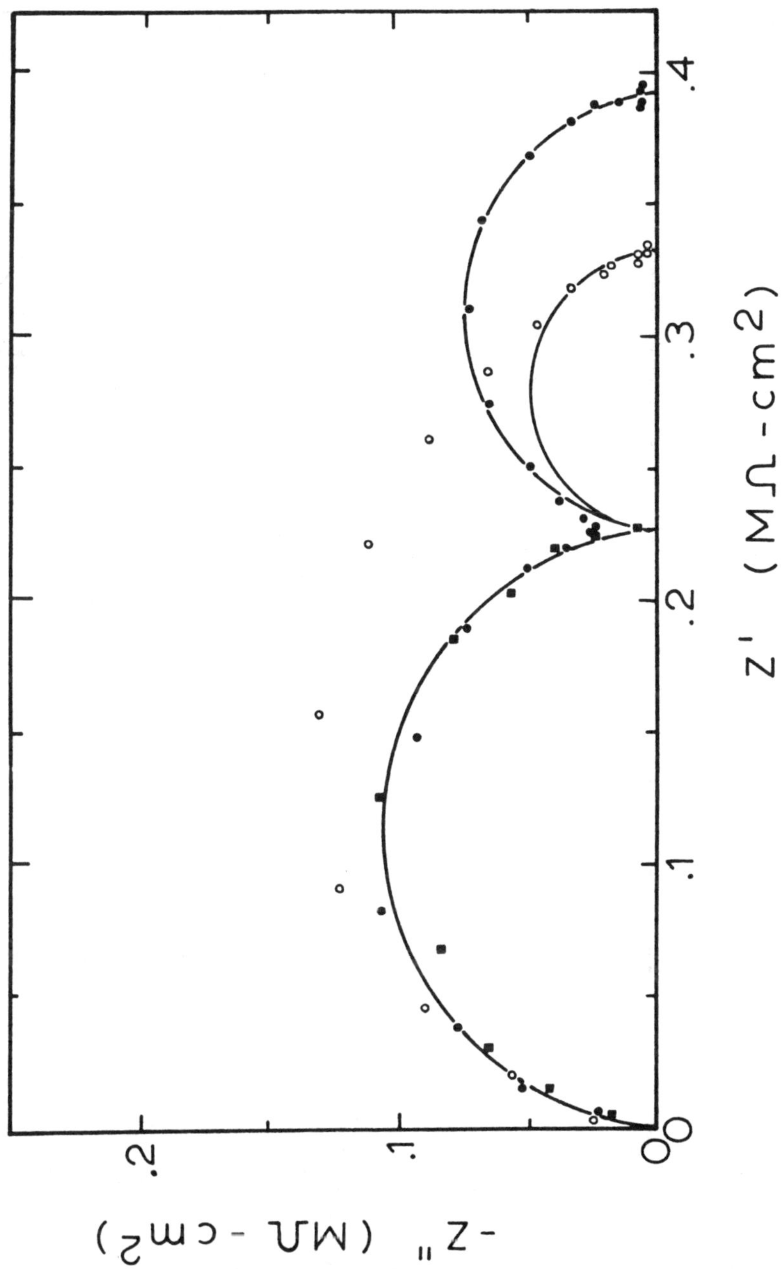

Figure 15b. Impedance plane diagram of 0.4 cm thick AR grade nitrobenzene liquid membrane as affected by constraining polymer. (■) dialysis membrane; (O), 10μ polypropylene; (●) acrylic coated polypropylene. (Ref. 51).

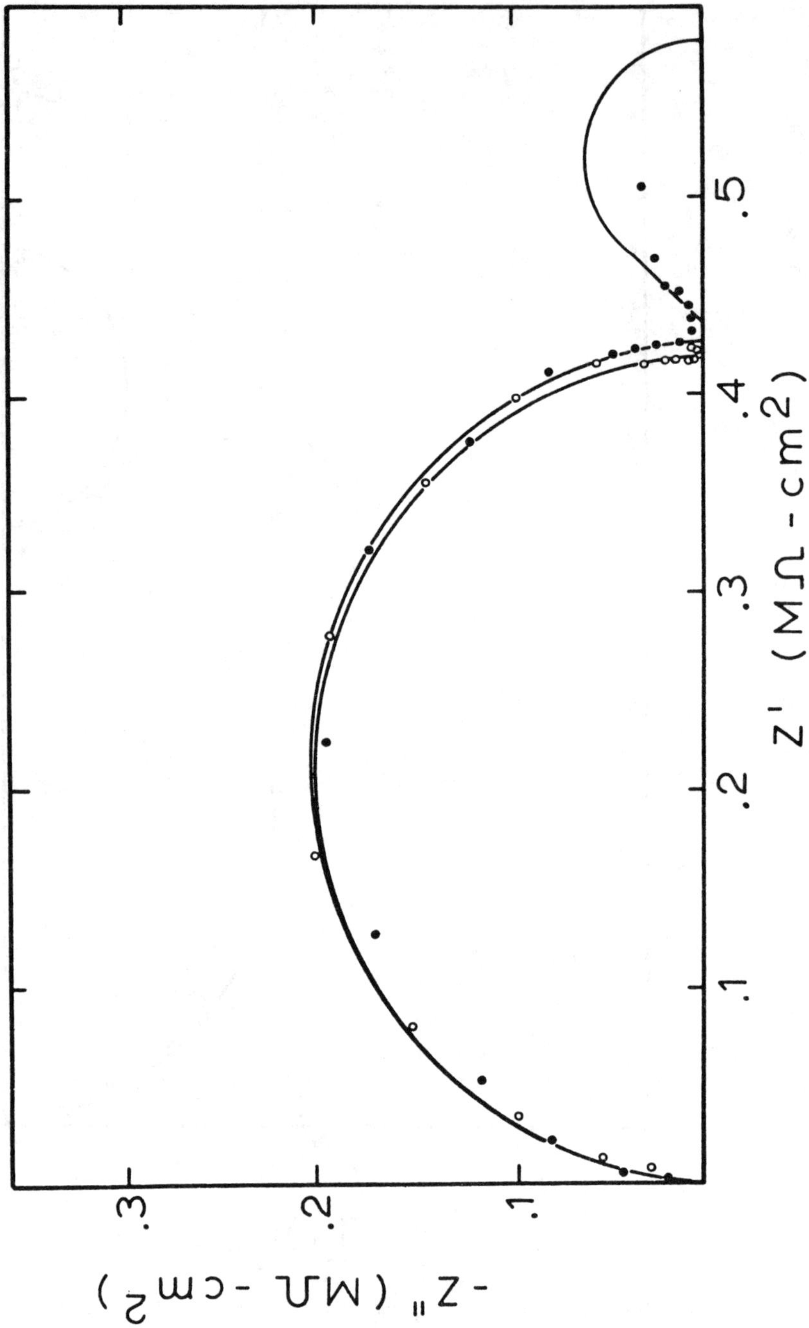

Figure 15c. Impedance plane diagram of 0.4 cm thick membrane of 1×10^{-5}M Aliquat nitrate in nitrobenzene. (O) platinum cell resullts; (●) reversible contact measurement with dialysis membrane constraint. (Ref. 51).

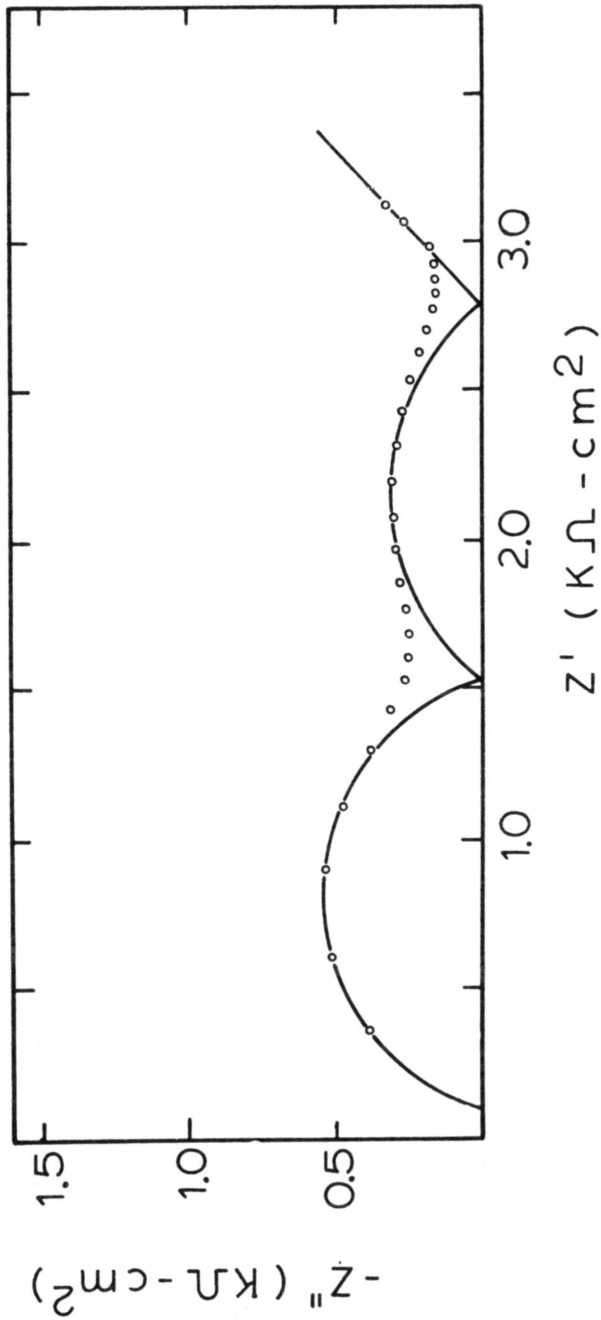

Figure 16a. Reversible contact impedance plane diagram of 0.01 cm thick membrane of 8×10^{-4}M Aliquat nitrate in purified nitrobenzene in a 0.45μ PVC-acrylonitrilecopolymer membrane. (Ref. 51).

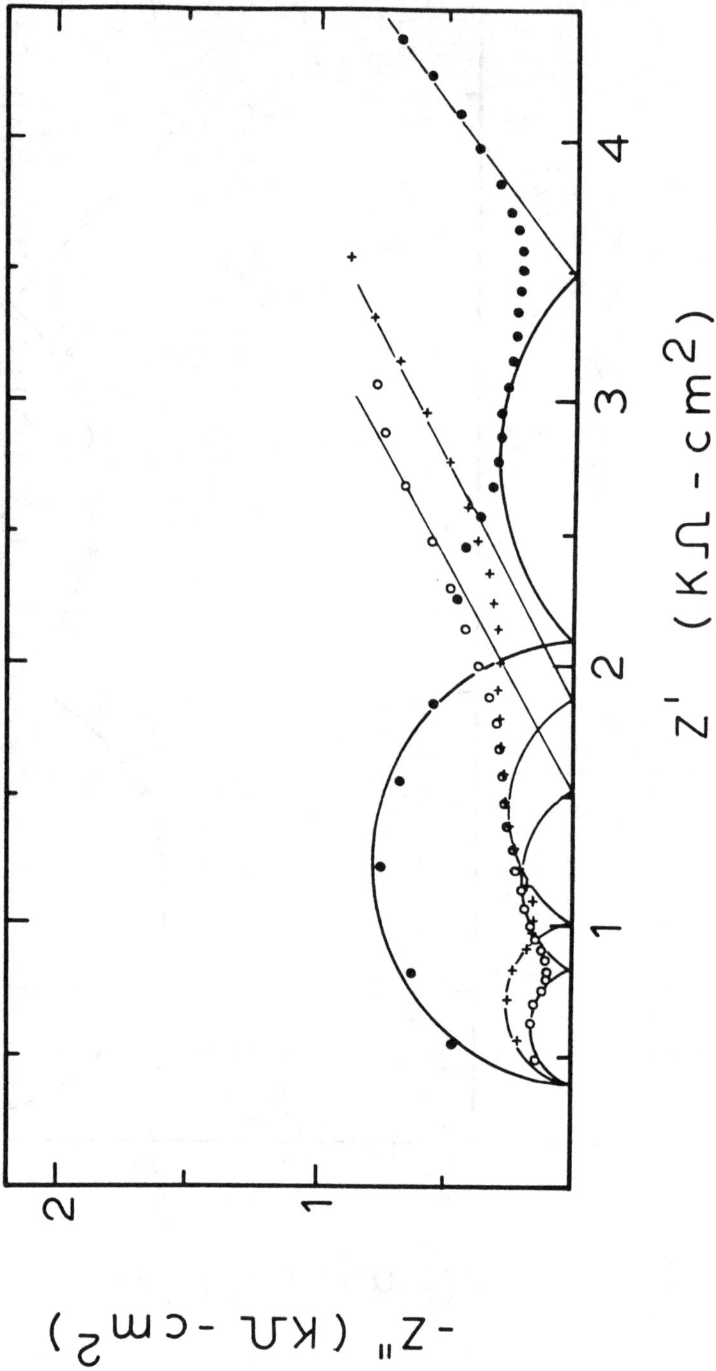

Figure 16b. Reversible contact impedance plane diagram of a cell as a function of time after fabrication. (●) new; (+) 24 hs.; (O) 48 hs. Cell membrane system was 0.01 cm thick, 0.45μ hole size, PVC-acrylonitrile containing 8x10^{-4}M Aliquat nitrate in nitrobenzene. (Ref. 51).

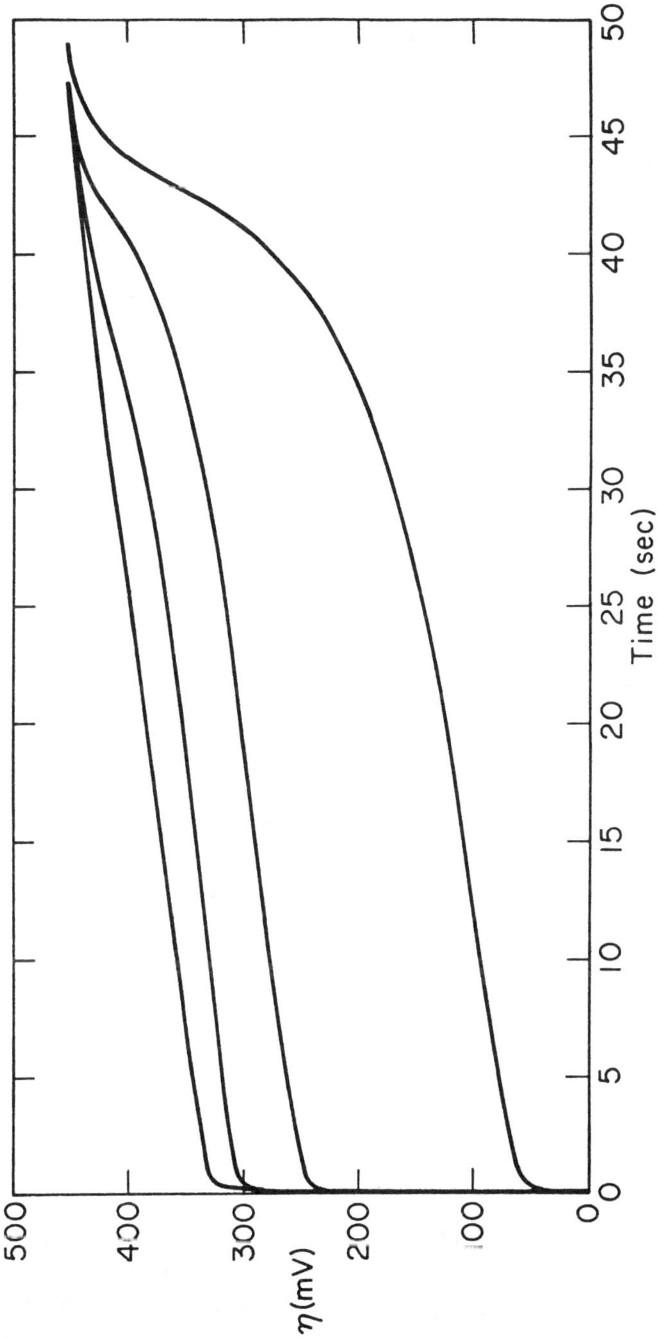

Figure 17. Experimental Log Analysis Plot for Transport of Tetrabutylammonium Chloride from water to nitrobenzene in the absence of supporting electrolyte C = 5.7 x 10^{-4} M $TBA^{+}Cl^{-}$; $\bar{C}$ = 0.0198 M $TBA^{+}TPB^{-}$. TBA = tetrabutylammonium; TPB = tetraphenyl borate; solid line: I = 15 μA; dashed line: I = 20 μA; slopes 122 mV/decade or α = ∿0.5. (Ref. 28).

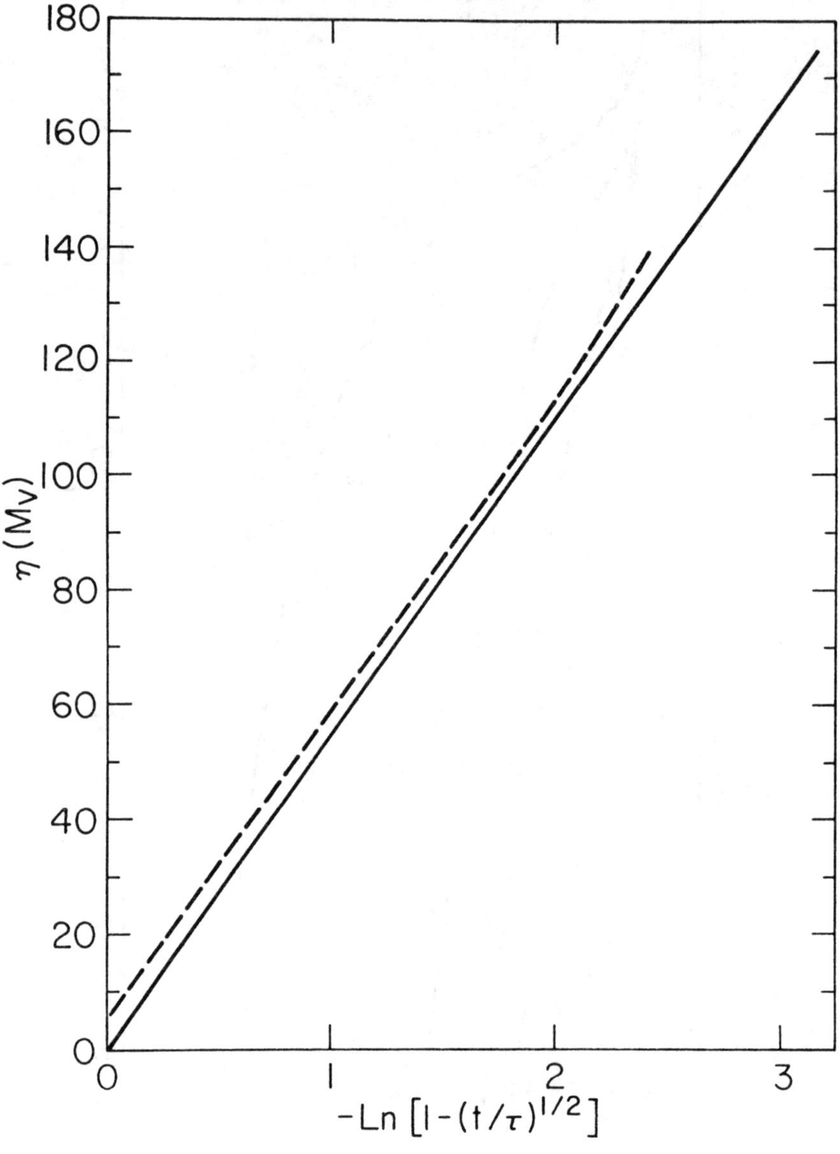

Figure 18. Overvoltage-time profiles for various TAA^+ ions from water to nitrobenzene, Aqueous solution M LiCl, organic solution 2 x 10^{-2}M TAATPB; current density: 62.5μA/cm^2 curve a) 1 x 10^{-3}M TBACl, b) 1 x 10^{-3}M tetrapropylammonium chloride, c) 1 x 10^{-3}M tetraethylammonium chloride and d) 1 x 10^{-3}M tetramethylammonium chloride. (Ref. 28).

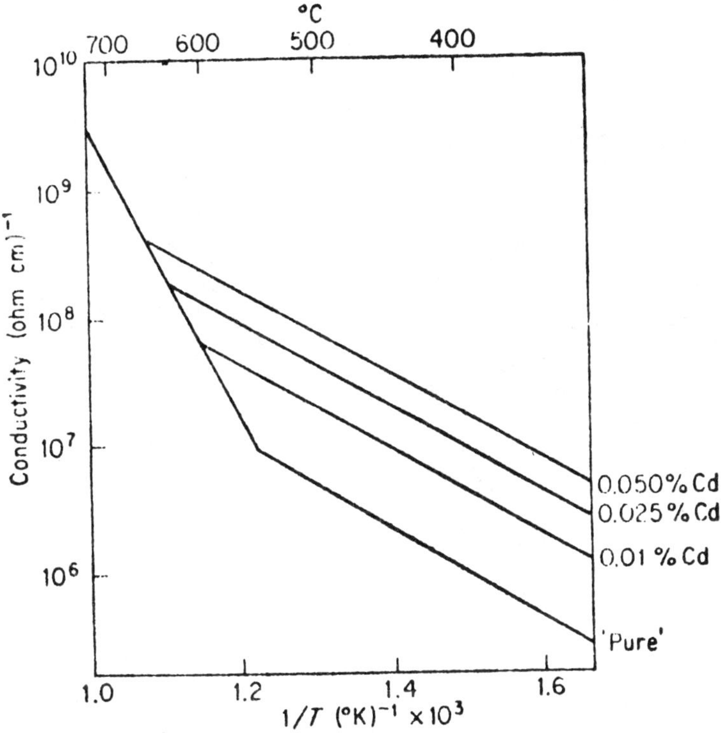

Figure B-1 Log σ vs. 1/T for NaCl crystals doped with varying atom per cent of $CdCl_2$. P.G. Shewmon, Diffusion in Solids, McGraw-Hill, New York, 1963.

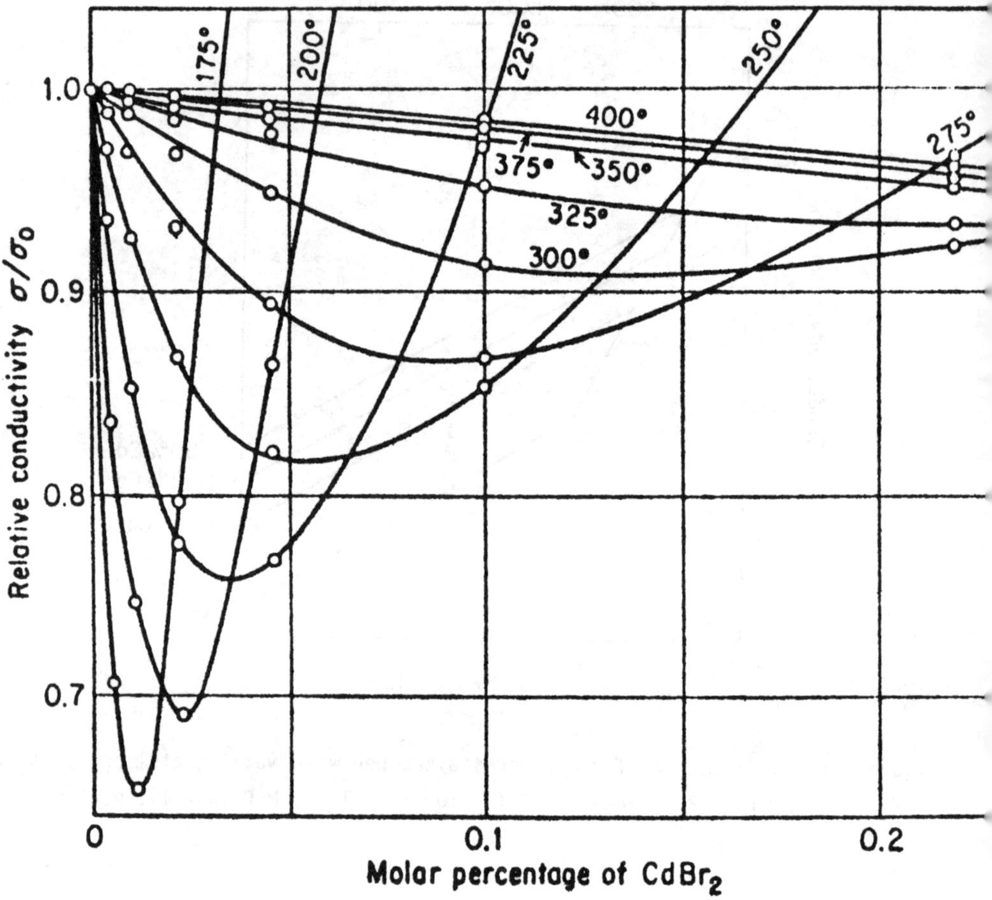

Figure B-2. Relative conductivity (σ/σ_0) of AgBr vs. atom fraction of $CdBr_2$ at several temperatures. After J. Teltow, Ann. Phys., 5, (1949) 63.

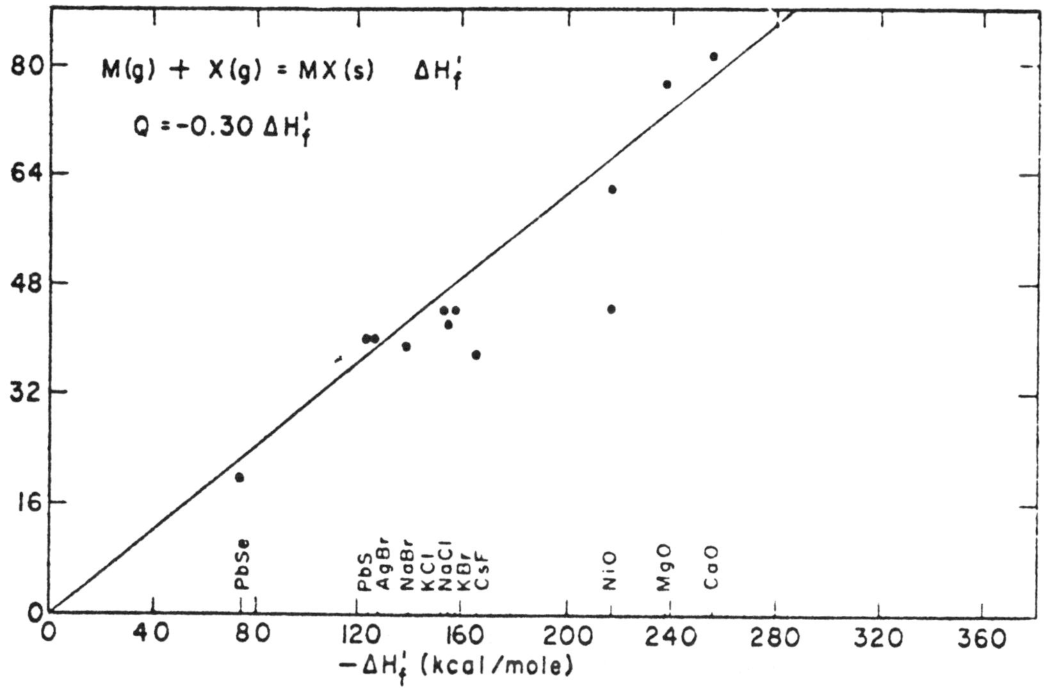

Figure C-1. Relation between activation energy, Q, and modified heat of compound formation $\Delta H'_f$ for cation diffusion in crystals with the NaCl structure. R.A. Swain, Thermodynamics, Vol. II. Proc. Symp. Vienna, July 22-27, 1965, IAEA and IUPAC, 1966.

NERNST-PLANCK OR NO?

J.C.R. Turner

Professor of Chemical Engineering,
University of Exeter,
Exeter EX4 4QF, England.

ABSTRACT

The Nernst-Planck equation for a species represents the effect of different driving forces on the flux of that species. It is a simplification in that the concentration gradient of that species, together with the gradient of electrical potential, are alone regarded as important in the flux equation; the ionic 'cross-coefficients' are ignored.

For simple situations, such as isothermal diffusion in dilute solutions, the Nernst-Planck equations are both easy to handle and useful in their conclusions. This paper leads on to other situations, of greater complexity - for example exchange within macro-porous ion-exchange resins. Here it may not be profitable to attempt detailed solutions, but the physical meaning lying behind the Nernst-Planck equations should not be lost sight of.

1. INTRODUCTION

The Nernst-Planck equation, applied to a species i, relates the flux of that species, J_i, to the 'forces' acting upon it. When the species i is charged, then one of the relevant forces is the electrical field, and it is that field which gives rise to those relationships which are implied under the heading of 'Nernst-Planck theory'.

2. THE FORCES

The forces can be regarded, for our purposes in isothermal ion-exchange, as being of two types - those due to concentration gradients, and that due to the electrical field. When we examine the terminology used to describe these forces, we find them more closely defined. Firstly it is the gradient of chemical potential, not of concentration, which is the basic force (consider uncharged species for the moment). It is not easy to obtain a gradient of chemical potential without a gradient of concentration (1) but it is thermodynamically more satisfying to refer to the gradient of chemical potential, and in certain rather unusual circumstances (2) the experimental results lean in this direction, too.

The gradient of electrical potential, ∇E, is the 'electrical force', which acts together with the gradient of chemical potential, $\nabla\mu$. It is of course necessary to ensure that these forces are expressed in the same units, and this will be mentioned again in the next section.

Some authors combine these two forces by considering the gradient of the 'electrochemical potential', $\nabla\mu'$, where $\mu' = \mu + zFE$. This is clearly a procedure equivalent to considering $\nabla\mu$ and ∇E separately.

Finally in the terminology of irreversible thermodynamics the forces are categorised as X_j, the effect of each force on species i being given (in the linearised theory) by coefficients L_{ij}, such that the flux $J_i = \sum_j L_{ij}X_j$.

3. THE COEFFICIENTS

Given a chosen terminology for the forces, the coefficients by which the forces must be multiplied to give the fluxes then follow. If we choose a concentration gradient frame (and relate the gradient of electrical potential to these units) then the appropriate coefficient is D_i, the (isothermal) diffusion coefficient. In some branches of electrochemistry it may seem more appropriate to base the flux on the potential gradient (volts/m) and to adjust the concentration force to these units. In such cases the mobility, λ_i, is the appropriate coefficient. We have already mentioned the L_{ij} terminology, where L_{ij} will have dimensions depending on those used to define the X_j.

For dilute solutions of electrolytes two simplifying assumptions may be acceptable.

i) $\nabla\mu_i = RT\,\nabla(\ell n f_i c_i) \quad RT\,\nabla(\ell n c_i)$. We take the activity coefficient, f_i, to be constant. In dilute solutions f_i may be acceptably close to unity, and even in more concentrated solutions (e.g. in ion-exchange resins) containing ions all of the same valency the ionic strength may be constant, and thus one could argue that f_i (which is primarily a function of ionic strength) is acceptably constant.

ii) We assume that those 'cross-coefficients' L_{ij} which relate to the effect of concentration gradients of species j on the flux of species i are negligible. These coefficients can be determined, and they are not zero, but to leave them in the general expressions leads to great complications.

4. ELECTRONEUTRALITY

If we therefore restrict our attention to only two forces on each species - that relating to its own concentration gradient, and that due to the electrical field, then we have a set of linear equations giving the fluxes, J_i, in terms of these two forces.

If we now assume electroneutrality in every element of volume, then the gradients of concentration of all the species obey the relationship $\sum_i z_i \nabla c_i = 0$. This relationship, combined with the presence of the ∇E term in each equation leads to the 'coupling' of the flux equations with the results now to be described. We must mention, however, that when ∇E is high, then significant 'charge separation' may occur, and the relations obtained using the principle of electroneutrality may break down.

5. EXAMPLES OF COUPLED FLUXES

i) Isothermal Diffusion of a Single Binary Electrolyte

Here the requirements of electroneutrality demand that ions of opposite charge move in the same direction at the same speed. When they do not possess the same mobility, this will require the presence of an electrical field, which is set up by an initial (usually in other respects negligible) charge separation. This electrical field is the 'diffusion potential' which is measurable, and is of particular interest in two situations:

a) the 'membrane potential', when diffusion is occurring across a membrane.

b) the 'liquid-junction potential', encountered in concentration cells. The aim of the electrochemist's 'salt bridge' is to eliminate the liquid-junction potential.

ii) Concentration Polarisation in Electrolysis

In elementary texts on electrochemistry, frequently used in metallurgy courses, attempts are made to divide the measured flux of an ion (e.g. the current-density at an electrode where a metal is being deposited) into two parts - the 'current flux' and the 'diffusion flux'. When considering results obtained with, and without, the presence of an excess of 'supporting electrolyte', it is not explained why the relevant diffusion coefficient changes from that of the ion to that of the (unsupported) salt containing the ion. These two values will in general be different; the proper use of the Nernst-Planck equation, in which the flux is related to forces (and not itself divided into two types of flux) leads naturally to the correct expressions for both limiting cases.

iii) Ion-exchange - Liquid Film Processes

Here we have at least three mobile ions. Two have the same sign of charge and are moving in opposite directions; one is entering the ion-exchanger, the other leaving it. To keep electroneutrality, when these ions do not have the same mobility, requires an electric field. In the presence of this field the ion of opposite sign of charge will also tend to move. If we assume a steady state, within a stagnant liquid film, then we have three flux equations which are subject to the following relations:

a) equal and opposite fluxes of the exchanging ions.

b) zero flux (due to the balancing of concentration and electric-field forces) of the non-exchanging ion of opposite charge.

The fluid mechanics of the liquid-film will depend upon the flow situation. The stagnant-film model is not the only possible one; one can use a developing-boundary-layer model.

In either case it becomes possible, via the Nernst-Planck equations, to describe the composition-dependence of the ion-exchange mass-transfer rate. For the stagnant-film model, it is necessary to estimate, or measure, another parameter, the stagnant film thickness. Given that parameter, the rate at any composition follows.

iv) Electrodialysis in Desalination

Here the rate of desalination may become limited by diffusion processes in the dilute liquid film close to the ion-exchange membrane. Potential gradients may become very high, if only over very short distances.

Ionic transports higher than Nernst-Planck would predict are obtained, even if the 'water-splitting' (generation of H^+ and OH^- ions from water) near to the membrane surface is allowed for (3). Rubinstein (4) has considered the effect of deviation from local electroneutrality, and gives a discussion at this NATO ASI.

v) Ion-exchange inside Ion-exchange Resins

Within practical ion-exchange resins the ionic strength is high, and the neglect of activity coefficients may be seriously objected to, particularly when the exchange of ions of different valency is being considered. While electroneutrality is an acceptable simplification (itself a consequence of high ionic strength) one may well query the assumption of a homogeneous phase. However, provided the resin is of gel type, and not too highly cross-linked, the Nernst-Planck equations give relations which are reasonably well obeyed in practice. As an instance, in the exchange of H^+ with Na^+ in 8% cross-linked polystyrene sulphonic resin the strong composition dependence of the exchange rate is explained. For example, the different shapes of forward and backward exchanges covering substantial changes in composition are described by Helfferich (5), while the measured diffusion coefficients for differential exchanges also fit the model (6).

For macroporous ion-exchange resins, the assumption that the resin phase is homogeneous is obviously even more suspect. Even differential exchanges cannot be described by a simple (composition-dependent) diffusion coefficient. A bidisperse model, based on Ruckenstein's treatment of porous catalysts (7), can be made to work reasonably well (8), but with one more parameter to be chosen a better fit should of course be obtained.

6. CONCLUSIONS

There seems little doubt that the Nernst-Planck equations are correct, and give useful results, for dilute electrochemical solutions. Their application to the more complicated situations of ion-exchange is more difficult, and may require more information, and further complexity, which the value of the results does not justify. Ion-exchangers are not precisely-defined materials. Different beads in a given batch have different physical (and ion-exchange) properties, while variation between batches is also significant. Manufacturers have specifications to meet; they may exceed them, but should not fail to reach them.

What the chemical engineer, using industrial ion-exchange materials, needs is a physically-sensible model, of reasonable simplicity. The Nernst-Planck equations give a physical reason for the marked composition dependence of ion-exchange rate. The engineer should be aware of this, even if he may regard some of the electrochemical arguments on matters of detail as being theological. If to describe a situation more accurately one must have available an ever-increasing number of quantities or parameters, the engineer, given the variability of the substances he has to deal with, may quite sensibly choose to stick with a simple model.

REFERENCES

1. Turner, J.C.R. Chemical Engineering Science 30 (1975) 151.
2. Turner, J.C.R. Chemical Engineering Science 30 (1975) 1304.
3. Makai, A.J. and J.C.R. Turner. Transactions of the Institution of Chemical Engineers, London, 60 (1982) 88.
4. Rubinstein, I. Journal of the Chemical Society, Faraday Transactions II, 75 (1979) 936.
5. Helfferich, F. Ion-Exchange (McGraw-Hill, New York, 1962).
6. Ruckenstein, E., A.S. Vaidyanathan and G.R. Youngquist. Chemical Engineering Science 26 (1971) 1305.
7. Patell, S. and J.C.R. Turner. Journal of Separation Process Technology 1 (2) (1980) 31.

APPLICATION OF THE STEFAN-MAXWELL EQUATIONS TO MULTICOMPONENT ION EXCHANGE

E. Earl Graham

Department of Chemical Engineering
The Pennsylvania State University
University Park, PA 16802 U.S.A.

ABSTRACT

Theoretical and experimental studies have shown that the Stefan-Maxwell relations may be used to describe binary ion exchange and will reduce to the Nernst-Planck equations only under limiting conditions where the self-diffusion coefficient of the Nernst-Planck equations are taken as the limiting tracer diffusion coefficient of each ion measured in an ion-exchange resin completely in the competing ion form. However, when the Stefan-Maxwell relations are extended to ternary ion exchange, it can be shown that no such simple limiting case exists and it is suggested that the Stefan-Maxwell relations should be used for multicomponent ion exchange. In effect this replaces the concept of a single ionic mobility for each ion as required by the Nernst-Planck equations, by combinations of ion-solvent and ion-ion interaction coefficients.

INTRODUCTION

The Nernst-Planck equation has become well established for the description of diffusion of ionic species and electrolytes. This is particularly so for reasonably dilute solutions for which activity coefficients may be neglected.

However, for concentrated electrolyte solutions such as exists in the interior of strongly basic or acidic ion-exchange resins, not only is it reasonable to suspect that activity coefficients may become significant, but also that ion-ion interactions may be important. The possibility of ion-ion interactions and the framework for handling such interactions is well established in the theory of nonequilibrium thermodynamics [see, for example, DeGroot (1)]. It is the purpose of this paper to review briefly previous work on the application of the theory of nonequilibrium thermodynamics for the description of binary ion exchange in ion-exchange resins and then extend the analysis to ternary and multicomponent ion exchange in general.

THEORY

Many formulations describing diffusion exist that take into account ion-ion interactions and are consistent with the theory of nonequilibrium thermodynamics. A review of these formulations has been made by Graham and Dranoff (2) with the conclusion that for ion-exchange resins for systems for which pressure gradients and the flux of solvent can be neglected, the following formulation of the Stefan-Maxwell equations written for n-ionic species has a number of important advantages.

$$-x_i \nabla\mu_i' = \frac{RT}{C} \sum_{j=0}^{(n)} \frac{(x_j J_i^o - x_i J_j^o)}{a_{ij}} \qquad i = 1 \ldots (n) \tag{1}$$

where $\nabla\mu_i'$ is the gradient of electrochemical potential (mole basis)

a_{ij} are the Stefan-Maxwell interaction coefficients in which 'o' denotes the solvent and 'n' the n-ionic species.

The other symbols have standard definitions and are given in the Symbols List.

It was shown by Miller (3,4,5) that for this formulation the Onsager reciprocal relations will hold even though the ionic species are no longer independent (due to electroneutrality requirements). For equation (1) we may then assume $a_{ij} = a_{ji}$.

Graham and Dranoff (2) also showed, using experimental data of Miller (5) for diffusion of mixed electrolytes in free diffusion, that the a_{ij}'s of the Stefan-Maxwell relations written for ionic species are independent of other ionic species that may be present and are insensitive to changes in total ionic concentration and the relative concentrations of each ionic species. For a given type of ion-exchange resin, the a_{ij}'s should therefore be constants. Another major advantage of the Stefan-Maxwell relations is that they are independent of the reference frame used. Therefore, for example, the J_i^o (flux relative to solvent velocity) of equation (1) may be replaced with N_i (flux with respect to fixed coordinates) and the a_{ij}'s will be the same.

A general form of the Nernst-Planck equations incorporating a convective term to account for transport by convection of the pore liquid can be written along the lines suggested by Helfferich (6) or Lightfoot and Scattergood (7). This may be written as

$$N_i = -\frac{d_i C_i \nabla\mu_i'}{RT} + x_i \sum_{j=0}^{n} N_j \tag{2}$$

In two special cases, equation (2) simplifies considerably. The first case is the case of no net molar flux so that $\sum_{j=0}^{n} N_i = 0$. The second case, where the ions diffusing are all of the same valence, the requirement of no current $\sum_{j=1}^{n} z_j N_j = 0$ reduces to $\sum_{j=1}^{n} N_j = 0$ and equation (2) becomes

$$N_i = -\frac{d_i C_i \nabla\mu_i'}{RT} + x_i N_o \tag{3}$$

In the case of ions of equal valence, one obtains the condition of zero molar flux if the solvent flux, N_o, is zero. In either case with zero net molar flux, equation (3) reduces to the usual Nernst-Planck equation.

$$N_i = -\frac{d_i C_i \nabla\mu_i'}{RT} \tag{4}$$

For this case, the general relation between the d_i of the Nernst-Planck equations and the a_{ij}'s of the Stefan-Maxwell relations reduces to the "intrinsic diffusivity" defined by Lightfoot and Scattergood (7) which also could be considered the "effective diffusivity" for the Nernst-Planck equations.

$$\frac{1}{d_i} = \sum_{j=0}^{n} \frac{x_j}{a_{ij}} \left(1 - \frac{x_i}{x_j}\frac{N_j}{N_i}\right) \tag{5}$$

The most important assumptions required in the application of equation (5) are that the net molar flux be zero or in the case of equal-valent exchange that the solvent flux is zero and that the a_{ij}'s are constants.

For binary ion exchange with 0 = water, a = active sites, 1 = counterion 1, 2 = counterion 2 and with the further assumption of no co-ions being present, equation (5) becomes

$$d_1 = \frac{1}{\dfrac{x_o}{a_{10}} + \dfrac{x_a}{a_{1a}} + \dfrac{x_2}{a_{12}}\left(1 - \dfrac{x_1}{x_2}\dfrac{N_2}{N_1}\right)} \tag{6}$$

Using the condition of no current ($z_1 N_1 + z_2 N_2 = o$) and electroneutrality ($z_a x_a + z_1 x_1 + z_2 x_2 = o$), equation (6) becomes

$$d_1 = \frac{1}{\dfrac{x_o}{a_{10}} - \dfrac{z_a}{z_2} x_a \left(\dfrac{1}{a_{12}} - \dfrac{z_2}{z_a}\dfrac{1}{a_{1a}}\right)} \tag{7}$$

A similar "effective diffusivity" may be defined for the other diffusing counterion, d_2, and within the restraints set by the assumption, d_1 and d_2 will be constant for a given binary ion exchange but will vary for each particular ion pair studied due to the interaction coefficient a_{12}.

It was shown by Graham and Dranoff (2) that the two Nernst-Planck equations using the "effective diffusivities" may be expressed in terms of concentration gradients by eliminating the gradient of electric potential, $\nabla\phi$, and using activity coefficients to express deviations from ideality. They in turn showed that as the concentration of either diffusing counterion goes to zero, the effect of deviations of ideality as expressed by activity coefficients also goes to zero. Therefore, as long as the "effective diffusivities" are used, the Nernst-Planck equations should give a good representation of binary ion exchange in resins even neglecting activity coefficients.

The "effective diffusivities" for the Nernst-Planck equations can be evaluated for binary ion exchange from isotopic exchange where only radioactive species 1* exchanges with nonradioactive 1 such that $N_o = N_a = N_2 = 0$ and $N_1 = -N_1^*$. Equation (5) becomes

$$d_1^* = \frac{1}{\dfrac{x_o}{a_{10}} + \dfrac{x_a}{a_{1a}} + \dfrac{x_2}{a_{12}} + \dfrac{x_1}{a_{11}}} \tag{8}$$

Then if $x_1 \to 0$; $x_2 \to -\dfrac{z_a}{z_1} x_a$ and (8) becomes

$$d_{1(2)}^* = \frac{1}{\dfrac{x_o}{a_{10}} - \dfrac{z_a}{z_2} x_a \left(\dfrac{1}{a_{12}} - \dfrac{z_2}{z_a}\dfrac{1}{a_{1a}}\right)} \tag{9}$$

and if $x_2 \to 0$; $x_1 \to -\frac{z_a}{z_1} x_a$ and (8) becomes

$$d^*_{1(1)} = \frac{1}{\frac{x_o}{a_{10}} - \frac{z_a}{z_1} x_a \left(\frac{1}{a_{11}} - \frac{z_1}{z_a} \frac{1}{a_{1a}} \right)} \tag{10}$$

Equation (8) is the general form of the tracer ion diffusion coefficient for species 1* = 1 while (9) and (10) are particular values. Equation (9) defines the limiting value for diffusion of '1' in an ion-exchange resin in which only '2' is present while equation (10) defines the value for the diffusion of '1' in an ion-exchange resin in which only '1' is present. The former is exactly the combination of Stefan-Maxwell interaction coefficients required for the "effective diffusivity" for the Nernst-Planck equation, that is, equation (7), and may be called the limiting tracer coefficient while the latter is the pure self-diffusion coefficient often used in the Nernst-Planck equation. As shown in the next section, the use of the limiting tracer coefficients for the "effective diffusivity" in the Nernst-Planck equations can explain their apparent success in describing binary ion exchange and at the same time explain the contradictions that result if the usual self-diffusion coefficients are used.

The foregoing analysis may be extended to multicomponent ion exchange in general, but specifically for a ternary system, 0 = water, a = active sites, 1 = counterion 1, 2 = counterion 2, 3 = counterion 3, equation (5) becomes

$$\frac{1}{d_1} = \frac{x_o}{a_{10}} + \frac{x_a}{a_{1a}} + \frac{x_2}{a_{12}} \left[1 - \frac{x_1}{x_2} \frac{N_2}{N_1} \right] + \frac{x_3}{a_{13}} \left[1 - \frac{x_1}{x_3} \frac{N_3}{N_1} \right] \tag{11}$$

For the restriction of electroneutrality and no current, this becomes

$$\frac{1}{d_1} = \frac{x_o}{a_{10}} + x_2 \left[\frac{1}{a_{12}} - \frac{z_2}{z_a} \frac{1}{a_{1a}} \right] + x_3 \left[\frac{1}{a_{13}} - \frac{z_3}{z_a} \frac{1}{a_{1a}} \right]$$
$$+ \frac{x_1 z_1}{z_2 N_2 + z_3 N_3} \cdot \left[\frac{N_2}{a_{12}} + \frac{N_3}{a_{13}} \right] - x_1 \frac{z_1}{z_a} \frac{1}{a_{1a}} \tag{12}$$

Also as before the case for isotopic exchange of radioactive 1* and 1 may be obtained from (5) as

$$\frac{1}{d_1^*} = \frac{x_o}{a_{10}} + \frac{x_1}{a_{11}} + \frac{x_2}{a_{12}} + \frac{x_3}{a_{13}} + \frac{x_a}{a_{1a}} \tag{13}$$

As with binary ion exchange the special cases of the general tracer coefficient given by equation (13) may be considered.

The usual self-diffusion coefficient can be obtained in the limit as $x_2 = x_3 = 0$ and $x_1 z_1 = -x_a z_a$ in which case equation (13) becomes

$$\frac{1}{d_1^*(1)} = \frac{x_o}{a_{10}} - \frac{z_a}{z_1} x_a \left[\frac{1}{a_{11}} - \frac{z_1}{z_a}\frac{1}{a_{1a}}\right] \tag{14}$$

This self-diffusion coefficient is clearly not the "effective diffusivity" required by equation (12) except for the case of all univalent counterions and the unlikely event that

$$a_{11} = a_{12} = a_{13} .$$

The limiting tracer coefficient defined by equation (13) where $x_1 = o$ will take on the same form as the "effective diffusivity" in the limit of an ion exchange process as x_1 goes to zero. In both cases, equation (12) and equation (13) become

$$\frac{1}{d_1^*(2,3)} = \frac{1}{d_1} = \frac{x_o}{a_{10}} + x_2 \left[\frac{1}{a_{12}} - \frac{z_2}{z_a}\frac{1}{a_{1a}}\right] + x_3 \left[\frac{1}{a_{13}} - \frac{z_3}{z_a}\frac{1}{a_{1a}}\right] \tag{15}$$

This limiting tracer coefficient, however, unlike that for binary exchange, is not a constant and will depend on the relative amounts of the 'other' counterions 2 and 3. This result strongly implies that if ionic interactions are important, that for ternary ion exchange it will be impossible to explain the exchange kinetics exactly using any single effective diffusivity for each ion. This would explain the highly variable results and general lack of agreement between theory and experiment found when the Nernst-Planck equations are used to describe ternary ion exchange.

RESULTS

For binary ion-exchange it is well known that the Nernst-Planck equations fit the experimental values as long as "effective diffusivities" that "best fit" the data are used. The value of the best fit coefficient for a given ion depends strongly on the other ion present and is not the same as the pure self-diffusion coefficients. The variation of a given "effective single-ion diffusivity" is often of the order of two or three fold as shown clearly by the studies of Hering and Bliss (8) and Morig and Rao (9). The variation in the values of the diffusion coefficients reported by these authors for the same ion-exchange resin can easily be explained using the concept of "effective diffusivity", defined by equation (7), which depends on the specific ion-ion interaction, a_{12}. Many other examples of such anomalies are reported by Graham and Dranoff (2).

The experimental part of the papers by Graham and Dranoff (2) give a detailed study of the Na^+-Cs^+ exchange on Dowex 50-X8. They show that the limiting tracer coefficients give much improved agreement over the pure self-diffusion coefficients when used to describe the Na^+-Cs^+ ion exchange using the "effective diffusivities" of equation (7). The best agreement for the Na^+-Cs^+ exchange in their study was obtained when values of the interaction coefficients were first obtained from the tracer diffusion coefficients and then an average value of the concentration of active sites and unhydrated water was used to calculate the "effective diffusivities" of equation (7). The almost exact agreement this method gave for both directions of exchange between Na^+ and Cs^+ may have been due to a compensating effect obtained which took into account the flux of water (solvent) which though small must have existed in their ion-exchange experiments.

The most thorough study of ternary ion exchange has been reported by Bajpai, Bupta and Rao (10). In their theory, they used a single self-diffusion coefficient for each ion and the Nernst-Planck equations. Even for the binary exchanges, the agreement between their experimental results and theory is not good as they used self-diffusion coefficients rather than limiting tracer coefficients or "effective diffusivities". For the ternary exchange, the deviation between theory and experiment is even more obvious with widely varying results being obtained depending on the ionic species present and on the relative amounts of the three ionic species present in the exchange resin.

CONCLUSIONS

The use of the Stefan-Maxwell relations as a theoretical basis has demonstrated that ion-ion interactions can be significant and are simply accounted for. In particular, for binary ion exchange it has been shown that it is incorrect to use the self-diffusion coefficients in the accepted Nernst-Planck equations but instead, if one uses the limiting tracer diffusion coefficients or two best-fit coefficients, the Nernst-Planck equations should (and do) describe ion exchange kinetics very well.

No such limiting conditions exist for ternary exchange and multicomponent exchange in general. It is proposed, therefore, that the Stefan-Maxwell relations themselves be used as the basis for ternary ion exchange using experimentally evaluated values of the interaction coefficients. This will involve twelve interaction coefficients (assuming the reciprocal relations apply). These coefficients and the consistency can be checked by simple radioactive tracer experiments. It is felt that these interaction coefficients will be constant for a given type of ion exchange resin and so once determined would have general application. Support for this is given by the fact that pure self-diffusion coefficients for the same resin as measured by different experimentors are generally nearly the same.

If the interaction coefficients can be shown to be constants, then once the six coefficients for one set of binaries were known only six new coefficients would be needed to completely explain a ternary system formed by adding a third ion. Furthermore, it is anticipated that as a number of systems are studied, simplified guidelines or approximate values of the interaction coefficients could be developed based on the theory and experimental results.

SYMBOLS LIST

a_{ij} Stefan-Maxwell interaction coefficient for ionic species, $m^2\ s^{-1}$

C total molar concentration including solvent, mol m^{-3}

C_i molar concentration of species 'i', mol m^{-3}

d_i an effective diffusivity for the Nernst-Planck equation, $m^2\ s^{-1}$

$d_{i(j)}$ a tracer ion diffusion coefficient for species i measured in an ion-exchange resin in the (j) form, $m^2\ s^{-1}$

J_i^o flux of species 'i' referred to the solvent velocity, mol $cm^{-1}\ s^{-1}$

N_i flux of species 'i' referred to fixed coordinates, mol $cm^{-1}\ s^{-1}$

R gas constant, J mol^{-1} K^{-1}

T absolute temperature, K (degree Kelvin)

x_i mole fraction of species 'i' in resin phase

z_i valence of ionic species 'i' in resin phase

Greek Letters

μ_i' electrical chemical potential (mole basis) J mol^{-1}

∇ gradient, m^{-1}

Superscripts

* tracer ion species

Subscripts

a active ionic sites

o solvent

i ionic species (counterions)

REFERENCES

1. S. R. DeGroot and P. Mazur, Non-Equilibrium Thermodynamics, North Holland, Amsterdam, 1962.

2. E. E. Graham and J. S. Dranoff, Application of the Stefan-Maxwell Equations to Diffusion in Ion Exchangers. Part I: Theory and Part II; Experimental, Ind. Eng. Chem. Fundam., to be published, 1982.

3. D. G. Miller, Thermodynamics of Irreversible Processes, Chem. Rev., 60 (1960) 15.

4. D. G. Miller, Application of Irreversible Thermodynamics to Electrolyte Solutions. I. Determination of Ionic Transport Coefficients ℓ_{ij} for Isothermal Vector Transport Processes in Binary Electrolyte Solutions, J. Phys. Chem., 70 (1966) 2639.

5. D. G. Miller, Application of Irreversible Thermodynamics to Electrolyte Solutions. II. Ionic Coefficients ℓ_{ij} for Isothermal Vector Transport Processes in Ternary Systems, J. Phys. Chem., 71 (1967) 616.

6. F. Helfferich, Ion Exchange, McGraw-Hill, New York, 1962.

7. E. N. Lightfoot and E. M. Scattergood, AIChE J., 11 (1965) 175.

8. B. Hering and H. Bliss, AIChE J., 9 (1963) 495.

9. C. R. Morig and M. G. Rao, Diffusion in Ion Exchange Resins: Sodium Ion-Strontium Ion System, Chem. Eng. Sci., 20 (1965) 889.

10. R. K. Bajpai, A. K. Gupta and M. G. Rao, Single Particle Studies of Binary and Ternary Cation Exchange Kinetics, AIChE J., 20 (1974) 989.

CONCENTRATION POLARIZATION AND MEMBRANE RESISTANCE EVALUATION BY CURRENT TRANSIENT AND NERNST-PLANCK EQUATIONS.

G. Tiravanti, G. Boari and R. Passino

Istituto di Ricerca sulle Acque, CNR, via De Blasio, 70123 Bari, Italy.

SUMMARY

An analysis of the Nernst-Planck equations in steady and transient conditions and the application to the description of concentration polarization phenomena is presented.
Interfacial concentration can be calculated both during current flow and after current interruption. In the first case it can be directly calculated, while in the second it is evaluated by extrapolating the data derived by overpotential measurements at zero time.
Careful evaluation of experimental data on cation membranes enables the validity of Nernst-Planck equations to be verified for "critical" interfacial concentrations in the order of 10^{-6} N KCl.
Different design criteria are formulated taking into account polarization phenomena in a full scale plant.
The same type of transient approach, based on the application of a step change in direct current and on the analysis of the potential variation subsequent to the application of the current step, is used for measuring electrical resistance in charged membranes.
Experimental results are in good agreement with those computed from the diffusion equation.

INTRODUCTION

Monitoring the behaviour of a complex membrane system, following the application of a current or a potential step, gives useful information in terms of physico-chemical description of the system itself. Furthermore, large use of solid state components in the monitoring system coupled with computer facilities makes these techniques handy and reliable. Excitation and propagation have been extensively studied especially in the field of biological membranes. The first physical formulation of a mechanism of excitation was reported by Nernst at the beginning of this century. He showed that ion concentrations close to a selectively ion-permeable membrane would change accordingly to the square root function of time after the application of a constant current [1].

In this paper, some applications of transient techniques are described and applied to charged membranes, making use of the classical Nernst-Planck equations, in order to better understand the phenomena of concentration polarization, which limit the rate of an electrodialysis process.

The application of a step change in direct current in a system with a membrane separating two aqueous solutions of the same electrolyte at the same concentration induces a potential difference, measured by reversible electrodes, which varies with time until steady state conditions are reached (Fig. 1). The analysis of the potential transient subsequent to the application of the current step allows the calculation of the membrane electrical resistance.

On the other hand, the current interruption technique, introduced by Cooke in the 60s [2-5] is very useful in making it possible to calculate interfacial concentrations and concentration boundary layer thickness under different operative and hydrodynamic conditions.

THEORETICAL

The system under study is constituted by a charged membrane separating two electrolytic solutions, i.e. when electric current is applied

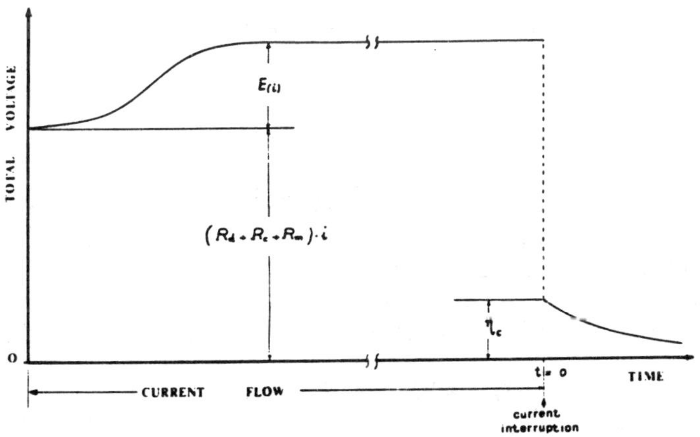

FIG. 1: Qualitative total voltage curve VS time.

or interrupted, the Fick diffusion equation in a bidimensional system holds valid:

$$\frac{\partial c}{\partial t} = D\left[\frac{\partial^2 c}{\partial x^2} + \frac{\partial^2 c}{\partial z^2}\right] - u\frac{\partial c}{\partial z} - v\frac{\partial c}{\partial x} \tag{1}$$

Assuming

$$v = 0$$

$$u \simeq 0 \qquad \text{for } x \leqslant \delta_c$$

$$\frac{\partial^2 c}{\partial z^2} \simeq 0$$

eq. (1) reduces to:

$$\frac{\partial c}{\partial t} = D\frac{\partial^2 c}{\partial x^2} \tag{2}$$

a) Current Application Step

When the concentration of the electrolytic solutions on both sides of the membrane is the same, the knowledge of boundary conditions allows the calculation of the membrane electrical resistance. Assuming that the probe electrodes are located at a known distance from the surface of the membranes and that they do not interfere with the current flow paths, the application of a constant current step by two working electrodes gives rise to a potential difference given, for short polarization times ($\leqslant 5$ s), by [6]:

$$\Delta V(t) = (2R_d + R_m + R_{b,t} - 2R_{b,o})\, i \tag{3}$$

where

$$R_{b,t}\, i = (R'_{b,t} + R''_{b,t})\, i \tag{4}$$

and ($R_{b,o}$ i) represent the ohmic potential drops within the boundary layers at time t and zero, respectively.
In the same system without membrane, the same current step should raise a potential difference given by:

$$\Delta V_o = 2R_d\, i \tag{5}$$

Assuming the membrane electrical resistance to be independent of i, the difference between the potential drops in the system with and without the membrane is given by:

$$\Delta V(t) - \Delta V_o = (R_m + R_{b,t} - 2\,R_{b,o})\,i \tag{6}$$

An "apparent" membrane resistance $R_{m,t}$ can be defined as:

$$R_{m,t} = R_m + R_{b,t} - 2\,R_{b,o} \tag{7}$$

When $t = 0$, $R_{b,t} = 2\,R_{b,o}$, thus eliminating the influence of concentration polarization phenomena within the boundary layers.
The electrical resistance of the solution within the dilute boundary layer is given by:

$$R'_{b,t} = \int_0^{\delta_c} \frac{x}{\Lambda_d c(x,t)}\,dx \tag{8}$$

Considering that a similar expression can be written for the concentrated boundary layer resistance, then $R_{b,t}$ can be evaluated by calculating the concentration distribution within the boundary layer, $c(x,t)$, assuming a one-dimensional diffusion perpendicular to the membrane surface.
The Fick equation (2) was solved [7] making use of the following boundary conditions as illustrated in Fig. 2:

$c = c_o$	$0 \leqslant x \leqslant \delta_c$	$t = 0$
$c = c'_m$	$x = \delta_c$	$t = \infty$ (steady state)
$c = c_o$	$x = 0$	

Of course for $t = \infty$ the concentration distribution within the boundary layer is linear, according to the classical Nernst equation in steady state conditions.
Fig. 3 gives the theoretical trend of the dimensionless electrical resistance of the boundary layer $\bar{R}_{b,t}$ VS dimensionless time $\bar{T}$, defined as:

$$\bar{R}_{b,t} = \frac{R_{b,t} - 2\,R_{b,o}}{R_{b,\infty} - 2\,R_{b,o}} \tag{9}$$

$$\bar{T} = \frac{t\,D}{\delta_c^2} \tag{10}$$

It is worth noting that $\bar{R}_{b,t}$ is a function of δ_c and c_o but, for short polarization times, ($t \leqslant 5$ s), the same curve is obtained for δ_c values up to 250 μm and for bulk concentrations up to 1 N NaCl solutions. Furthermore, anion and cation membranes having the same permselectivity give the same results. This can be explained considering that the diffusion equation does not take into account any interaction between different ions in solution and different fixed charges in cation and anion membranes.
The diagram in Fig. 3 shows a linear pattern for $\bar{T} \leqslant 0.2$. It follows

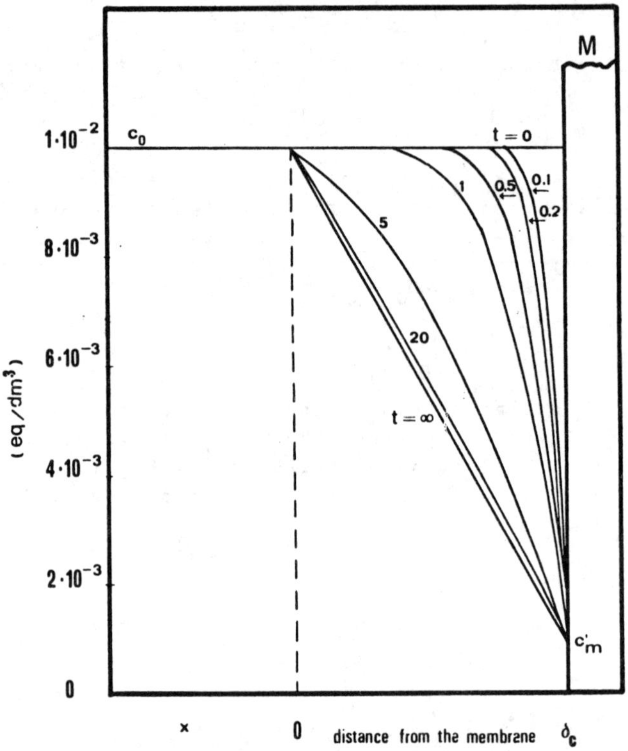

FIG. 2: Schematic representation of the concentration profiles within the diluted boundary layer at various times after the application of current step.

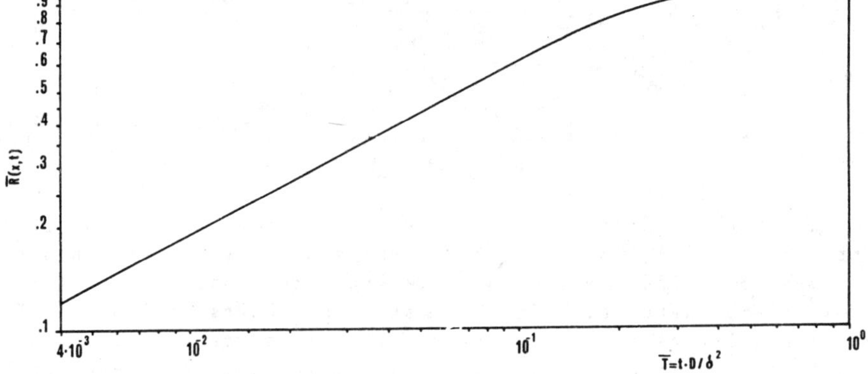

FIG. 3: Theoretical dimensionless boundary layer electrical resistance SV dimensionless time.

that the variation of the resistance through the boundary layers can be interpolated with good accuracy by the function:

$$R_{b,t} = 1.9 \; \overline{T}^{\;1/2} \tag{11}$$

Taking into account eq. (7), (9), (10), eq. (11) becomes:

$$R_{m,t} = R_m + \alpha \, t^{1/2} \tag{12}$$

Hence a simple model allows the calculation of the membrane resistance R_m by least square extrapolation at zero time of the $R_{m,t}$ values.
Experimentally, the $R_{m,t}$ values can be determined by eq:

$$R_{m,t} = \frac{\Delta V(t)}{i} - \frac{\Delta V_o}{i} \tag{13}$$

from the difference between the potential drops in the system with and without membrane at time t after the application of the current step.

b) Current Interruption Step

Another interesting application of Nernst-Planck equations in transient conditions is related to the evaluation of concentration polarization phenomena leading to the formulation of new design criteria for the electrodialysis process.
In fact, it is known that conventional plants operate at a fraction of the "limiting" current density, generally determined according to a Cowan diagram, in order to prevent polarization effects occurring at membrane/solution interfaces.
The calculation is based on the analysis of the differences of electrochemical potential during current flow and after current interruption, taking into account the hydrodynamic and mass transfer conditions in the compartments of a full scale plant.
The following simplifying assumptions are made:
-bidimensional (x,y) system;
-constant hydrodynamic (δ_h) and concentration (δ_c) boundary layer thickness along the flow path;
-fully developed turbulent flow;
-constant velocity field outside δ_h;
-constant velocity gradient within δ_h;
- $\delta_h / \delta_c = Sc^{0.33}$;
-negligible diffusive transport along the stream velocity direction.
The set of equations to be applied to any subsystem, each constituted by a portion of the stack half cell limited by two contiguous turbulence promoters, were derived developing the Nernst-Planck, Ohm, Kirchhoff equations. In steady state conditions (Fig. 1), the boundary layer overpotential E(i) represents the potential difference

across the two boundary layers, excluding the term due to the membrane resistance:

$$E(i) = \Delta V'_b + \Delta V''_b + E'_d + E''_d + E_m \tag{14}$$

Expressing each term in eq. (14) as a function of the interfacial concentration and of electrolyte and membrane parameters leads to:

$$E(i) = K_d \ln\left\{\frac{(N_1 + 1)\, c_d - c'_m}{c'_m}\right\} + \frac{B}{\Lambda_d}\left\{\ln\left[(N_1 + 1)\, c_d - c'_m\right]^{\lambda - 1} - \ln N_1^{\lambda}\, c_d^{\lambda - 1}\right\} + e_2 \ln N_1 \tag{15}$$

Developing eq. (15) with respect to c'_m one obtains the interfacial concentration during current flow:

$$c'_m = \frac{(N_1 + 1)\, c_d}{1 + \left\{[N_1 + 1]^{-\frac{B(\lambda - 1)}{\Lambda_d K_d}}\right\}\left\{N_1^{\frac{B\lambda - e_2 \Lambda_d}{\Lambda_d K_d}}\right\} \exp\left(E(i)/K_d\right)} \tag{16}$$

When current is interrupted, ohmic potentials drop to zero and the residual potential difference is constituted by concentration overpotential, defined as:

$$\eta_c = E'_d + E''_d + E_m \tag{17}$$

In this case, the expression of each term in eq. (17) gives the interfacial concentration determined at and after the current interruption:

$$c'_{m,t} = \frac{(N_1 + 1)\, c_{d,t}}{1 + [N_1^{-e_2/e_o}] \exp(\eta_{c,t}/e_o)} \tag{18}$$

and the concentration overpotential:

$$\eta_c = e_o \ln \frac{(N_1 + 1)\, c_d - c'_{m,o}}{c'_{m,o}} + e_2 \ln N_1 \qquad (19)$$

η_c cannot be measured directly, due to some electric interfering phenomena in the electronic circuit and to probe polarization so that it must be extrapolated at zero time. In order to overcome these difficulties, eq. (2) has been solved numerically [9] (finite differences method), in order to find out a physically consistent extrapolation expression for interfacial concentrations. η_c at the instant of the current interruption can be then calculated by using the value of $c'_{m,o}$ extrapolated according to the relation:

$$c'_{m,o} = c'_{m,t} - \alpha' t^{1/2} \qquad (20)$$

The boundary layer overpotential E(i) can be experimentally determined as:

$$E(i) = \Delta V - \Delta V_o + \frac{i\ \delta_c}{c_d \Lambda_d} \left(1 + \frac{\lambda}{N_1} \right) \qquad (21)$$

δ_c can be calculated by:

$$\delta_c = \frac{(R_m + R_d + R_c)\ B\ \Delta c}{\Delta V - \eta_c - \frac{B}{\Lambda_d} \left\{ \ln \frac{c_d}{c_d - \Delta c} + \lambda \ln \frac{N_1 c_d + \Delta c}{N_1 c_d} - \frac{\Delta c}{c_d} \left(1 + \frac{\lambda}{N_1}\right) \right\}} \qquad (22)$$

Eq. (21) can be rearranged as:

$$\Delta V = E(i) + \left\{ R_m c_d + \left[\frac{d - \delta_c}{\Lambda_d} \right] \left[1 + \frac{\lambda}{N_1} \right] \right\} \frac{i}{c_d} \qquad (23)$$

Combining the Nernst equation:

$$\frac{i}{c_d} = \frac{B}{\delta_c} \left(1 - \frac{c'_m}{c_d} \right) \qquad (24)$$

and eq. (16) in eq. (23) the following relation is obtained:

$$\Delta V = E(i) + f(R_m, c_d, E(i)) \qquad (25)$$

EXPERIMENTAL

Electrical resistance measurements were carried out making use of a cell described elsewhere [6]. NaCl stagnant solutions in the range 0.01 to 1.0 N were used. The cell and the solutions were mounted in a thermostat-controlled (25 $\pm$ 0.01°C) air bath; this corresponded to an uncertainty in the resistance measurements of about $\pm$ 0.02 %.
Cation membranes Ionics 61 CZL-183 and anion membranes Ionics 111 BZL-219 completely converted into Na^+ and Cl^- form, respectively, were used. Ag/AgCl probe electrodes, made according to Ives and Janz [10], were connected to the cell solutions by Luggin capillaries as a liquid junction.
Overpotential measurements with the current interruption technique were performed on a regular scale ED tortuous path Ionics Mark III stack, made up by 26 cell pairs, fed with 0.05 N KCl solutions at 25 $\pm$ 0.5 °C in both diluted and concentrated compartments [11]. The characteristics of the stack and of Ag/AgCl electrodes, the electric and flow operating conditions are given in Table 1.
A constant DC supply was used in the experiments.
In both cases, electric parameters were measured by a HP data acquisition system previously described [8].
The procedure for measuring the membrane electrical resistance was as follows. The asymmetry potential of the probes was measured at the start of the experiments; then the desired constant current density was selected. The circuit was closed to the cell by means of a scanner, making use of a relay having less than 1 ms switch time. Potential differences across the cell were measured by probes every 43 $\pm$ 0.5 ms for 5 sec.
Data also obtained by reversing either the membrane in the cell or the current flow direction were averaged.
Tests conducted on the Mark III stack with the current interruption technique were started by recirculating the feed solutions through the system at the desired flow rate. Potential differences between the cell pairs were continuously measured by probes placed at different locations along the flow path. The data were recorded, both during the transient of the electric current until steady state conditions were reached, and after current interruption.
The electrical resistances of the solutions and of membranes were calculated from potential measurements taking into account corrections due to membrane and probe electrode potentials.

RESULTS AND DISCUSSION

The current step method was applied to the mentioned commercial ion-exchange membranes and the results compared with classical AC methods [12-13] at a frequency of 1000 Hz.
The results showed less than 2 % standard deviation while the spread

TABLE 1
Characteristics of the experimental system for the current interruption runs.

STACK	
number of desalting cell pairs	26
number of neutral compartments	2
spacer type	tortuous path
spacer thickness	10^{-3} m
length of flow path	3.3 m
width of flow path	0.04 m
membrane dimensions	0.46 x 0.51 m^2
active membrane area	55 %
cation membranes	61 AZL-183
anion membranes	111BZL-219
neutral compartments membranes	61 AZL-183
membrane resistance	0.002 $\Omega\,m^2$
number of membranes per neutral compartments	2
neutral compartments feed concentration	0.05 N KCl
counterion transport number in both membranes	0.95
cathode working electrode	stainless steel
anode working electrode	Pt plated Ti
working electrode feed conc: 0.1N $FeCl_2$+ 0.1N $FeCl_3$+0.2N KCl(pH=1.5)	
FEED	
total diluted flow rate	3.6 10^{-4} m^3/s
total concentrated flow rate	"
ion transport number in solution	0.5
mean flow velocity	0.175 m/s
feed concentration in both compartments	0.05 N KCl
equivalent conductivity	0.0141 $m^2/eq\Omega$
ELECTRIC CONDITIONS	
total applied voltage	140 V
maximum applied current	40 A
MEASURING ELECTRODES	
tipe of probes	electrolytic Ag/AgCl
probe active surface	3 10^{-3} x 2 10^{-3} m^2
number of probes	3 sets of 3
probe thickness	1.5 10^{-4} m
probe location along the flow path	
1st set	0.4 m
2nd set	1.8 m
3rd set	3.2 m

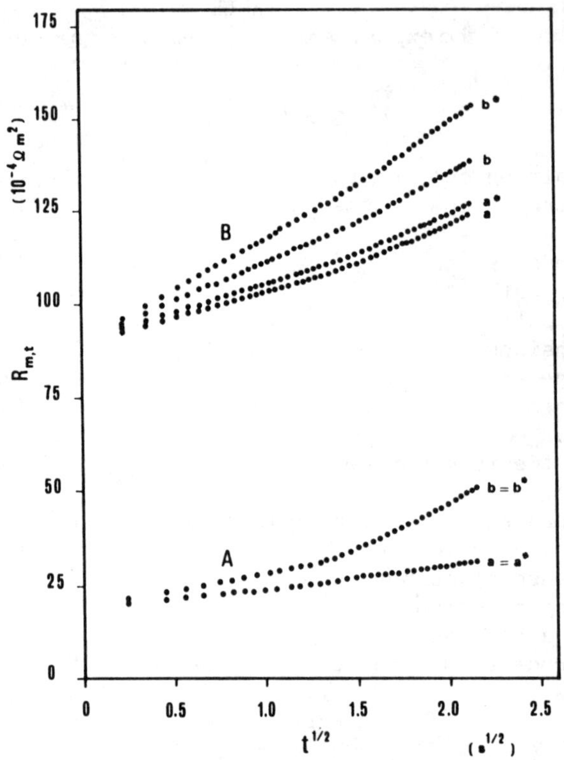

FIG. 4: Apparent electrical resistance/time curves of the anion membrane Ionics 11 BZL-219 in 0.01 N NaCl VS $t^{1/2}$ for direct* and reverse current flow direction.
A = clean membrane; B = fouled membrane; a = 15 A/m^2; b = 30 A/m^2.
* membrane with the fouling deposited on the surface facing the cathode.

of results was about 5 % for AC methods [6].
An interesting application is the investigation of the different response of the voltage/current curves in asymmetrical membranes. Membranes of this kind may spontaneously form during electrodialysis of waters containing colloidal organic and/or inorganic particles which, being negatively charged, migrate towards the anode and precipitate on the dilute side of anion membranes as well as within the membrane. Fig. 4 reports the apparent electrical resistance/time curves for the clean and silica fouled anion Ionics BZL-219, taken from an industrial plant, as a function of $t^{1/2}$ at different current densities and for direct (fouling facing the cathode) and reverse

flow directions. The membrane specimens were equilibrated for 20 minutes with 0.01 N NaCl solutions before the measurements.
As expected, the clean membrane gave the same response by varying the current direction, at a given current density while the curve slope increased with increasing the applied current density. On the contrary, the fouled membrane evidenced current rectification, defined as the change of electrical resistance by changing current direction. It is wotth noting that the increase of the apparent membrane electrical resistance with time becomes significant just few milliseconds after the application of current flow, as a consequence of concentration polarization.
The extrapolation to zero time of the set of homogeneous data in Fig. 4 gave the same values, independently of current density and flow direction. The membrane resistance value of the fouled membrane was about 4 times higher than the corresponding one of the clean membrane. Considering that the concentration distribution in the solution near the membrane is almost independent on concentration boundary layer thickness δ_c for short polarization times [14], the different response of the fouled membrane is to be ascribed to depletion phenomena occurring in the membrane itself.
On the other hand, when steady state conditions are reached in the system, the transient phenomena develop by suddenly interrupting the current flow (Fig. 1). This technique provides additional information on parameters which limit the mass transfer through the membranes.

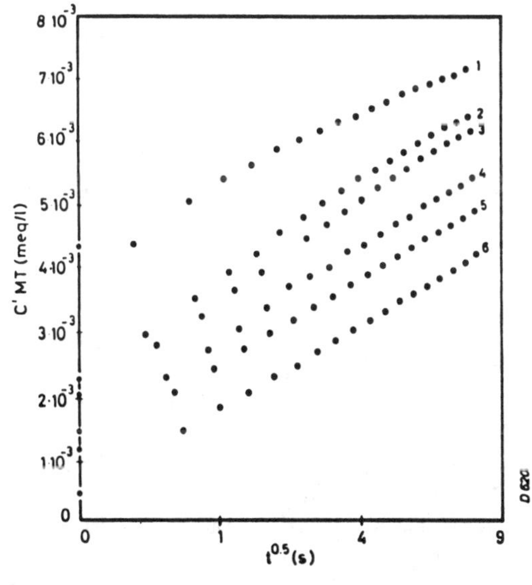

FIG. 5: Interfacial concentration relaxation as a function of time after the current interruption. Cation membrane Ionics 61 CZL-183. c_o = 0.009 N NaCl; 1-6=probe couples located along the flow path.

According to the theory described above, Fig. 5 shows an example of interfacial concentration relaxation, calculated after current interruption, VS $t^{1/2}$ for several couples of probes along the flow path.
The interfacial concentration obtained during current flow and at current interruption, working with the cation exchange membranes, were compared. A good agreement was found independently of the bulk solution concentrations (10^{-3} to 10^{-2} N) used. The current densities giving rise to interfacial concentrations as low as 10^{-6} N are taken as "critical" values and are a function of the membrane length.
For higher polarization conditions, because of water dissociation equilibrium and electroneutrality, there is a limiting K^+ flux through the membrane and the equations used for calculations do not hold. The disagreement is to be ascribed to H_3O^+ transport due to the Bethe-Toropoff effect, competitive with K^+ transport through the cation membrane.
Of course, limiting current densities which are a consequence of concentration polarization should be avoided in the ED plants, because of some unwanted effects, mainly membrane fouling and scaling.
Correlations found between operating parameters and boundary layer overpotential were applied to a full scale plant, in order to formulate different design criteria.
The direct analytical solution of eq. (25) for a regular size stack is complex because c_d is also a function of E(i). Consequently the solution was performed for each of the subsystems, by assuming that c_c and c_d entering the (i+1)th subsystem were identical to those leaving the ith subsystem.
The δ_c value was determined by eq. (22) making use of electric parameters only and its mean value was found to be

$$\delta_c = 70 \pm 20\ 10^{-6}\ \text{m}$$

working with a mean flow velocity of 0.175 m/s. It is in agreement with other values [15] derived by hydrodynamic considerations, for the same flow path characteristics and the same flow velocity.
Membrane and elecrolyte characteristics, as reported in Table 1, and a $\delta_c = 70\ 10^{-6}$ m were used for the calculations.
The theoretical correlation E(i) vs V for the half cell under constant applied voltage was determined for different flow path lengths, as reported in Fig. 6. The agreement between experimental data and the calculated ones, assuming equal initial bulk concentration in the dilute and concentrate compartments, is satisfactory.
It was observed that, at current densities lower than 200 A/m^2, E(i) reaches higher values for the anion membrane. This confirms the higher polarization of anion membranes, as already remarked by Spiegler [16]. The Nernst-Planck model described by Helfferich [17] does not allow this phenomenon to be quantified.
Fig. 6 also shows that the slope of the E(i) lines increases with the voltage applied. Generally, the flow path length required to obtain a given desalting factor is inversely proportional to the constant voltage applied, but E(i) will increase and c'_m decrease with the voltage, and a maximum "safe" value of the applied voltage

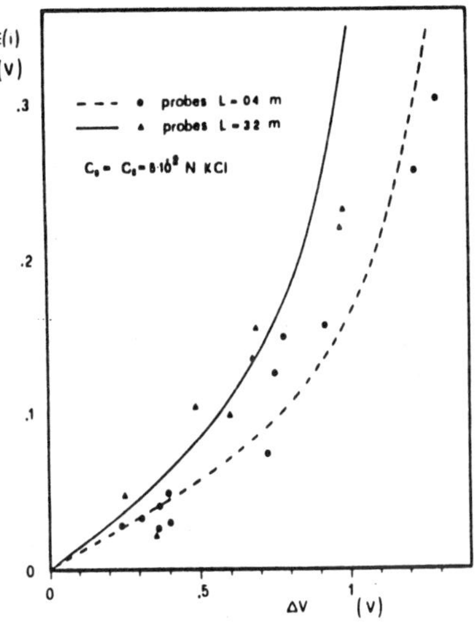

FIG. 6: Theoretical and experimental boundary layer overpotential E(i) as a function of applied voltage.

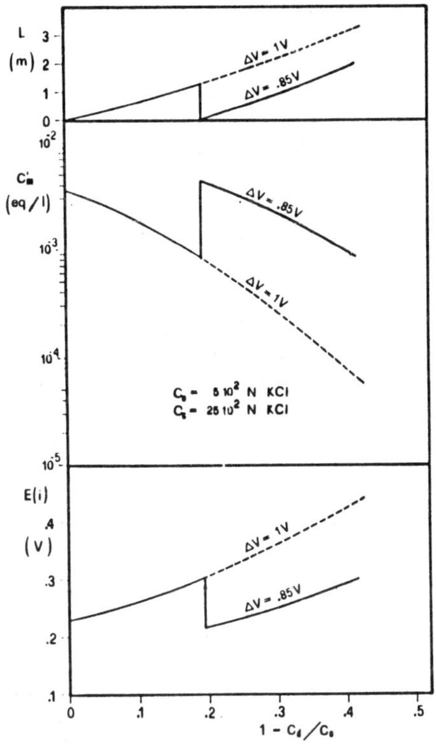

FIG. 7: Calculated stage length, interfacial concentration and boundary layer overpotential VS desalting factor for one stage and two stage ED plant.

can be used as design criterion for given hydrodynamic and feed concentration conditions.
The equation developed were used to find out whether a better control of c'_m, by varying the applied voltage along the flow path, could represent a better design criterion.
As an application example, a two stage operation was calculated, for which $c'_m \geq 8\ 10^{-4}$ N; $E(i) \leq 0.3$ V. The desired minimum value of c'_m is reached with an applied voltage $\Delta V = 1$ V after a flow path length in the first stage L = 1.3 m; the voltage applied in the second stage is $\Delta V = 0.85$ V, using which voltage the minimum for c'_m is almost reached at the end of the residual 2 m of flow path, as shown in Fig. 7. The values of c'_m when the applied voltage is kept unchanged (one stage operation) are also reported as dashed lines. For the same membrane length, the desalting factors in the two stage and in the one stage operations are 0.416 and 0.427, respectively, but the performance should be much safer in the first case, in which c'_m is one order of magnitude higher than reached in the one stage operation.

CONCLUSIONS

The data presented demonstrate the utility of analyzing the response of ion-exchange membrane/solution systems in transient conditions. The Nernst-Planck equations can be successfully applied in stagnant solutions for short times ($\leqslant$ 5 sec) after the application of a constant current step allowing the calculation of the membrane electrical resistance by a DC method and taking into account corrections due to concentration polarization phenomena.

The current interruption technique and the set of equations developped can be applied to a full scale ED plant making it possible to calculate the concentration boundary layer thickness and the boundary layer overpotential.

The correlation of these parameters with the interfacial concentration and the applied voltage allows for a different approach to process design, according to which the minimum value of c'_m can be fixed "a priori" and, consequently, the applied voltage varied along the stages.

LIST OF SYMBOLS

B	$B = F D/(\bar{t}_i - t_i)$, A m^2/eq
c_i	Electrolyte concentration, defined by subscripts, eq/dm^3
c_m	Interfacial concentration, eq/dm^3
$c_{m,o}$	Interfacial concentration at the instant of current interruption, eq/dm^3
c_o	Bulk concentration on dilute compartment, eq/dm^3
D	Electrolyte diffusion coefficient, m^2/s
d	Compartment depth, m
E_d	Diffusion potential, V
E(i)	Potential difference across the boundary layers, excluding the term R_m i, V
E_m	Membrane potential, V
e_o	$e_o = (2RT/F)(\bar{t}_i - t_i)$, V
e_1	$e_1 = (RT/F)(1 - 2t_i)$, V
e_2	$e_2 = e_1 + (RT/F)$, V
F	Farady constant, coul/eq
i	Nominal current density defined as total current/active membrane area, A/m^2
K	$K = B/\Lambda_d + e_o$, V
L	Membrane flow path length, m
N_1	$N_1 = c_c/c_d$
R	Gas constant, J/°K mol
$R_{b,o}$	Electrical resistance of the solution between the membrane surface and the concentration boundary layer, at time zero, Ωm^2
$R_{b,t}$	Electrical resistance of the solutions between the membrane surfaces and the concentration boundary layers, at time t, Ωm^2

$\bar{R}_{b,t}$	Dimensionless electrical resistance defined by eq. (9)
R_i	Electrical resistance per unit area, defined by subscripts, Ωm^2
R_m	Membrane electrical resistance, Ωm^2
$R_{m,t}$	Apparent electrical resistance of the membrane, Ωm^2
Sc	Schmidt number, D/ν
t	Time, s
T	Absolute temperature, °K
t_i	Counterion i transport number in solution
$\bar{t}_i$	Counterion i transport number in membrane
$\bar{T}$	Dimensionless time defined by eq. (10)
u	Mean linear velocity in z direction, defined as flow rate/channel section, m/s
v	Flow velocity in x direction, m/s
x	Coordinate of the membrane compartment depth, m
y	Coordinate of the membrane compartment width, m
z	Coordinate of the membrane flow path length, m

Greek letters

α	Constant in eq. (12), $m^2 s^{-1/2}$
α'	Constant in eq. (20), eq $dm^{-3} s^{-1/2}$
Δc	Difference in concentration, eq/dm^3
ΔV	Difference in potential, defined by superscripts and subscripts, V
δ_c	Concentration boundary layer thickness, m
δ_h	Hydrodynamic boundary layer thickness, m
η_c	Concentration overpotential, V
λ	$\lambda = \Lambda_d/\Lambda_c$
Λ	Equivalent conductivity of the electrolyte, defined by subscripts, m^2/Ω eq
ν	Kinematic viscosity, m^2/s

Subscipts

b	Diffusion boundary layer
c	Concentrate
m	Membrane
o	Referred to time zero
t	Referred to time t
x	Referred to coordinate x

Superscripts

'	Referred to the dilute side of the membrane
"	Referred to the concentrate side of the membrane

REFERENCES

1. Nernst, W., Zur theorie der elektrishen reizung, Nach. K. Gesell., (1899) 104.
2. Cooke, B.A., Concentration polarization in electrodialysis, I. The electrometric measurement of interfacial concentration, Electrochimica Acta, 3 (1961) 307.
3. Cooke, B.A., Concentration polarization in electrodialysis, II. Systems with natural convection, Electrochimica Acta, 4 (1961) 179
4. Cooke, B.A. and S.J. Van der Walt, Concentration polarization in electrodialysis, III. Practical electrodialysis systems, Electrochimica Acta, 5 (1961) 216.
5. Cooke, B.A., Some phenomena associated with concentration polarization in electrodialysis, First Symp. on Water Desalination, Washington, D.C., 2 (1967) 219.
6. Tiravanti, G., The direct current method for measuring charged membrane conductance, J. Membrane Sci., 9 (1981) 229.
7. Crank, J., The mathematics of diffusion, Clarendon press, Oxford, 1975.
8. Passino, R., G. Boari, A. Rozzi and G. Tiravanti, Overpotential phenomena on charged membranes in flow conditions. Transport mechanism in polarization conditions with KCl and HCl solutions and analysis of boundary layer overpotentials, Desalination, 24 (1978) 55.
9. Passino, R., A. Rozzi and G. Tiravanti, Membrane/solution polarization in dynamic conditions, in R. Passino (Ed.), Study Week on Biological and Artificial Membranes, Pontificia Academia Scientiarum, Elsevier Sci. Publ. Co., Amsterdam, 1976.
10. Ives, D.J.G. and G.J.Janz, Reference Electrodes, Academic Press, New York, 1961.
11. Passino, R., C. Marra, A. Rozzi and G. Tiravanti, Performance evaluation and process design of an electrodialysis plant by overpotential measurements, Desalination, 41 (1982) 181.
12. Hills, G.J., A.C. Jacubovic and J.A. Kitchener, Determination of electrical conductivity of crosslinked polyelectrolytes, resins and gels, J. Polym. Sci., 19 (1956) 382.
13. Subrahmanyan, V. and N. Lakshminarayanaiah, A rapid method for the determination of electrical conductance of ion exchange membranes, J. Phys. Chem., 72 (1968) 4314.
14. Forgacs, C., J. Leibovitz, R.N. O'Brien and K.S. Spiegler, Interferometric study of concentration profiles in solutions near membrane surfaces, Electrochimica Acta, 20 (1975) 555.
15. Forgacs, C., N. Ishibashi, J. Leibovitz, J. Sinkovic and K.S. Spiegler, Polarization at ion-exchange membranes in electrodialysis, Desalination, 10 (1972) 181.
16. Spiegler, K.S., J. Leibovitz and J. Sinkovic, Polarization at membrane/solution interfaces, in R. Passino (Ed.), Study Week on Biological and Artificial Membranes, Pontificia Academia Scientiarum, Elsevier Sci. Publ. Co., Amsterdam, 1976.
17. Helfferich, F., Ion Exchange, Mc Graw-Hill, New York, 1962.